NATO Science for Peace and Security Series C: Environmental Security

NATO Science for Peace and Security Series
Sub-Series C. Environmental Security

Volumes in the Environmental Security Sub-Series deal with scientific and technological aspects of innovative and promising projects related to environmental security, including regional environmental problems and natural and man-made disasters. Floods, droughts, landslides, desertification, earthquakes and industrial accidents are examples of topics covered, as are regional environmental problems related to transboundary water resources or transfrontier air pollution. Other areas covered relate to climate change, environmental chemistry, disaster prevention and forecast, remote sensing and detection, and decision support systems/risk assessment.

Special offer: For all clients with a print standing order we offer free access to the electronic volumes of the Series published in the current year.

Tünde Anna Kovács · Igor Fürstner
Editors

Critical Infrastructure Protection: Advanced Technologies for Crisis Prevention and Response

Springer

Editors
Tünde Anna Kovács
Bánki Donát Faculty of Mechanical and Safety Engineering
Óbuda University
Budapest, Hungary

Igor Fürstner
Subotica Tech—College of Applied Sciences
Subotica, Serbia

ISSN 1874-6519 ISSN 1874-6543 (electronic)
NATO Science for Peace and Security Series C: Environmental Security
ISBN 978-94-024-2310-5 ISBN 978-94-024-2308-2 (eBook)
https://doi.org/10.1007/978-94-024-2308-2

Proceedings of the Advanced Training Course on Security-Related Advanced Technologies in Critical Infrastructure Protection, with a View to Prevent Crises Belgrade, Serbia 10–16 June 2024

This Springer imprint is published by the registered company Springer Nature B.V.
The registered company address is: Van Godewijckstraat 30, 3311 GX Dordrecht, The Netherlands

Organization

Scientific Guarantors

Prof. Dr. Ing. Zoltán Rajnai, Bánki Donát Faculty of Mechanical and Safety Engineering, Óbuda University, Hungary
Dr. Ing. Igor Fürstner, Subotica Tech—College of Applied Sciences, Serbia

International Scientific Commitee

Prof. Zoltán Rajnai, Ph.D., Óbuda University, Hungary
Ing. Igor Fürstner, Ph.D., Subotica Tech—College of Applied Sciences, Serbia
Prof. Tünde Anna Kovács, Ph.D., Óbuda University, Hungary
Prof. Tamás Berek, Ph.D., Ludovika University of Public Service, Hungary
Prof. Petrica Vizureanu, Ph.D., Gheorghe Asachi Technical University of Iasi, Romania
Prof. Dražan Kozak, Ph.D., University of Slavonski Brod, Mechanical Engineering Faculty in Slavonski Brod, Croatia
Prof. Alexis Rusinek, Ph.D., Lorraine University, France
Prof. Sanja Maravić Čisar, Ph.D., Subotica Tech—College of Applied Sciences, Serbia
Prof. János Besenyő, Ph.D., Óbuda University, Hungary
Ing. Petar Čisar, Ph.D., University of Criminal Investigation and Police Studies, Serbia
Robert C. Castel, Ph.D., Israel National Parks Authority, Israel
Ing. Zlatko Čović, Ph.D., Subotica Tech—College of Applied Sciences, Serbia
Ing. Róbert Pintér, Ph.D., Subotica Tech—College of Applied Sciences, Serbia
Ing. János Simon, Ph.D., Subotica Tech—College of Applied Sciences, Serbia
Ing. László Gogolák, Ph.D., Subotica Tech—College of Applied Sciences, Serbia
Livia Sedmina, Ph.D., Subotica Tech—College of Applied Sciences, Serbia

Piroska Stanic Molcer, Ph.D., Subotica Tech—College of Applied Sciences, Serbia
Zoltán Pap, Ph.D., Subotica Tech—College of Applied Sciences, Serbia
Ing. Zoltán Nyikes, Ph.D., Milton Friedman University, Hungary
Ing. László Tóth, Ph.D., Óbuda University, Hungary
Ing. Jovan Trajkovski, Ph.D., University of Ljubljana, Slovenia
Ing. Norbert Daruka, Ph.D., Óbuda University, Hungary
Darko Trifunovic, Ph.D., Institute for National and International Security (INIS), Serbia
Ing. Branislav Todorovic, Ph.D., Institute for National and International Security (INIS), Serbia
Géza Gulyás, Ph.D., Ludovika University of Public Service, Hungary
Darjana Sredić, Ph.D., PIM University, Faculty of Philosophy, Bosnia and Herzegovina
Ljubica Janjetović, Ph.D., PIM University, Faculty of Philosophy, Bosnia and Herzegovina

Preface

The protection of critical infrastructure stands as one of the most pressing challenges of our era. In a world marked by rapid technological advancements, increasing interdependence, and unforeseen crises, the resilience and security of essential systems have become paramount. As societies grow more complex, the vulnerabilities of critical infrastructures to both natural and human-made threats demand innovative and robust solutions. This book seeks to address these challenges by bringing together diverse perspectives and cutting-edge research to provide a comprehensive understanding of the field.

This publication is a part of the NATO Science for Peace and Security Series, a platform dedicated to advancing collaborative research and innovation in service of global stability and safety. By integrating knowledge from different fields such as mechatronics, informatics, management, and related fields, this book offers a holistic view of the strategies and technologies required to enhance resilience and response capabilities.

The contributions presented in this volume are the result of dedicated research efforts by an international group of authors. These researchers hail from a variety of countries, with a significant representation from Central Europe and the Balkans, regions that have historically faced unique challenges in infrastructure protection. Their work reflects a diverse array of methodologies and perspectives, demonstrating the strength of international collaboration in addressing global issues. By bringing together expertise from different cultural, geographical, and disciplinary backgrounds, this book embodies the spirit of cooperation that is essential for advancing scientific understanding and practical solutions.

The international scope of this book also reflects the global nature of critical infrastructure challenges. While specific vulnerabilities and threats may vary across regions, the underlying principles of resilience, adaptability, and proactive planning are universal. The authors' contributions provide valuable insights into how different countries approach these issues, offering opportunities for cross-border learning and the sharing of best practices.

The workshop held in Belgrade—Serbia in 2024 that inspired this book was not only a venue for presenting research but also a forum for fostering collaboration and building networks among experts. The discussions and exchanges that took place during the workshop laid the groundwork for the chapters that follow, each of which represents a unique piece of the puzzle in the quest to secure critical infrastructure.

In assembling this book, the aim has been to provide readers with a resource that is both intellectually rigorous and practically relevant. The chapters span a wide range of topics, from advanced sensor technologies and predictive maintenance techniques to crisis management frameworks and policy recommendations. Each contribution is grounded in thorough research and reflects the latest developments in its respective fields. Together, they offer a multifaceted view of the challenges and opportunities in critical infrastructure protection, making this volume an indispensable resource for researchers, practitioners, and policymakers alike.

Hopefully, this book will serve not only as a repository of knowledge but also as a catalyst for further research and collaboration in the field of critical infrastructure protection. The challenges are great, but so too are the opportunities to make a lasting impact. By working together and embracing the power of interdisciplinary and international collaboration, we can build a future where critical infrastructure is not only protected but also empowered to support the needs and aspirations of societies worldwide.

Subotica, Serbia
January 2025

Igor Fürstner

Contents

Chapter 1
Exploring the Intersection of Human Movement and Collaborative Robotics: Challenges and Opportunities in Computational Modeling

Nada El Yasmine Aichaoui and Tünde Anna Kovács

Abstract By focusing on the challenges and promise of collaborative robotics (cobotics), the paper poses the communicative dimensions of human movement, introducing the concept of mastery and examining the psychological aspects of collaborative scenarios. Transitioning to computational modeling, it discusses classical methods that provide technical insights into approximately modeling human movement, which provides visions into utilizing these models to enhance the intelligibility of robots during their interactions with humans. This work includes results of applied experiments, detailing the evaluation of walking through step estimation and foot position detection during assisted ambulation, bridging theoretical discussions with real-world applications. Blending history, theory, and practical insights provides a comprehensive exploration of the dynamic realm of human–robot interaction and computational modeling.

Keywords Collaborative robotics · Human movement · Computational modeling · Human–robot interaction

1.1 Introduction

Robotics is a relatively young science compared to other disciplines. A robot is an assembly of technologies such as motors, sensors, and computers. Robotics was born when these technologies became mature enough to be combined. The field of robotics has the unique characteristic of relying on technological developments to propose and

N. E. Y. Aichaoui (✉)
Doctoral School on Safety and Security Sciences, Obuda University, Budapest, Hungary
e-mail: nada.aichaoui@phd.uni-obuda.hu

T. A. Kovács
Bánki Donát Faculty of Mechanical and Safety Engineering, Óbuda University, Budapest, Hungary

T. A. Kovács and I. Fürstner (eds.), *Critical Infrastructure Protection: Advanced Technologies for Crisis Prevention and Response*, NATO Science for Peace and Security Series C: Environmental Security, https://doi.org/10.1007/978-94-024-2308-2_1

study new applications [1]. This dependency makes the subject exciting because it is constantly evolving. The maturity of robotics for specific tasks in an environment is undeniable. The industry greatly exploits these technologies to improve productivity [2].

Among the current challenges in robotics development, situations where the robot physically interacts with humans are very promising. The common use of robots in industrial settings involves protecting people through a cell isolating the device [3]. There are still many tasks to automate, and humans remain operators capable of handling complex situations that require responsiveness that robotics cannot fully address. For this reason, cobotics are emerging [4, 5]. It allows for scenarios where humans collaborate with robotic partners. Although very promising, this type of robotics is still in its early stages. The issues raised by such interactions are related to the safety of the human operator around the robot.

Surprisingly, the fundamental problem of this type of robot is older than robotics itself. It originates from the theory of cybernetics. The goal is to propose systems capable of predicting and considering the human reaction to the action of the robotic system [6]. Cybernetics was invented during World War II, with Norbert Wiener's creation of an automatic pilot system called the "predictor," an antiaircraft firing device that anticipated the pilot's evasion trajectory in the path of the defense projectile. Norbert Wiener then theorized and created this research field with the foundational work in 1948, "Cybernetics or Control and Communication in Animals and the Machine," in which he announced a theory of control and communication for both animals and machines. It is in this theory that research establishes itself for robotics in interaction with humans [7].

The concept of dexterity is a performative prism. The limitation of this approach is that it diverges from the reality of human–human cooperation. Indeed, the motivations behind human movement are not always aimed at precision or efficiency. For instance, an artistic dance gesture may be motivated by an aesthetic choice that the dancer wishes to convey to the audience. Feth et al. [8] studied the role of haptic feedback in interaction, observing an indication of communication with an energy expenditure higher than necessary for the movement. Thus, there seems to be a communication goal that overlays the task to be performed. Humans appear capable of transforming a performative movement to give it an informative dimension during collaboration [9]. One particularity is the ability of human–human collaborations to dynamically evolve during interaction. Similarly, the social context can change the gesture; [10] demonstrated that one could alter their trajectory to convey an information signal ("signaling") about the ongoing task. The example of "signaling" shows that it is possible to modify the movement to add objectives to it. However, it has been shown that human–human collaboration strategies involve a dynamic change in the degree of involvement/control of each person in the task. Thus, in cases where guiding and being guided roles are necessary, humans can naturally shift from being guided (follower) to guiding (leader) to optimize collaboration [11, 12].

One effect of collaborations is the sense of collective agency (we-mode), which presupposes a movement prediction system involving humans. The sense of agency is defined by [13] as the feeling of controlling one's actions and their consequences on the world ("controlling one's actions, and, through them, events in the outside world"). Indeed, [14] shows that collective agency (we-mode) is observed when an agent interacts with a human but not observed when it involves an artificial agent. When observing one's own movement, it is an actor, and this movement is clearly human, so the cognitive load is low. An image of the latter is measured by the subjective estimation of time spent in action. This measure is called "intentional binding." It evolves significantly based on the degree of involvement in the movement. In other words, when one is active, the perceived time is shorter than when the movement is endured. The hypothesis is that there are movement predictions that are very accurate when it comes to our own gestures [15]. In collective agency [14], the prediction system seems to be engaged in interaction with another person. These predictions also prove to be very accurate when it comes to a joint action with another human. This leads to the conclusion that the predictor is built on the human property of movement, making it a form of movement prediction system that creates motor imagery and allows prediction, thereby reducing mental load.

The uniqueness of this research paper lies in its attempt to identify the results of methods that can operate under realistic conditions for users the possibilities of modeling are limited by the condition of the measured subjects. Thus, individuals with pathological movement may not be able to perform a large number of experimental trials/recordings. Possible models are those capable of being effective despite a small amount of data.

1.2 Computational Modeling of Movement

This paper aims to define methods that deepen classical modeling by closely approximating the observed movement. Developing such models relies on evaluating their ability to perform human-like tasks [16], thus, such an approach also allows for selecting quality criteria. This work proposes methods that are generally capable of adapting to the modeling of pathological movement based on collective experiences. Particular attention is given to the ability of the proposed methods to be applied to movement modeling with a limited amount of data.

1.2.1 Classical Computational Models Operating with Few Data

Generative methods. Provide information derived from movement by comparing it with its measurement. Principal Component Analysis (PCA) is an approach based on linear algebra. It assumes that the measured data is governed by a linear function that can be represented by a matrix. The method first involves calculating the parameters of this matrix that make this function representative of the relationship between inputs and outputs [17]. The PCA strategy involves arranging the eigenvectors of this matrix by the influence they have on the accuracy of the function. Accuracy is obtained by measuring the error resulting from applying the vector to the input data compared to the true output data [18]. The eigenvector that describes the majority of the data behavior will be the first component (the first vector), and so on for the subsequent components. It is thus possible to reduce the dimension of the function by keeping a finite number of principal components [19].

Moreover, a computational model can be generated using hidden Markov models. The system is represented as having a probability given the current state [20]. The goal here is to find the stochastic parameters of transition between different states, which can be referred to as a hidden Markov Model (HMM) because it assumes that the decision is memoryless, meaning that the information at the moment determines the transition, represented by a probability [21]. With this assumption, the complexity of probabilities is reduced to an estimate based on current data, and the states are defined one after the other, thus speaking of a hidden Markov chain [22].

Discriminative methods. The Heuristic discriminative methods (state machines) determine the category of the data. A simple approach to determining the category of a data vector can be through a simple heuristic [23]. Thus, it is possible to define an "algorithmic reasoning" resulting in the membership of a dataset in one category or another. There are different ways to represent the algorithm. One can consider describing the processing by a flowchart (data processing diagram) or by a state machine. This algorithm performs tests and measurements on the provided data to determine the category to which the data belongs [24, 25].

The following table highlights comparative aspects of each computational method, offering a clear overview of their purposes, strengths, and limitations. These models provide varying approaches to handling data, with generative models like PCA and HMM focused on data generation and state transitions, and discriminative models like state machines determining categorical memberships (Table 1.1).

1.2.2 Evaluation of Walking Through Step Estimation

Several strategies are possible to use to predict movement. Firstly, understanding the possibilities related to human movement is highly useful. There is a frequency property. Human movement has limited dynamics. The walking frequency is reported to be between 0.5 and 5 Hz according to [26], while in other experiments, maximum frequencies are observed to increase up to 12 Hz. Having a maximum limit in natural frequencies implies limits in accelerations and possible limb controls. Thus, this assumption was used in working on foot detection and step length estimation [27]. The goal is to reliably detect the position of the feet during assisted ambulation with a walker. The first experiment Fig. 1.1 involves installing a depth camera (ADS, Active Depth Sensor) on a passive walker and testing it with elderly individuals in a geriatric ward, as well as evaluating it with subjects present in the laboratory.

Table 1.1 Comparison of classical computational models operating with few data [18, 21, 23]

Type	Method	Key feature	Advantages	Limitations
Generative	PCA	Reduces dimensionality by identifying principal components	Simplifies data by focusing on significant variance	Assumes linearity; less effective with non-linear data
	HMM	Models system behavior with probability transitions between states	Captures temporal and sequential patterns in data	Assumes memoryless transitions; complex for large state spaces
Discriminative	Heuristic State Machines	Classifies data into categories based on a predefined heuristic	Simple to implement and interpret	Lacks flexibility; may not generalize well with more complex data

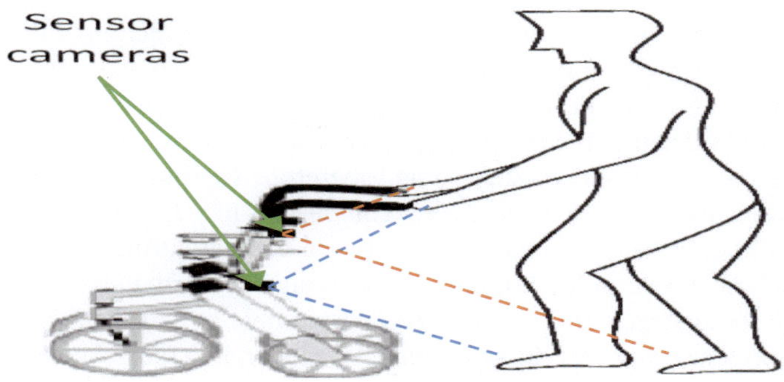

Fig. 1.1 Sensor-based gait analysis through sensor cameras (author)

From this usage [28–30], it is possible to observe the direction taken by the user as well as their speed while recording the 3D map of the foot area. Based on these observations, the robust proposal is to estimate the position and orientation of the foot when it is placed. Using this information precisely, it is possible to accurately estimate the direction taken by the user. To obtain these positions and orientations of both feet during walking with a passive walker, image processing involves segmenting the movement, examining the silhouette of the feet, and identifying the orientation of the feet. To maintain stable conditions, the foot must be at ground level. This work assumes that the ground is flat. It is thus possible to identify the parameters of the ground initially. Once this condition is validated, a data zone is chosen parallel to the ground a few centimeters above. This zone will be where the foot is supposed to appear when placed on the ground during walking with the walker.

Next, the data from this zone are segmented, and the two most voluminous segments are selected. Thus, by filtering the data to obtain the "blobs" of the feet.

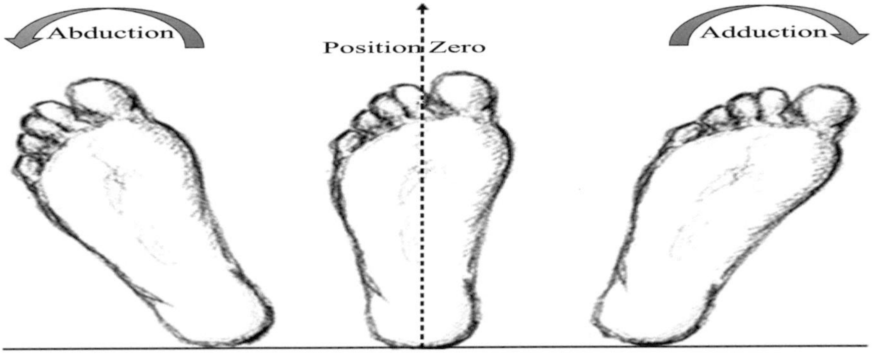

Fig. 1.2 Movement adduction/abduction of the relative angle of the foot (author)

The processing to identify the orientation of the feet involves a principal component analysis of the data [31, 32]; Fig. 1.2 shows the projection of the first component (the first vector) onto the ground plane gives an axis in the direction of the foot. This axis is the relative angle of the foot concerning the robot.

A limitation of this work is the need to handle cases of joined feet and the loss of a foot. An inspired heuristic from tracking involves considering that a too-significant change is an error (an inconsistency) and therefore waiting for coherent data to reappear to resolve these specific cases. This results in a foot position evaluation system. It allows measuring the distance between the feet during walking, which is one of the clinical criteria for assessing the walking of elderly individuals [33, 34].

Through these stages of study, it is indeed possible to explore the modeling of human movement using specific properties. Moreover, it is more likely to conclude that these models have thus allowed, starting from simple models, to represent complex movement in a detailed manner in terms of force and position. Furthermore, modeling with variable impedance, in particular, can be defined as a modeling approach that explores the power of using Gaussian noise in the person's movement to obtain system parameters [35]. Indeed, Gaussian noise being independent of time, it can allow examining data from a movement phase across multiple movements. However, introducing Gaussian noise into the movement of a robot physically attached to a subject requires a high level of assurance regarding user safety. Although it is more considered for its ability to work with patients, the risk still exists and acts as a deterrent to such detailed modeling [36].

Basically, modeling tools require the ability to adapt parameters according to the property. Thus, variable parameters based on the phase, detection, and interpolations according to the classical walking model have been explored. Finally, it is always possible to move towards non-linear tools such as neural networks [37, 38].

1.2.3 Computational Modeling of Human–Robot Kinesthetic Interaction

Understanding the dynamics of human–robot interaction is truly essential. To achieve this, it is necessary to conduct experiments comparing physical human–human interactions and physical human–robot interactions [39]. However, to construct the human–robot interaction program, it is crucial to have analyzed human–human interactions. For clarity, the results of these experiments can be identified through the comparison of human–human and human–robot interactions. It is important to note that the parameters and design of the virtual partner were determined through preliminary human dyads before the presented results [40, 41].

The setup for conducting a human–human study implements a teleoperation scheme [42]. The virtual partner is a state machine designed to be a faithful representation of the observed negotiation. The first phase of this state machine involves deciding whether the partner takes the initiative. To do this, it must conduct a study on various cumulative criteria that best characterize the interaction [43], considering geometric parameters such as position, speed, or average acceleration, examining the time at which it is possible to start the movement, integrating the minimum jerk property into the generated movement, and finally, considering a threshold criterion, Fig. 1.3 shows the initiate stage chart for this first phase. All the mentioned criteria were studied in terms of their ability to predict the final choice. It turns out that the criterion of the first crossover, defined as a switch that determines the correct side because the movement started in that direction first, is significant. As a result, the experiment defined a dead zone where if the cursor displayed on the screen exceeds this zone, the final destination is defined as the side of the cursor, and the ability to predict the final result is over 94% in this case [44, 45].

This approach demonstrates the potential of the simplified question of kinesthetic communication in one direction. The experiment aims to gain a deeper understanding of the effect of movement on humans. Knowing that it is conducted in an environment

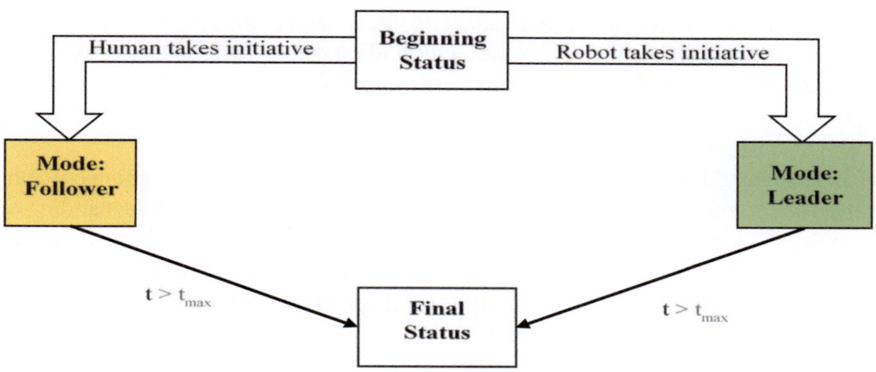

Fig. 1.3 Virtual partner state machine for left–right negotiation (author)

that is sufficiently precise, fast, and controlled to conclude the impact these artificial systems have on humans. An important result to note is the virtual partner's ability to act as effectively as a human.

1.3 Conclusion

This paper has explored the intricate dynamics of human movement within the context of collaborative robotics, focusing particularly on the concept of dexterity as both a performative and communicative tool. By examining dexterity through a dual-lens—one that considers both the mechanical precision and the subtle communicative gestures inherent in human motion—we have endeavored to bridge the gap between human–human cooperation and and human–robot interaction.

Our analysis began with a critical examination of existing models in human movement and computational modeling, identifying their limitations in fully capturing the complexity of human dexterity. We argued that traditional approaches, often centered around efficiency and task completion, fail to account for nuanced, non-verbal communication that is pivotal in human–human collaboration. This led us to propose an enriched understanding of dexterity, not just as a technical skill but as a medium through which humans convey intentions, adapt to their robotic counterparts, and co-create a collaborative environment.

In developing this argument, we drew upon insights from both robotics and psychology, highlighting how human movements in a collaborative setting often serve multiple purposes—such as signaling intentions, responding to the robot's actions, and facilitating a more intuitive partnership. This perspective challenges the prevailing notion that human–robot collaboration should aim to replicate human precision and instead suggests that the success of such collaboration hinges on the robot's ability to interpret and respond to the communicative aspects of human movement.

Through case studies and experimental data, we demonstrated that enhancing robotic systems to recognize and adapt to these subtle movements can significantly improve collaboration outcomes. Our findings underscore the need for future research to focus on developing algorithms and systems that can understand and replicate the communicative nuances of human dexterity, thus moving beyond a purely mechanistic approach to robotics.

In summary, this paper calls for a rethinking of our approach to designing collaborative robots, encouraging the field to adopt a more comprehensive understanding of dexterity—one that combines both functional and communicative aspects. This shift would pave the way for future human–robot partnerships that are not only more efficient but also more seamless and intuitive, resulting in more successful and balanced interactions across different environments.

References

1. Nguyen TH, Kovacs L, Nguyen TD (2024) The development of robotic manipulator for automated test tube. Acta Polytech Hung 21(9)
2. Sabanovic S (2010) Robots in society, society in robots: Mutual shaping of society and technology as a framework for social robot design. Int J Soc Robot 2(4):439–450
3. Marecek-Kolibisky M, Janik S, Mikva M, Szabo P, Czifra G (2024) Human-machine co-working for socially sustainable manufacturing in industry 4.0. Acta Polytech Hung 21(2)
4. Simoes AC, Pinto A, Santos J, Pinheiro S, Romero D (2022) Designing human-robot collaboration (HRC) workspaces in industrial settings: a systematic literature review. J Manuf Syst 62:28–43
5. Treusch P (2020) Robotic knitting: re-crafting human-robot collaboration through careful coboting, transcript Verlag, , p 166
6. Suciu DA, Dulf EH, Kovacs L (2024) Low-cost autonomous trains and safety systems implementation, using computer vision. Acta Polytech Hung 21(9)
7. Wiener N (2019) Cybernetics or control and communication in the animal and the machine. MIT Press
8. Feth D, Groten R, Peer A, Buss M (2009) Control-theoretic model of haptic human-human interaction in a pursuit tracking task. In: RO-MAN 2009-the 18th IEEE international symposium on robot and human interactive communication, IEEE, pp 1106–1111
9. Kovari A (2024) Industry 5.0: generalized definition, key applications, opportunities and threats. Acta Polytech Hung 21(3)
10. Pezzulo G, Donnarumma F, Dindo H (2013) Human sensorimotor communication: a theory of signaling in online social interactions. PLoS ONE 8(11):e79876
11. Reda A, Benotsmane R, Bouzid A, Vásárhelyi J (2023) A hybrid machine learning-based control strategy for autonomous driving optimization. Acta Polytech Hung 20(9)
12. Jarrasse N, Sanguineti V, Burdet E (2014) Slaves no longer: review on role assignment for human–robot joint motor action. Adapt Behav 22(1):70–82
13. Haggard P, Chambon V (2012) Sense of agency. Curr Biol 22(10):R390–R392
14. Obhi SS, Hall P (2011) Sense of agency in joint action: influence of human and computer co-actors. Exp Brain Res 211:663–670
15. Kinany N, Khatibi A, Lungu O, Finsterbusch J, Büchel C, Marchand-Pauvert V, Doyon J (2023) Decoding cerebro-spinal signatures of human behavior: application to motor sequence learning. Neuroimage 275:120174
16. Stergiou N, Decker LM (2011) Human movement variability, nonlinear dynamics, and pathology: is there a connection? Hum Mov Sci 30(5):869–888
17. Garello L, Lastrico L, Rea F, Mastrogiovanni F, Noceti N, Sciutti A (2021) Property-aware robot object manipulation: a generative approach. In: 2021 IEEE international conference on development and learning (ICDL). IEEE, pp 1–7
18. Nishimura Y, Nakamura Y, Ishiguro H (2020) Human interaction behavior modeling using generative adversarial networks. Neural Netw 132:521–531
19. Hansson A (2020) Understanding people movement and detecting anomalies using probabilistic generative models
20. Tanwani AK, Lee J, Thananjeyan B, Laskey M, Krishnan S, Fox R, Calinon S (2020) Generalizing robot imitation learning with invariant hidden semi-Markov models. In: Algorithmic foundations of robotics XIII: proceedings of the 13th workshop on the algorithmic foundations of robotics, vol 13. Springer International Publishing, pp 196–211
21. Manouchehri N, Bouguila N (2023) Human activity recognition with an HMM-based generative model. Sensors 23(3):1390
22. Nasfi R, Amayri M, Bouguila N (2020) A novel approach for modeling positive vectors with inverted Dirichlet-based hidden Markov models. Knowl-Based Syst 192:105335
23. Li S, Su W, Huang R, Zhang S (2020) Mobile robot navigation algorithm based on ant colony algorithm with A* heuristic method. In: 2020 4th international conference on robotics and automation sciences (ICRAS). IEEE, pp 28–33

24. Pernkopf F, Bilmes JA (2010) Efficient heuristics for discriminative structure learning of Bayesian network classifiers. J Mach Learn Res 11:2323–2360
25. Xu Y, Fern A, Yoon SW (2007) Discriminative learning of beam-search heuristics for planning. In: IJCAI, pp 2041–2046
26. Godfrey ACRMDOG, Conway R, Meagher D, ÓLaighin G (2008) Direct measurement of human movement by accelerometry. Med Eng Phys 30(10):1364–1386
27. Yao Y, Pan L, Fen W, Xu X, Liang X, Xu X (2020) A robust step detection and stride length estimation for pedestrian dead reckoning using a smartphone. IEEE Sens J 20(17):9685–9697
28. Lee SH, Kim J, Lim B, Lee HJ, Kim YH (2023) Exercise with a wearable hip-assist robot improved physical function and walking efficiency in older adults. Sci Rep 13(1):7269
29. Khan MS, Mandava RK (2023) Design of dynamically balanced gait for the biped robot while crossing the ditch. Acta Polytech Hung 20(7)
30. Tlalolini D, Chevallereau C, Aoustin Y (2010) Human-like walking: optimal motion of a bipedal robot with toe-rotation motion. IEEE/ASME Trans Mechatron 16(2):310–320
31. Rohith KK, Varma N, Sudheer AP, Joy ML (2020) Mathematical modeling and comparative study of 12-DoF biped robot using screw theory and Denavit–Hartenberg convention. In: Innovative product design and intelligent manufacturing systems: select proceedings of ICIPDIMS 2019. Springer Singapore, pp 979–989
32. Emken JL, Benitez R, Reinkensmeyer DJ (2007) Human-robot cooperative movement training: learning a novel sensory motor transformation during walking with robotic assistance-as-needed. J Neuroeng Rehabil 4:1–16
33. Fried LP, Tangen CM, Walston J, Newman AB, Hirsch C, Gottdiener J, McBurnie MA (2001) Frailty in older adults: evidence for a phenotype. J Gerontol A Biol Sci Med Sci 56(3):M146–M157
34. Sobhani A, Fadayevatan R, Sharifi F, Kamrani AA, Ejtahed HS, Hosseini RS, Mortazavi S (2021) The conceptual and practical definitions of frailty in older adults: a systematic review. J Diab Metab Disord 1–39
35. Davids K, Bennett S, Newell KM (2006) Movement system variability. Hum Kinet
36. Seehafer L, Morrison S, Severin R, Ness BM (2021) A multi-systems approach to human movement after ACL reconstruction: the cardiopulmonary system. Int J Sports Phys Ther 17(1):60–73
37. Liu Y, Bi Q, Zang X, Li Y (2020) Human-like walking of a biped robot actuated by pneumatic artificial muscles and springs. In: 2020 IEEE 16th international conference on automation science and engineering (CASE). IEEE, pp 1395–1400
38. Larson D, Vien V, Ness BM, Wellsandt E, Morrison S (2022) A multi-systems approach to human movement after ACL reconstruction: the musculoskeletal system. Int J Sports Phys Ther 17(1):27–46
39. Bartneck C, Belpaeme T, Eyssel F, Kanda T, Keijsers M, Sabanovic S (2024) Human-robot interaction: an introduction. Cambridge University Press
40. Thomaz A, Hoffman G, Cakmak M (2016) Computational human-robot interaction. Found Trends® Robot 4(2–3):105–223
41. Huber M, Rickert M, Knoll A, Brandt T, Glasauer S (2008) Human-robot interaction in handing-over tasks. In: RO-MAN 2008-The 17th IEEE international symposium on robot and human interactive communication. IEEE, pp 107–112
42. Ong KW, Seet G, Sim SK (2008) An implementation of seamless human-robot interaction for telerobotics. Int J Adv Rob Syst 5(2):1
43. Markkula G, Lin YS, Srinivasan AR, Billington J, Leonetti M, Kalantari AH, Merat N (2023) Explaining human interactions on the road by large-scale integration of computational psychological theory. PNAS Nexus 2(6):163
44. Castro A, Silva F, Santos V (2021) Trends of human-robot collaboration in industry contexts: handover, learning, and metrics. Sensors 21(12):4113
45. Fong T, Rochlis J, Currie N, Mishkin A, Akin DL (2013) Space telerobotics: unique challenges to human–robot collaboration in space. Rev Hum Factors Ergon 9(1):6–56

Chapter 2
Serverless Approaches to Incident Response in Critical Networks

Azar Mamiyev

Abstract Serverless computing offers the opportunity for improving incident response (IR) support within critical infrastructure networks, and this paper explores the maximisation of that potential. However, traditional IR approaches based on large amounts of manual processes and on-premises infrastructure lack the ability to scale with the increasing scale, velocity, and sophistication of today's cyberattacks. The paper posits that the elasticity, automation, and pay per use nature of serverless computing in comparison is a more agile, scalable as well as a cost-effective proposition. It shows how serverless functions are used to implement automated threat detection and response pipelines alongside existing Security Information and Event Management (SIEM) systems. The paper in particular shows how serverless functions can simplify data ingestion, threat intelligence enrichment, live feed analysis, auto responding and reporting. Moreover, the paper further demonstrates the application of serverless capabilities to post incident forensics in critical infrastructure environments, specifically showing the use of serverless to build event driven forensic systems which enable rapid, comprehensive and automated investigations. An example of real-world application is introduced through a case study of an energy plant. Integrating serverless architectures helps organisations get faster incident response, less downtime and better protection of important services.

Keywords Serverless computing · Incident response · Critical infrastructure · SIEM · Automation · Security · Forensics · Cybersecurity

2.1 Introduction

Today, critical infrastructure systems such as electric power grids, water treatment facilities and financial and healthcare institutions among others are more interconnected than ever, spanning greater distances globally and with more vulnerable

A. Mamiyev (✉)
John Von Neumann Faculty of Informatics, Obuda University, Budapest, Hungary
e-mail: azarmamiyev@stud.uni-obuda.hu

T. A. Kovács and I. Fürstner (eds.), *Critical Infrastructure Protection: Advanced Technologies for Crisis Prevention and Response*, NATO Science for Peace and Security Series C: Environmental Security, https://doi.org/10.1007/978-94-024-2308-2_2

control points being exposed to increasingly advanced cyber-attacks. These essential services are being attacked by attackers using advanced tactics: ransomware, supply chain compromises and advanced persistent threats (APTs) to break in, with sometimes dire results. A reminder that vulnerability comes in the form of Colonial Pipeline's 2021 attack, whereby disrupting fuel supplies across the southeastern US and incurring millions of dollars in ransom. Unlike other attacks, today's attacks are not only about stealing our data, but they can also disrupt critical operations, undermine public safety, and cost billions of dollars [1]. But traditional incident response (IR) methods, relying on manual processes and protected-on premise infrastructure, are quickly becoming unable to keep up. Manual analysis and remedy is inherently slow, error prone, and difficult to scale in response to the quickly evolving threat landscape. The reality is that on prem security infrastructure is both up front and ongoing investment, constraining the agility and scale. Additionally, processing the sheer volume, velocity, and variety of security data produced by modern critical networks is challenging, and more importantly, obviating timely and effective response is severely constrained.

Serverless computing is proposed in this chapter as a transformative approach for incident response in critical network environments to address the critical need for more agile, scalable, and cost friendly security solutions. This means that serverless platforms by nature are elastic and charge based on the pay per use model, allowing us to quickly deploy and scale security functions on demand and from the comfort of not having to worry about the underlying infrastructure [2]. Serverless computing's event driven nature makes responding to security events almost an automated process, dramatically reducing response times and mitigating the effects of attacks. However, serverless computing provides an attractive solution to this problem through the ability to design highly scalable and automated incident response pipelines. In particular, security logs from a variety of sources can be ingested and analysed, correlated around events to detect security compromises and potentially threats, and automated responses invoked based on pre-defined rules, and even used in post incident forensic investigations using serverless functions. This granular, function-based approach enables organisations to configure and scale their incident response resources in a way that matches their unique requirements and adjusts in reaction to evolving threat environments. Serverless computing's agility and cost effectiveness make it particularly well positioned to address security operations in critical net-work environments, which are inherently dynamic and resource intensive. The limitations of traditional SIEM systems and an examination of the role are then discussed.

2.2 Security Information and Event Management (SIEM)

Organisations focussed on enhancing their cybersecurity posture are now relying on Security Information and Event Management (SIEM) systems. SIEM systems supply the machinery to centralise security collection and analysis so that organisations can better find and regulate security threats [3]. But traditional SIEM architectures are struggling to keep up with the changing threat landscape and the ever-increasing demands of the complex critical networks [4]. This explores the core functionality of SIEM systems, and examine their limitation for tackling the current day cyber threats, and the possibility of serverless computing to enhance SIEM capabilities and aid in a more agile and effective incident response.

2.2.1 Core Functionalities of SIEM Systems

Security Information and Event Management (SIEM) systems are fundamental to security operations centres (SOCs), which are a mechanism for aggregating, analysis and correlation of security logs from disparate sources across an organisation's network. In traditional incident response there are several key functions that SIEM systems perform [3]. All of these include log aggregation from various sources i.e. from firewalls, intrusion detection systems (IDS), servers, endpoints etc. Data would be normalised to a common format out of which analysis and correlation could be made efficiently [5]. Rule based correlation engines in SIEM systems are used to detect such patterns and anomalies indicative of security incidents. With this correlation, it generates alerts that will notify security personnel of a potential threat. Additionally, many SIEM platforms also integrate threat intelligence feeds to fill voids in the SIEM capabilities and provide context to events [3]. Firstly, SIEM systems also help with compliance auditing as it generates reports and dashboards to ensure security posture from a compliance perspective and that this requirement's compliance is met. Traditionally, SIEM systems have been crucial to the visibility into security posture, identifying malicious activity, and beginning the process around how to response to an incident.

2.2.2 Limitations of Traditional SIEM in Critical Networks

Yet, the evolution of the cybersecurity space filled with critical network environments has exposed some key limitations in existing SIEM deployments. The ever-increasing volume, velocity, and variety of data generated by modern networks, often referred to as the "3Vs" of big data, pose a formidable challenge. In its handling of such massive data influx, traditional SIEM architectures, based on on-premises infrastructure and relational databases, struggle to scale efficiently. As a result, this scalability bottleneck causes the performance degradation, delayed alerts, and eventually less time available for the timely detection and response to incidents [4]. At the same time, traditional SIEM systems are prohibitively expensive to scale, as it is very expensive to buy hardware, software licences, and create specialist personnel. In addition, the integration of diverse and relatively often proprietary log formats provides data from a multitude of security tools proves to be a major challenge. To ensure reliable and efficient ingestion of data at scale, there are complex parsing and transformation of the data [5]. More critically, traditional SIEM systems don't have the real time processing and automation capabilities needed to handle today's dynamic threat environments. It makes it difficult for rapid response whenever an attacker has a long dwell time; they had time for an impact beyond measures we can take. In most cases, basic automated alerting is provided but the activation of more complex automated response actions such as shutting off infected systems or blocking malicious traffic is often complex and difficult to deploy and manage within traditional SIEM platforms. The lack of automation requires a significant amount of manual intervention that increases response times as well as placing an exceedingly high burden on security personnel. Security Orchestration, Automation and Response (SOAR) platforms seek to conquer this gap by combining and automating several Security tools and process, however, integrating these with typical SIEMs can become a headache. While SOAR is essentially a software automation solution, serverless computing, with its inherent automation capabilities and seamless integration with other Cloud native services, is well suited for enhancing or even replacing some solution areas of SOAR functionality. SIEM alerts can in turn be used to trigger serverless functions which respond to automatic actions, speeding incident workflows and automating out manual inputs. Combining serverless computing with SIEM systems gives much advantage over relying on the conventional SOAR systems. serverless functions provide a more granular and flexible approach to automating in your organisation which enables organisations to build a custom workflow based on that organisation's needs. The ability to configure granular control of incident response tasks enables more precise and efficient automation of incident response tasks and minimises risk posed by an overly broad, automated action. Additionally, given that serverless computing is entirely scalable and cost efficient, it becomes a compelling option for organisations with limited resources, and might not be able to afford the licencing and infrastructure expenditure required to set up a full fledge SOAR deployment. Some serverless functions, combined, or instead, can augment or even replace aspects of SOAR functionality, giving a faster, cheaper way to automate incident response.

2.2.3 The Advantages of Integrating Serverless with SIEM

SIEM system integration with serverless computing brings benefits to it when contrasted to solely relying on conventional SOAR platform or solely relying on conventional SIEM. Serverless functions offer a more granular and flexible automation mechanism which allows organisations to build custom, event-based flows tailored to their requirements [6]. This granular control provides much greater control over automation of incident response tasks compared to broad force automation, reducing the risk user can retaliate against themselves with unwanted broad automation of incident response tasks. In addition, serverless computing provides the scalability and cost-effective element where your organisation can be tight in resource constraints without having to run the licencing and infrastructure costs of running full-fledged SOAR, scaling a traditional SIEM infrastructure [7]. A SOAR capability rich enough to address all security incident needs can be too much for some situations to manage, but Serverless functions can fill gaps or completely replace these parts of SOAR functionality to push automated incident response into a more agile, more cost-effective path. Furthermore, serverless helps alleviate scaled and cost constraints of traditional SIEM deployments by way of providing on demand compute power, using a pay per use model.

2.2.4 The Need for a New Approach

The limitations of these approaches, combined with the benefits inherent within serverless integration, make the need for a new approach to incident response in critical networks more pressing: one that addresses the mountains of scale and speed, and the ensuing automation issues. With such inherent scalability, elasticity, event driven architecture, serverless computing is a powerful solution. Serverless functions have an on-demand nature for the data volume, which means it can automatically scale up or down based on the data volume vs. the typical scalability bottlenecks that SIEMs have. In addition, the serverless platform pay-per-use pricing model reduces the cost and complexity of maintaining and scaling on-premises infrastructure [8]. Since the event-driven architecture of serverless computing lends itself very well to automating real time responses to security events, organisations are that can react quickly and effectively to developing threats. By employing serverless technologies, organisations can enhance their own incident response capabilities to levels previously only dreamed of within traditional SIEM deployments.

2.3 Serverless-Based Threat Detection and Response Pipelines

Current security architectures typically cannot handle the dynamic environment of modern cyber threats well. Having solid scalability, flexibility and low cost, serverless computing becomes a perfect alternative when constructing robust and responsive threat detection and response pipelines. In this section, we look at the fundamental concepts of serverless computing to understand how the capabilities can be used to build automated incident response workflows.

2.3.1 Serverless Computing Fundamentals

Serverless computing is a cloud computing execution model which enables developers to build and run applications without thinking about servers. Computation resources are allocated dynamically to the cloud provider. By deploying code as individual functions that are triggered by events and executed on demand, developers are no longer provisioning and managing servers. A defining characteristic of serverless computing, this event driven architecture is really well suited for runtime security eventing, providing a real time response to security events as they occur. One of the two other key benefits of serverless computing is automatic scaling [8], and pay per use pricing [8]. Serverless functions automatically scale with fluctuating workloads, so that resources are intentionally and automatically allocated as demand warrants [9]. It eliminates the need for manual capacity planning, and even during peak loads, offers consistent performance to handle bursts of activity in security incident. Organisations pay for the compute time their functions consume, thus paying only for what they consume [8], resulting in lower operational costs than running always-on servers and thus a particularly attractive advantage for security operations, which can be resource intensive.

2.3.2 Architecting Serverless Incident Response Pipelines

Serverless functions can be orchestrated to build threat detection and response pipelines that scale to solve the limitations of legacy security architecture. Incident response would consist of these linked functions that serve the specific functions in the overall incident response workflow. Say, a serverless incident response pipeline could start with a function, which ingests security logs from a firewall, for example. The function would trigger another function, which would enrich the log data with threat intelligence. The enriched data could then be fed to a subsequent function which would analyse the events to find correlations amongst events to help identify potential security incidents. If an incident is seen, another function could execute automatically to handle it, such as perhaps isolating an affected system or blocking the malicious traffic. Finally, another function could produce notifications and alerts for security personnel. The flexibility of such a modular approach offers the ability to customise the pipeline to their unique requirements. What's more, serverless platforms also come with tools and services enabling orchestration of these functions, managing data flow between them, monitoring overall health and performance of that pipeline.

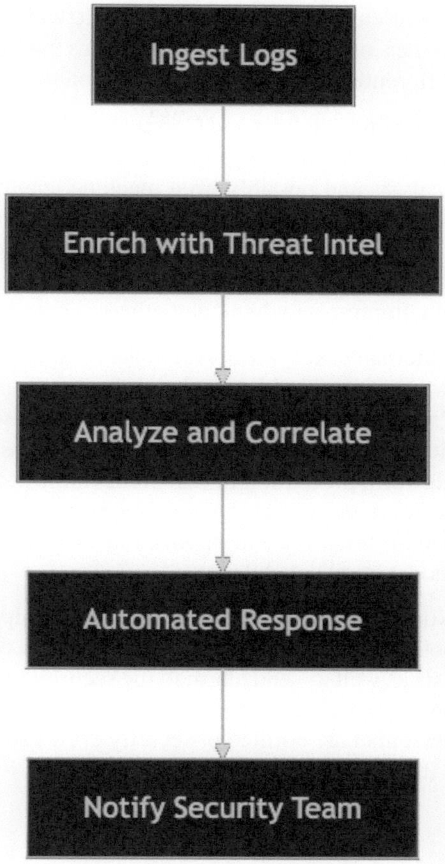

Fig. 2.1 Incident response pipeline with serverless functions

2.3.3 Key Components of a Serverless Incident Response Pipeline

Because of its lightweight nature, serverless functions can be orchestrated to form automated threat detection and response pipelines that can overcome the restrictions of typical security architectures. This set of pipelines consist of a series of functions which are executed in sequence to perform the overall incident response workflow. A typical serverless incident response pipeline might follow this architecture:

Ingestion and Enrichment: For the ingestion of security logs from the various sources (cloud platforms, security appliances, endpoint agents, SIEM systems and more), the pipeline starts with serverless-functions. These are the entry point that raw data has for security. Enrichment functions immediately following ingestion further enrich the raw logs with contextual information rich in threat intelligence feeds. It could be an enrichment of the seen activity to add known malicious IP address, domains, file hashes, or vulnerabilities. It provides this added context which enables the system to differentiate between benign and potentially malicious events with greater accuracy.

Real-time Analysis and Correlation: Functions perform real time analysis and correlation of enriched logs that then flow [5]. Through rule-based correlation, statistical analysis and possibly even machine leaning algorithms, these functions will identify patterns and deviations of malicious activity. Here real time processing is key to enable fast responses to emerging threats and to minimise dwell time. This stage additionally gives rise to proactive threat hunting ways of finding more subtle indicators of compromise which may otherwise not be discovered.

Automated Response: Automated response functions are triggered when a security event is identified during the analysis stage, which satisfies a set of predefined criteria. Pre-authorised actions are executed on those functions, and they depend on the security policies defined. Automated responses include isolating infected systems from network, prevention of malicious traffic at the firewall, disabling compromised user account, or even automating malware analysis. Automated responses save security people time, prevent application of inconsistent policies and frees them to tackle more complicated threats.

Notification and Reporting: Finally, our pipeline contains functions serving to create notifications and reports. It can also create detailed reports for investigation and analysis as well as send alarm to the security personnel through various channels (email, SMS, and chat platform). This final stage ensures a loop is closed as substantive personnel are notified of security events and records are kept keeping in compliance with and prepared for forensic purposes.

2.3.4 Benefits of Serverless for Threat Detection and Response Pipelines

Using serverless computing to build threat detection and response pipelines could provide attractive benefits, particularly in critical infrastructure environments. SOAR platforms have an incredible amount of automation but may not always be the ideal solution because of the cost and scalability limitations. A more granular, more flexible, and less expensive means of building automated response workflows is offered by serverless functions. This allows real time security event response by enhancing agility and responsiveness. In serverless, there are no servers to provision when needed—every time there is a new security log, threat intelligence update, or an anomaly detected, your function will be called instantaneously to begin processing those logs, updates, and anomalies. It is this real time responsiveness that is necessary to respond quickly to threats and minimise the effect it can have in critical infrastructure where downtime can potentially be devastating. Additionally, serverless provides significant scalability and elasticity required to cope with the large volumes of security data typically generated by mission critical infrastructure systems. Serverless functions are automatically scaled, so that the pipeline can process large datasets without compromising in performance anywhere in the pipeline, from peak load times to security incidents. The pay-per used model of serverless platforms also reduces cost and complexity to a great extent. Infrequent peak loads are now accommodated without having to pay for the expensive hardware and software needed to support them.

2.4 Serverless Integration with SIEM Systems

Building serverless security pipelines from scratch is possible, however, many organizations want to learn how to use their existing serverless functions within their already deployed Security Information and Event Management (SIEM) systems. In this hybrid approach, organizations benefit from the advantages of serverless computing yet continue spending on the existing secure-ty investments. We explore this large space of possible integrations for serverless functions with SIEM systems, showing compelling use cases for having both.

2.4.1 Integration Strategies

There are several strategies involved in integrating serverless functions with SIEM systems. A common way is to use functions triggered by SIEM. Lots of modern SIEMs out there have APIs or integration hooks to allow external systems (like serverless functions) to be fired in response to the sorts of predefined rules (and/or detected events). As an example, if the SIEM is configured to trigger a serverless function to isolate a potentially compromised host when a high severity alert is generated automatically. A function-initiated data enrichment is another strategy. Before the SIEM system stores the data, serverless functions that enrich the data can be added [6]. If you can invoke a serverless function to query a threat intelligence API and add context to incoming log entries, you can greatly improve the accuracy and speed of subsequent analysis in the SIEM. Real time streaming integration has a powerful solution for scenarios of high-volume, real-time data streams. You can use streaming platforms like Apache Kafka or Amazon Kinesis, and integrate them with your serverless functions to process and philtre the data before it gets to the SIEM, thereby reducing the load and allowing real time analysis. The approach is especially rewarding for working with large volumes of log data coming from cloud environments or high-speed network sensors. Finally, certain SIEM platforms allow you to extend the SIEM with the SIEM extension development as some give platforms support its extension development which means organisation can develop customised extension or apps using serverless functions. Tailoring the functionality of

SIEM with custom logic and integrations is possible to specific needs like automating a specific incident response workflow within SIEM environment. Some examples of these integrations are using AWS Lambda functions as Webhooks [10] with Splunk, so Splunk can trigger Lambda functions based on events. In the same way, Azure Sentinel can be integrated with Azure Functions via Logic App [11] to automate workflows via Sentinel's alerts to Azure Functions for specific actions.

2.4.2 Use Cases for Serverless SIEM Integration

Several compelling use cases show how serverless functions can be useful in conjunction with SIEM systems. One such use case is real time threat intelligence enrichment which is crucial for real incident response. As the SIEM is processing alerts, in real time serverless functions can also query the threat intelligence APIs to enrich security events with the most current information. It provides security analysts with immediate context that can be used to evaluate severity and relevance of alerts much more quickly and accurately with less information required. If applied towards automated malware analysis, it also is another valuable application. A Suspicious file detected by the SIEM can kick off a serverless function to send file (or sometimes other data) to cloud based malware analysis service. The analysis results are fed back into the SIEM and added to records to help with the investigation and response. The automation throws light on the malware analysis process and reduces the time for evaluating a possible threat. The final contribution addresses a critical challenge of existing SIEM deployments—SIEM capacity scaling. In case of high activity or security incidents, the SIEM can ingest extremely high amounts of data which can thoroughly surpass its capacity. And serverless functions can be used to dynamically scale the SIEM's processing power on demand, to be able to cope with loads without the performance degradation or delays in alert generation that would occur from either smaller instances or hardcoding such decisions. It gives much needed flexibility and cost efficiency in managing fluctuating workloads.

2.5 Serverless Framework Solutions for Automated Post-Incident Forensics

The dynamic and distributed nature of modern critical infrastructure make traditional digital forensics methods difficult for it to reliably support. The ephemeral and interconnected nature of these serverless deployment configurations makes manual, instance-centric approaches ill suited. Post incident forensics has come at a price, not only from the inefficiencies of on-call databases and page hits but also the time spent provisioning additional computers, numerous network firewalls, and the hardware itself. A vicious cycle of compatibility and instrumentation requirements keeps follow up forensics an unplanned investment. This critical shift applies strongly to critical infrastructure, where response and comprehensive analysis must occur quickly and comprehensively in order to minimise damage and maintain operational continuity. In addition, serverless computing's inherent scalability and cost effectiveness makes it an attractive solution for low resource critical infrastructure environments. In this section we explore how serverless frameworks enable this shift into automated forensics, and are used in the building of automated forensics tools geared for the special challenges of critical infrastructure protection.

2.5.1 *Event-Driven Forensics for Critical Infrastructure*

Traditional forensics often focuses on analysing individual systems—an "instance-centric" approach. This approach breaks in serverless environments due to the ephemeral nature of functions. While looking at individual instances fails to capture attacks spanning across several function or service boundaries. Currently, serverless frameworks require us to think in terms of event driven forensics, tracing function execution flow and analysing the triggering events throughout the entire system. With this paradigm shift, we find critical infrastructure work crucial because it enables investigators to trace execution flow across interdependent systems and to see security events as a whole rather than as disjointed pieces, as is the case with methods instance centrically focused on distributed systems. It makes it possible to understand much better the complex attacks.

2.5.2 *Event-Based Tracking and Monitoring for Enhanced Security*

Comprehensive logging and metadata capture is required to resolve many cases Ranger has worked on using effective event driven forensics. Audit trails are important in critical infrastructure, and need to be detailed. Logging capabilities are available with serverless platforms, but the configuration of serverless logging should be forensically relevant. It also covers function execution logs (start/stop times, resource usage), inputs and outputs (payloads, API calls), env vars, and network connexions (source/destination IPs, and port and protocol). With serverless functions able to automate this data capture, we get a very complete and very readily available activity record, extremely important for reconstructing attacks, identifying vulnerabilities, meeting compliance requirements, and in many cases more effectively than by hand keeping the logs. By automating, this removes some of the drag on security teams, and further allows for faster response time.

2.5.3 *Building Serverless Forensics Tools for Critical Infrastructure*

The power to automate forensic tasks with serverless functions is quite strong. These functions automatically collect and store evidence, analyse logs, and reconstruct attack timeline, creating faster response times than average manual processes based on trigger events or proactively. Fast containment is very important in critical infrastructure.

Automated System Isolation: If suspected of security compromise, serverless functions can automatically isolate the affected systems or subparts to isolate further damage or lateral movement. In sensitive environments, it's critical to isolate attacks really quickly and really automatically, minimising downtime, minimising cascading effects.

Real-time Anomaly Detection: Critical infrastructure components of servers in distributed environments are monitored by serverless functions that analyse real time data streams, using machine learning to raise an alarm of malicious behaviour. As an active method, it allows for early detection and early response to threat before incidents evolve.

Immutable Log Storage: Forensic evidence integrity has to be preserved. Immutable storage for logs from processes that run on servers is a crucial automation that automates storing logs to make tampering and chain of custody undetected and totally out of bounds from credible forensic investigations and then regulatory compliance requirements.

2.5.4 *Example Architectures: A Serverless Forensic Pipeline for Industrial Control Systems*

Real time analysis, along with an automated response, can be achieved with an industrial control system pipeline instead of a serverless pipeline which can reduce downtime and speed up incident resolution. Ingested by a serverless function, the data streams of the ICS sensors are produced in real time. This function triggers a second function, when suspicious activity is detected, to isolate the affected component. Meanwhile another function grabs all of the detailed logs and metadata and immutably store = root in a checkpoint. With this automated pipeline, damage minimization can be achieved in a rapid, evidence preserving manner. This RT capability is a major advantage to traditional forensic approaches that often take weeks to complete.

Example: Responding to a Cyberattack on an Energy Plant with Serverless Forensics

A modern energy plant based on serverless functions. The SCADA system of a targeted cyberattack uses malware and unauthorised access to try to disrupt power distribution. A serverless forensic pipeline is instrumental in responding:

Anomaly Detection and Initial Response: A serverless function is fed with real time SCADA data streams and uses machine learning models to detect anomalous behaviour (for example, abnormal voltage levels, control commands). When it detects attack, then it activates the separate functions for isolation and its evidence collection. This automatic response cuts substantially down on the time it takes to respond to an attack on critical infrastructure, which is incredibly important.

Automated System Isolation: Isolation function works with the plant's net-work infrastructure e.g. via API calls, to isolate affected SCADA components, inhibit lateral intrusion and minimise operational impact, thereby mitigating the spread of the power grid disruption. This automated isolation is much faster than manual intervention, it limits the attacker's access and cause of damage.

Automated Evidence Collection and Preservation: The evidence collection is where we capture the network traffic logs, create the memory snapshots on affected system and store all the collected evidence immutably (AWS S3 with object lock, Azure Blob storage with immutable storage policies etc.) If needed, this ensures data integrity and chain of custody (important to investigations and compliance). The automation helps collect the evidence quickly and consistently in the dynamic of a cyberattack.

Post-Incident Analysis: Once started, the analysis phase starts with another serverless function correlating the collected data. You can see it using tools or scripts to fingerprint malware signatures, analyse attacker TTPs and identify the entry point, compromise scope. Through a dashboard the results are presented. This analysis phase is automated, which in turn accelerates the investigation process and give security teams timely insight to help guide their response and recovery activities.

Recovery and Remediation: Recovery and remediation are informed by forensic analysis. This detailed information provides security teams with information for patching vulnerability, restoring systems, as well as implementing stronger security controls to prevent future incidents. They support compliance reporting and show that you are following regulatory guidelines.

2.6 Conclusion

During this exploration of serverless computing, its transformative power for incident response in the critical network environment has been revealed. But modern cyberattacks, aiming at our most vital systems, are becoming more sophisticated and more frequent, and traditional methods don't measure up. However, serverless computing presents as a highly compelling alternative to offer highly scalable, automated, and cost-effective solutions [6].

Examined were the limitations posed by traditional SIEM systems in dealing with the volume of data that is generated by modern networks. Serverless computing solves such limitations with on demand scaling, real time processing and at the same time optimised pricing. The integration of serverless functions with existing SIEM deployments should be the simplest and most practical way to extend security without ripping up existing infrastructure.

It documented how serverless functions can orchestrate automated threat detection and response pipeline. By decomposing incident response workflows into granular, event driven functions then, we are able to introduce new agility and responsiveness to our security operations. Resources are always scaled up dynamically, protecting critical systems during the peak loads or large-scale incidents.

Not only is this true of serverless frameworks, but they mean that the start of post-incident forensics is being revolutionised. Moving to event driven and service level forensics helps investigators analyse complex attacks across distributed systems in a more effective way, from an instance centric to a event centric. Evidence collection and log analysis tasks are automated and speed investigations as dashboards, log and case comparisons are reduced to a few mouse clicks.

The applications of these ideas to real world problems, in particular protecting an energy plant, were demonstrated. Automated isolation of the system and product evidence collection minimises system damage and preserves vital data.

Advanced analytics and machine learning can be used further in future research for serverless incident response. Promising avenues are integrating threat intelligence and developing smart automation workflows. With cyber threats becoming more innovative, serverless computing's agility, scalability, cost-effectiveness will become even more critical for securing lives-critical networks.

References

1. Marquardt EP, Zachary Cohen, Alex: First on CNN: US recovers millions in cryptocurrency paid to Colonial Pipeline ransomware hackers | CNN Politics, https://www.cnn.com/2021/06/07/politics/colonial-pipeline-ransomware-recovered/index.html. Accessed 23 Dec 2024
2. Sosinsky B (2010) Cloud computing bible. Wiley. https://doi.org/10.1002/9781118255674
3. Bhatt S, Manadhata PK, Zomlot L (2014) The operational role of security information and event management systems. IEEE Secur Priv 12:35–41. https://doi.org/10.1109/MSP.2014.103
4. Cinque M, Cotroneo D, Pecchia A (2018) Challenges and directions in security information and event management (SIEM). In: 2018 IEEE international symposium on software reliability engineering workshops (ISSREW), pp 95–99. IEEE, Memphis, TN. https://doi.org/10.1109/ISSREW.2018.00-24
5. González-Granadillo G, González-Zarzosa S, Diaz R (2021) Security information and event management (SIEM): analysis, trends, and usage in critical infrastructures. Sensors 21:4759. https://doi.org/10.3390/s21144759
6. Nupponen J, Taibi D (2020) Serverless: what it is, what to do and what not to do. In: 2020 IEEE international conference on software architecture companion (ICSA-C), pp 49–50. IEEE, Salvador, Brazil. https://doi.org/10.1109/ICSA-C50368.2020.00016
7. Serckumecka A, Medeiros I, Bessani A (2019) Low-cost serverless SIEM in the cloud. In: 2019 38th symposium on reliable distributed systems (SRDS), pp 381–3811. IEEE, Lyon, France. https://doi.org/10.1109/SRDS47363.2019.00057
8. Rajan AP (2020) A review on serverless architectures—function as a service (FaaS) in cloud computing. TELKOMNIKA Telecommun Comput Electron Control 18:530. https://doi.org/10.12928/telkomnika.v18i1.12169

9. Lynn T, Rosati P, Lejeune A, Emeakaroha V (2017) A preliminary review of enterprise serverless cloud computing (Function-as-a-Service) platforms. In: 2017 IEEE international conference on cloud computing technology and science (CloudCom), pp 162–169. IEEE, Hong Kong. https://doi.org/10.1109/CloudCom.2017.15
10. Amazon: What is AWS Lambda?—AWS Lambda, https://docs.aws.amazon.com/lambda/latest/dg/welcome.html. Accessed 23 Dec 2024
11. Microsoft: Azure Functions documentation, https://learn.microsoft.com/en-us/azure/azure-functions/. Accessed 24 Dec 2024

Chapter 3
Use of UAVs in Aerial Radiation Surveying in Radiological Contamination Detection

Tamás Berek

Abstract In the event of breakdowns at nuclear industrial installations and in the event of any radiological incidents, it is important to ensure rapid and accurate information on the radiological situation to ensure that safety measures are taken as accurately as possible. The method of radiological aerial survey has long been used to control contamination of large areas. Over time, the development of radiological detectors made it possible to detect the boundaries of the contaminated site and locate point sources. Advances in UAV control quickly made them suitable as an addition to aerial surveys with piloted aircraft. However, this technology also has some peculiarities that present challenges in its application.

Keywords Aerial radiological reconnaissance · UAV · Radiological detector · Aerial survey

3.1 Introduction

In addition to chemical, biological radiological and nuclear weapons and equipment, the state of the security environment in the coming decades may also be worsened by the criminal use of insufficiently "guarded" chemical, biological or nuclear components of peaceful industrial or research capacities, among other determinants.

Whether we are talking about CBRN incidents, accidental or intentional, including terrorist acts, it can be argued that, although there are not too many such incidents worldwide, the possible consequences are particularly serious. Early detection of CBRN incidents and rapid and effective response, which require detection and identification of the threatening agent, are essential to mitigate its effects. To this end, the European Union, like the rest of the world, needs to respond to these challenges. The ability to detect is an indispensable complement to prevention. The detection

T. Berek (✉)
Ludovika University of Public Service, Budapest, Hungary
e-mail: berek.tamas@uni-nke.hu

T. A. Kovács and I. Fürstner (eds.), *Critical Infrastructure Protection: Advanced Technologies for Crisis Prevention and Response*, NATO Science for Peace and Security Series C: Environmental Security, https://doi.org/10.1007/978-94-024-2308-2_3

of CBRN materials is key to ensuring that appropriate action is taken on CBRN incidents, as it is not possible to determine which substances are involved in an incident without detection. This requires the development of a CBRN threat detection system, equipped with a detector park based on standardised minimum standards that adequately responds to future CBRN threats and other application challenges.

The equipment of CBRN reconnaissance organisations is therefore a critical factor that fundamentally determines the capability of subunits. The capacity of instruments for detecting and identifying chemical, biological and radioactive substances is a rather sensitive factor in this respect and should not be overlooked. It is difficult to equip CBRN detection subunits, but also to equip chemical detection teams of any other organisation (e.g. disaster management), to develop a detector park suitable for the purposes of application of the organization [1].

The widespread use of autonomy, transport vehicles with high automation or some level of autonomy can be observed in both military and civilian use. In the case of the latter, as a result of digitalization and the use of new communication technologies, one of the main aspects is to increase transport safety, and efficiency and reduce pollutant emissions. During military use, aspects such as the conservation of human resources, the improvement of operational efficiency, observation and perception of conditions, and environment, processing of information, fast and accurate decision-making, and the execution of the given step also appear [2].

The widespread use of UAVs and the great development of control technology and control technology, together with the increasing efficiency of CBRN detection detectors, make it possible to use UAVs in CBRN detection.

Radars are more difficult to detect their presence and fly at near-ground altitudes, so under the influence of the environment, the coverage of the sensors is minimal, the possibility of counteraction success is significantly reduced [3].

Purpose of CBRN detection

Detection provides information collected visually and by other means on the existence of CBRN strikes, emissions and emerging threats. Reconnaissance shall include the collection of meteorological data necessary for the assessment. Organizing and evaluating diverse CBRN information from the field of operations is a complex task. Even instrumental detection data from the same area may differ due to operator activity, the heterogeneous nature of the contamination, and visual observation results.

Operations carried out in urbanised environments are also at risk of escaping hazardous industrial materials from industrial facilities, including as a result of their destruction by conventional weapons. Information about the CBRN situation is also essential for conducting rapid and high-intensity operations in urban combat, and let us not forget that the urban combat environment, characterised by fires, blockades and thick smoke, also makes the activities of CBRN reconnaissance subunits more difficult.

3.2 Application of UAV in Aerial Survey

The informality of aerial surveys makes it possible to detect areas of high radius level or areas that are difficult to carry out with ground-based radiation reconnaissance and with losses. Basically, aircraft and helicopters can be used for this task, the flight speed of which is proportional to the technical possibilities of the existing metrology devices.

Its advantage over ground reconnaissance is that it allows reconnaissance of large areas, search for facilities, maneuvering depending on the combat situation and weather conditions, quick evaluation of measured data in the air or immediate transmission.

Its disadvantage is that meteorological conditions highly influence it, so the aircraft has limited flying tactical capabilities, so the risk associated with their use must be considered.

Aerial radiological reconnaissance should normally be carried out at an altitude of 50 to 200 m and the boundaries of the contaminated site should be established at low altitudes of 25 to 100 m. During reconnaissance, the altitude of the flight must be changed in proportion to the decrease or increase in radiation. It is necessary to pay attention to the fact that if the radiation level is very large, then it is advisable to increase the flight altitude. Airborne and chemical reconnaissance should be carried out at a ground altitude of 5–10 m, by helicopter, hovering, and occasionally by landings, which is quite complicated and burdensome for the crew.

In a radiological survey using a UAV, if the drone is controlled manually and 1 drone is used similarly to a piloted aircraft, the detection method should follow the techniques used in aerial radiation assessment with manned aircraft. These are route, direction and point survey.

In route assessment, the pilot follows a route between two checkpoints, taking advantage of dominant terrain features.

In the case of a direction survey, the pilot flies in a straight line between two turning points. In both techniques, the readings of the measurement data are taken at time intervals corresponding to equal distances. In the case of direction and route techniques, the pilot flies to the starting point of the survey. The patrol begins the survey, maintaining a constant height and speed. The operator records the altitude and time of the aircraft, then takes the measurement every 10 s and records the time and dose rate. The pilot gives a signal and information to the operator at the end point of the survey, where the last measurement takes place. The point technique is used to determine the ground dose rate at points relevant to operations, which provides an opportunity to obtain more accurate information. In the case of a point survey technique, the aircraft shall land near the point to be measured. The operator moves out, approaches the selected point and takes the measurement. Suppose the high dose rate does not allowed, in the above case. In that case, the measurement should be carried out from the air, and the CBRN Center should calculate the radiation level using the air-ground attenuation coefficient [4].

When planning an aerial survey, an appropriate scanning detection method must be established, along with the minimum measurement point density and scanning distance. A maximum scan height must also be determined.

In some cases, the aerial radiological detection method follows the method of a search and rescue (SAR) flight developed with the help of test flights and applied in practice, which can ensure the maximum density of the necessary measurement points, and for this, before performing the flight, the corners of the reconnaissance area, the scanning distances must be determined, which depends on the distance from the source of radioactive contamination. In practice, two types of conversion are used for processing data generated during aerial radius surveying. The conversion of the irregular set of points formed by the measurement data into a raster (possibly pixelated) point network or a Delanuay triangle network. In the case of searching for a point source, the former can be used, and when searching for a territorial source, the latter can be used [5].

Based on the developed methodology, during the mapping of large-scale radiation pollution, the radiation level of each area can be determined by calculation, taking into account flight altitude, atmospheric and ground conditions. During the delimitation of point radioactive sources and the measurement of hot spots, the geographical coordinates of point sources and hot spots can be determined on the basis of dose intensity points different from the dose rate conditions of the environment [6].

One of the great advantages of using UAVs compared to conventional aircraft for aerial radiology survay is that drones can be deployed more easily near contaminated terrain and reconnaissance patrols are much less costly. There is no need to take into account the radiation dose received by the crew, and UAVs can be used in environmental and topographical conditions when piloted aircraft are not permitted from the point of view of flight crew safety. They can more easily approach objects that would be risky by other means, which can provide more accurate radiological information. They can get much closer to hot spots and stay near radioactive sources for longer. UAVs can be successfully used in several cases where ground reconnaissance poses an increased risk.

CBRN EOD team in the deactivation activity of the improvised explosive devices can also be particularly beneficial or walkable or impassable terrain [7].

In particular, radiological detection with UAVs is useful in areas devastated by fires, industrial disasters, and urban environments. UAVs can track terrain more efficiently and can descend, ascend, and quickly manoeuvre or change direction without taking into account crew loads.

Aircraft do not come into contact with the ground during the survey, can fly at different altitudes following the terrain, travel faster than ground elements, control a larger area in a given time period, and can move over the most difficult terrain. Most UAVs are capable of rising quickly vertically and do not need long runways for takeoff and landing. They fly very low, hover and turn, move in tight spaces, are cheap and easy to handle [8].

UAV selection requirements:

When selecting unmanned aerial aircraft for aerial radiological reconnaissance missions, a number of circumstances should be taken into account. During radiological survay, radioactive contamination of the UAV must be taken into account. After survay, the drone must be decontaminated, so a UAV of such design must be obtained that can withstand the load associated with decontamination.

The implementation of decontamination is fundamentally determined by the degree of contamination and the value of the UAV, the degree of its depreciation and substitutability, as well as the cost of decontamination and maintenance. UAVs of higher value are decontaminated in case of higher contamination, even if the costs of decontamination are high [9].

It is necessary to fly in complicated conditions. In addition to highly eroded natural and artificial terrain elements, radioactive-contaminated terrain sections formed as a result of nuclear catastrophe or nuclear strike are characterized by extremely high dose intensities in places. Especially in urban environments, conditions can occur, especially caused by fires, which put a great strain on the performance of UAVs. The aircraft shall be selected in such a way that, in addition to withstanding the above loads, specified flight performance shall be achieved under different weather conditions.

UAVs used for radiological surveys shall be equipped with a navigation device for rapid identification of measurement points. Accurate height measurement is also important, and in addition to radiation detectors, ambient temperature and humidity sensors must also be included in the sensor package. It is not necessary for radiation detection, but it is definitely useful to install visible spectra and thermal cameras to visually identify individual hot spots. Aircraft capable of lifting and transporting the mass of this sensor package should also be used when planning to operate with one or two drones. By using swarms of drones, of course, these sensors can be diversified between UAVs.

Propulsion systems play a crucial role in enabling efficient and reliable operation of UAVs. Choosing the ideal propulsion system depends on many factors, such as task requirements, flight duration, payload capacity and environmental considerations. While internal combustion engines dominated in the past, electric and hybrid systems are gaining ground due to their advantages. Among UAV types, the multirotor is one of the machines with the fastest energy consumption. Most of them have a battery life of less than 60 min [10].

The choice of UAV is also determined by the nature of the radiological survey task: scanning of large areas, fixed-wing drones capable of long flight distances can be used to determine the boundaries of the contaminated area, however, the flight performance of drones capable of point assessment and landing and take-off can be used in industrial and urban environments is much more modest. The use of drones in agricultural land management faces similar challenges. Some of the multirotor UAVs used in agriculture are work drones capable of lifting loads in addition to performing precision flight operations.

There, rotary-wing drones are preferred due to ease of operation, slower speed, better maneuverability, relatively low cost. These benefits provide greater opportunities for gathering information about plants through imaging, which is one of the uses of UAVs in production. In cases where the observation area is relatively large, preference is given to fixed-wing aircraft, which allow the entire territory to fly over in a short time [11].

Among the types of drones suitable for aerial CBRN reconnaissance of small areas from low altitudes, battery-powered and hovering multirotor UAVs are suitable, for which battery capacity must be taken into account, which determines and even primarily limits flight time and route, thus the size of the terrain that can be explored.

In the case of drones currently used in agriculture that can carry payloads of up to 50 kg, a flight time of about 20–30 min with a charged battery is expected when powered by battery power [12].

3.3 Detector Selection

Data on the CBRN situation are part of the commander's critical information needs in his decision-making process. The task of CBRN detection is to obtain information. Refuting the presence of danger in a particular area (negative meaning) is just as important, but often means more than confirming it (positive meaning). A clear demonstration of a harmless status is significant, given the danger of an implied—possibly false—sense of security knowing it. The primary condition for confident refutation of CBRN contamination, the presence of toxic warfare agents, is the use of detectors operating on different principles and with appropriate sensitivity.

The equipment of reconnaissance organizations is a critical factor that fundamentally determines the capability of subunits. The capacity of instruments for detecting and identifying chemical, biological and radioactive substances is a rather sensitive factor in this respect and should not be overlooked. It is difficult to equip CBRN reconnaissance subunits, but also to equip CBRN assessment teams of any other organisation (e.g. disaster management), to develop a detector park suitable for the purposes of the organisation's application.

This requires examining the concept of aerial radiological detection corresponding to the expected application of the detector to be used for aerial radiological surveying. The possible hardware and software design of the air-radiological detection module that can be fitted to the carrier vehicle shall also be examined, as the component components determining the size of the module shall be explored. Aerial radiological assessment requires an efficient combination of sensors and detectors, where a significant limiting factor is the payload weight that the drone can carry. It is also important to define the concept of the hardware and software structure of the radiological assessment module and to delimit the information content and structure of the operator's graphical interface using the system.

The UAV payload includes optical, radiological, chemical, biological, and meteorological sensors that need to be optimized for the given task. Accelerometers, gyroscopes, pressure sensors, and GPS devices help you control and navigate. For radiological detection, GM tubes used for dose rate measurement are the most common, and scintillation detectors are suitable for spectrum recording [13].

Therefore, assessing the characteristics of possible detectors by comparing sensitivity, operating range, mass, and temperature data is an important task when compiling the detector set. Determining the optimal operating voltage of the GM detector, comparing the dead time of GM tubes with the capacity and determining the recovery time are part of detector optimisation.

In case of detection of hot spots, the use of a scintillation detector has been successful, and in the case of an area survey, data from GM detectors provide information [5].

The cooling requirements of semiconductor detectors, which are currently most common for gamma spectrometry measurements, result in a high mass so scintillation detectors can be used here for measurement purposes. However, due to the dependence of the gamma spectrum measured by the scintillation detector on temperature changes, scintillation detectors require energy recalibration before environmental and outdoor measurements. Research has already shown that this method can be turned into an algorithm and burned into the instrument's EPROM to perform in situ measurements, where gamma spectra, which can be easily and quickly recorded, can provide vital information about hotspots, the main isotopes determining the radiation field and the determination of sampling sites [14].

However, scintillation detectors are vulnerable to mechanical effects and their measurement is influenced by many factors. The examination of the sensitivity of ideal detectors suitable for accurate and reliable measurements in field conditions to changes in air pressure, temperature, humidity and electromagnetic, ionizing radiation is an essential task, as well as the exploration of the sensitivity of components installed in detectors to external, physical influences.

The temperature dependence of the sensitivity of detectors used under field conditions is of decisive importance from a metrological point of view. This complex temperature dependence needs to be determined and tested under laboratory conditions. Here it is possible to distinguish between the temperature dependence and temperature behavior of the light emission of the scintillator, the photoelectron multiplier tube, and the electrical components.

It must be covered with a cover to prevent outside light from interfering with the measurement. The detector shall be made resistant to vibrations and other mechanical stresses by means of mechanical absorbers. Temperature changes cause significant errors in the measurement result, so it is necessary to apply temperature compensation in the scintillation detector, which is based on temperature measurement and automatic calibration using standard sources [15].

It is also necessary to assess gas-filled detectors suitable for an aerial radiological survey module, taking into account the specifics of the detectors that can be used. Due to the stability of the Geiger-Müller tube detector, it is well suited for this purpose, but determining its optimal operating voltage, comparing sensitivity, operating range, mass and operating temperature data is an essential task.

The instrumentation of the radiological survey module is therefore of great importance, as the module equipped with a GM tube detector performs reliable measurements at lower speeds and from a lower height than the version equipped with a scintillation detector. The UAV can measure at any location, practically at a very low altitude, which is a very big advantage for hot spot analysis [16].

In addition to radiological detectors, it is essential to monitor the relative humidity, temperature and barometric air pressure value of the ambient air, and it is also important to measure the temperature of the scintillation detector with the help of a scintillation crystal envelope temperature sensor.

An important condition for the placement of meteorological sensors is that the sensor works properly during the assessment of the environment so that the operation of the drone does not interfere with it. In addition to reliability, weight is also an important consideration to maximize flight uptime. The dynamism of flight conditions in a UAV requires the operability of meteorological sensors in environments with high and low temperatures, as well as high humidity [17].

Overall, the requirements for a drone for aerial radiological survey are fast deployability, simple guidance and programmable control, radiation-resistant sensor system and electronics, long flight time, wide-range radiation detector and speed [13].

This is an idealised set of requirements that requires compromises on many of its elements in order to comply with it.

Control of UAVs:

One of the drawbacks of CBRN surveys carried out 20–30 years ago with unmanned aircraft was that it can only detect areas to which it has been guided or programmed, and the detection process lacks initiative.

It should also be borne in mind that the Manual control of UAV devices requires physical activity from the operator.

According to biological research, the operator is classified as a "moderate" activity in terms of muscle function and energy expenditure, during which a significant neuropsychic load is placed on the body, which can significantly affect the efficiency of its activity. There is also a significant load on the senses. As a result of the physical and neuro-psychological strain that occurs, the ability to work decreases, mental fatigue plays a prominent role in this [18].

Long flight times and mission distances are desirable for aerial radiological surveys, but technical possibilities are limited.

In practice, the distance between the starting point of detection and the operator is relatively small, i.e. it has to be located close to the terrain section to be explored, which is very important in case of contaminated terrain [16].

This means that this form of UAV control still has to take into account the vulnerability and potential for error by personnel and human factors.

Various drone control algorithms provide solutions to the above problems. The results of recent research are promising and can be used to improve the performance of drones developed for aerial radiological surveying. The newly developed drones have achieved significant mobility through route planning algorithms. Their manoeuvrability and obstacle-avoidance capability allow them to be used in terrain where other methods could not carry out aerial or ground radiological surveys.

Self-driving UAVs can quickly reach their intended destination, and their hovering capabilities allow them to respond instantly to usage needs. In addition, UAVs easily penetrate confined spaces or dangerous terrain that would be difficult to penetrate. They can move around built-up environments, ravines, forests and other hard-to-reach places [19].

RGB and thermal cameras mounted on UAVs also allow the movement of ground-based radiological survey UGVs to be coordinated. Ground equipment is much slower and energy efficiency is also important when using drones on the ground. Aerial survey images can be used to support ground-based unmanned route planning effectively.

The radiological survey with a piloted aircraft is well complemented by UAVs. Drone measurements can ensure higher spatial resolution and refinement of information revealed during aircraft measurements through their lower speed and low-altitude flight. Using air data generated by the UAV, the UGV can pinpoint the sources. The results of this experiment have already been published by the research team [20].

One of the new, developing areas of autonomous vehicle control, including UAVs, is cooperative control, in which the task is to coordinate the movement of vehicles in order to achieve a predetermined goal. For example, to deliver the UAV to a predefined location at a specified time, such as to the observation site or to the reconnaissance site [21].

What algorithm is used depends on the task the drone needs to perform. When flying over open terrain or tracking routes without distractions, it is unnecessary to use an algorithm that can handle dynamic changes in the environment, especially if it only increases the computational demand. The obstacles in space determine how complex an algorithm we have to use [22].

In the case of a scanning aerial radiological survey, the routing of drones can be considered an optimization problem, the solution of which is to find a reconnaissance route from the starting point of the radiological survey to the end point of detection that ensures radiological survey of the terrain. However, in the case of a more detailed radiation assessment or search for sources, several other factors must be taken into account. Given that aerial radiation detection does not always take place in open areas, but often in crowded environments full of obstacles (e.g. urban areas, indoor spaces), it is important to take these requirements into account.

The drone must use route planning algorithms that guarantee a collision-free and optimal route. Route planning refers to a set of waypoints that bypass obstacles from the initial position to the final destination, while collision avoidance takes a specific waypoint assignment as a local destination to avoid obstacles [19].

This is a seemingly simple program design task on a flat terrain that is not too extensive, but in reality, on complex terrain, it is an extremely complex task.

In fact, we usually have minimal information about individual obstacles, especially about dynamic elements such as people in motion, cars, other drones. In order for the drone to detect these obstacles, various cameras and distance sensors must be used in order for the drone to recognize the movement of the object. Add to this the fact that atmospheric disturbances constantly hinder keeping the drone on track, and the uncertainty of the state of the drone is also a disturbing factor. This already requires support for AI. Such a method used in practice when operating a drone in an unknown environment is an algorithm that breaks down cells and cells [22].

Sampling-based methods, RRT (Rapidly-exploring Random Tree) and PRM (Probabilistic Roadmap) are randomly sampled route planning algorithms primarily used for navigating unfamiliar or complex environments. They are adaptable to dynamic environments, but have high computational costs as they generate many route patterns and do not guarantee an optimal solution.

RRT algorithm specifically designed to solve non-holonomic and kinematic routing problems. RRT avoids approaches requiring point-to-point convergence, which is an advantage for non-holonomic constraints [23].

The advantage of graph-based methods (e.g. Dijkstra, A, Voronoi graph) is that they are relatively simple and easy to implement, but they are not always suitable for efficient route planning in complex, obstacle-ridden environments. Biologically inspired algorithms are versatile, applicable to a wide range of problems, including multipurpose routing, and adaptable to constraints and complex environments. However, the disadvantage of biologically inspired algorithms is that they can fall into local minima, which does not guarantee finding the optimal route, which can cause problems with passage in tight spaces [19].

Biologically inspired routing algorithms have emerged with advances in machine learning and have high hopes, especially for controlling swarms of drones. These algorithms are based on the pattern of biological systems observed in nature (anthill, swarm of bees).

Another method of route planning for UAVs is the Potential Field Method. Its advantage is that it can move in unfamiliar territory by taking into account its current situation and the possibilities of its situation. The basis of this method is that the attractive forces are given by the goals and stations, and the repulsive forces are given by the obstacles. The danger is that the drone may end up in a trap position, and even though it could fit between two obstacle elements, its algorithm will choose to avoid [24]. In this case, obstacles can be defined using a map, RGB thermal cameras and LIDAR in real time.

Since in this method, the movement of the drone is determined by the resultant force of the potential field, the drone moves towards the minimum value of the potential field, thereby approaching the target and avoiding obstacles. The potential field method tends to fall into local minima, which means that the drone can get stuck at a point from which it cannot reach the target. This can especially happen when the target and obstacles are close to each other. Despite the simplicity and intuitiveness of the PF method, the problem of local minima and the lack of global optimality limit its practical applicability [19].

For these, it is absolutely necessary to address the issue of the positioning of drones.

For UAVs to solve tasks reliably, their positioning systems play an important role. During development, several possibilities arise, which, of course, bring different advantages and different possibilities for error.

Using low-cost3 hardware, these navigation devices are available at a lower cost, but the accuracy of low-cost MEMS4-based sensors is lower than that of tactical quality sensors. Accelerometer and angular velocity sensors are riddled with error phenomena such as bias, drift, scaling. Some research is investigating sensor fusion to reduce the effects of various faults, resulting in reliable navigation information that can be used for guidance. In the case of a loosely coupled solution, pre-processed solutions from different subsystems are combined. During tightly coupled integration, raw measurements of subsystems are integrated [25].

A radiological survey concept adapted to the metrological characteristics of the detector and the typical (expected) application task needs to be developed. The definition of the methodology of applied CBRN detection is important for the development of flight altitude, scanning distance and measurement point density. It is essential to determine the scanning distance that provides the point density necessary for measuring the radiation field and to determine the point density relevant to the success of data processing.

The large amount of data provided by cameras and sensors calls into question the relevance of human resource analysis. And when optimizing the UAV's flight path or the UGV reconnaissance route through aerial reconnaissance camera data, some kind of analysis algorithm must be used.

The efficiency of processing discovery data can be greatly improved by advances in automated data analysis methods and machine learning methods. On the one hand, this increases the autonomy of UAVs in target identification, route planning and decision-making, among others, and on the other hand reduces the risk of human error and the subjectivity of analyses [19].

In connection with the application of autonomous devices, it is essential to address the issue of artificial intelligence

Artificial intelligence

Advances in control technology, camera and sensor technology and rocket technology have created new opportunities for autonomous systems in space and the underwater environment. Whether we examine autonomous platforms from a functional point of view (machine operation, information gathering, target detection), the level of autonomy (autonomous, partially autonomous, remotely controlled) or lifetime (single-use or recurring), accuracy is a fundamental condition for operational safety. One of the reasons for ensuring accuracy is that unmanned systems, taking into account their extreme application environment and the physical limitations of piloted vehicle crews, go well beyond the capabilities of manned devices [26].

The military development and testing of drone swarms has also begun, during which a number of small, low-cost, cooperative and unmanned aircraft coordinate their activities, operating as a coherent unit. Unmanned devices should have adequate autonomy and intelligence [27].

Today, drone convoys are being tested in many countries as distributed collaboration systems consisting of many small, inexpensive, unmanned aerial vehicles. This type of autonomous, or "autonomous" weapon raises new strategic, ethical and legal questions. Swarming drones can bring real benefits, but they can also carry potential dangers. The researchers say the risks posed by swarming drones must be weighed before they develop destructive capabilities. Swarming systems typically consist of individual agents that interact with each other and their environment. Agents follow simple rules, but interactions between agents can lead to rather complex and sophisticated collective behaviors, including emerging intelligence. For example, a swarm can remain in shape while changing direction several times [28].

NATO's AI strategy states that AI applications are developed and used in accordance with national and international law, are developed and used with an appropriate level of discretion and care, and that clear human responsibility must be applied to ensure accountability. The document states that AI will be properly understood and transparent, including through the use of review methods, resources and procedures, and that the safety, security and robustness of such capabilities will be tested and proven within these use cases throughout their lifecycle [29].

Thanks to artificial intelligence, Reinforcement Learning (RL) also provides an opportunity for drone route planning, during which we assign specific actions to specific situations and evaluate them quantitatively. Learning has three main components: the environmental model (how we are positioned); the methods of procedure (go straight or turn right); and rewards (negative or positive depending on the outcome) [22].

3.4 Conclusion

Aerial radius survey is used to quickly survey large areas. The radiation dose suffered by personnel is less than in the case of ground survey. Aerial radiation assessment can be carried out in areas with high radius levels and impassable terrain, but for a long time it has been declared more inaccurate than ground reconnaissance and is also dependent on weather.

When performing an aerial radiological survey, it is necessary to explore the characteristics of the terrain determining the detection possibilities, as well as to examine the peculiarities of determining the dose intensity of the areas, taking into account flight altitude, atmospheric and ground conditions. The conditions and limitations of the technology for determining the qualitative, quantitative parameters of radioactive materials under field conditions must be analysed.

The comparison of the physical and metrological parameters of scintillation detectors optimized for different measuring ranges, as well as the examination of gamma dose rate measurement techniques and the examination of the characteristics of gamma dose rate transmitters are essential when determining the lower limit of discoverability. The study of the effect characteristics of scintillation detectors used in gamma specroscopic techniques should include the possibilities of scintillation measurement and the interference with the measurement (ambient light, vibrations, electromagnetic radiation coming from outside, temperature changes, irradiation with extremely high dose capacity.

When analysing the different properties of scintillators with different chemical compositions and determining their applicability, it is also necessary to compare the light yield of scintillators and their spectrum. The examination of the hygroscopicity of a scintillation crystal capable of producing a crystal with good resolution spectrum, which can be used for aerial radiological survey tasks, is essential in order to determine minimum requirements for protection against moisture.

When developing the sensor fleet required for a drone with significantly lower payload than piloted aircraft, the carrying capability of the drone is a strong limiting factor, which must be taken into account.

UAVs allow for more accurate area exploration, but there are battery capacity limitations, which significantly affect flight time. It is often necessary to perform radiological detection in an unfamiliar environment where GPS does not work. In this case, UAVs must follow a survey path that requires efficient route planning while minimizing survey time and cost. This requires UAVs to have real-time information about their surrounding environment.

This performance can be greatly improved by using swarms of drones and route planning and optimization algorithms.

References

1. Berek T (2020) Climate change as a factor influencing CBRN information collection and processing In: Földi L, Hegedűs H (eds) Climate change challenges and possible responses, pp 141–158. Ludovika Egyetemi Kiadó Budapest
2. Berek T, Répás J (2024) Opportunities and challenges of forensic testing of autonomous battlefield and transport vehicles In: Gőcze I, Padányi J (eds) Excerpts from the achievements of military engineering sciences IV, pp 7–20., Ludovika Egyetemi Kiadó Budapest
3. Kiss B, Palik M (2023) Military use of drones in modern military operations Repüléstudományi Közlemények 35(5):115–130
4. NATO 2112 Nuclear Biological and Chemical reconnaissance
5. Zelenák J, Csurgai J, Halász L, Solymosi J, Vincze Á (2009) Assessment of the applicability of aerial radiation reconnaissance capabilities to search for lost or stolen sources or to detect contaminated terrain Hadmérnök 4(1):46–62
6. Csurgai József: Drónra szerelhető radiológiai modul fejlesztési koncepció, hardveres felépítés és nukleáris méréstechnikai alapok Repüléstudományi Közlemények 34(3):181–202 (2022)
7. Ember I, Kovács Z (2020) Drones above EOD perators during their public duty. In: Beňovský M (eds) Zborník Prednášok Trhacia Technika 2020 Banská Bystrica, Szlovákia, Slovenská spoločnosť pre trhacie a vŕtacie práce, pp 90–97
8. Kovács Z, Ember I (2022) Landmine detection with drones Land Forces Acad Rev 27(1):84–92
9. Kiss B (2018) Evolution of UAV-s in the light of radiological detection Repüléstudományi Közlemények 30(3):15–26
10. Békési, B, Knapiczius A, Gajdács L (2023) Possible propulsion structures of drones. Repüléstudományi Közlemények 35(1):257–274. https://doi.org/10.32560/rk.2023.1.18
11. Hajnal K, Hegyi F, Békési B (2023) Multikopteres drón tervezése, fejlesztése és felhasználása. Repüléstudományi Közlemények 35(2):11–21

12. Kim J et al (2019) Unmanned aerial vehicles in agriculture: a review of perspective of platform, control, and applications. IEEE Access 7:105100–105115. https://doi.org/10.1109/ACCESS.2019.2932119
13. Manga L, Kátai-Urbán L, Vass Gy, Csurgai J (2017) Unmanned aircraft in radiological reconnaissance Védelem Tudomány 2(2):63–75
14. Csurgai J, Padányi J, Földi L (2020) Temperature dependence of NaI(Tl) radiation scintillation detectors' characteristics. Adv Military Technol 15(1):201–212
15. Petrányi J, Vass Gy, Csurgai J, Kátai-Urbán L (2021) Summary of smart detector application options Sugárvédelem 14(1):24–31
16. Csurgai J (2022) Drone mounted radiology module development concept, hardware architecture and nuclear metrology basics Repüléstudományi Közlemények 34(3):181–202
17. Bottyán Zs, Fekete Cs, Gyöngyösi A, Kardos P, Tuba Z, Vas T (2022) Development of a drone measurement network suitable for routine atmospheric vertical profile measurements Közlekedés és Mobilitás 1(1):55–65
18. Dunai P (2013) Energy consumption, analytical analysis of load parameters of circulatory and respiratory system during UAV operator work Repüléstudományi Közlemények 25(3):13–17
19. Lyu M, Yibo Z, Chao H, Hailong H (2023) Unmanned aerial vehicles for search and rescue: a survey. Remote Sensing 15(13):3266. https://doi.org/10.3390/rs15133266
20. Kochersberger K et al (2018) Unmanned aircraft applications in radiological surveys. 2018 IEEE international symposium on technologies for homeland security (HST), Woburn, pp 1–5. https://doi.org/10.1109/THS.2018.8574133
21. Szabolcsi R (2022) Cooperative control of unmanned aerial vehicles Repüléstudományi Közlemények 34(1):5–22
22. Mihályi G (2023) Challenges and possible solutions for UAV trajectory planning Repüléstudományi Közlemények 35(3):51–68
23. LaValle SM (1998) Rapidly-exploring random trees: a new tool for path planning, technical report 98–11, Computer Science Dept., Iowa State University
24. Borenstein J, Koren Y (1991) Potential field methods and their inherent limitations for mobile robot navigation. Proceedings—IEEE international conference on robotics and automation, pp 1398–1404. https://ieeexplore.ieee.org/stamp/stamp.jsp?tp=&arnumber=131810
25. Farkas M, Vanek B, Rózsa Sz (2018) Determination of unmanned aircraft position and orientation through tightly coupled integration of low-cost GNSS, inertial and magnetic sensors Repüléstudományi Közlemények 30(1):139–150
26. Jacquelyn Schneidera Julia Macdonaldb: Looking back to look forward: Autonomous systems, military revolutions, and the importance of cost. J Strategic Studies, 1–23. Online: https://doi.org/10.1080/01402390.2022.216457
27. Arash Heydarian Pashakhanlou: AI, autonomy, and airpower: the end of pilots?. Defence studies 19(4):337–352. Online: https://doi.org/10.1080/14702436.2019.1676156
28. Irving Lachow (2017) The upside and downside of swarming drones. Bulletin of the atomic scientists 73(2):96–101. Online: https://doi.org/10.1080/00963402.2017.1290879
29. NATO: Summary of the NATO Artificial Intelligence Strategy. 2023. április 25. Online: https://www.nato.int/cps/en/natohq/official_texts_187617.htm

Chapter 4
Prevention of the Terror Attack and the Social Aspects in the European Union

János Besenyő

Abstract Over the past decades, but especially in the aftermath of 9/11, the world's perception of security has been greatly affected by terrorist attacks against various targets, which both state actors and ordinary people are trying to counter and prevent. They are trying to do this by increasingly diverse means, such as legislation, giving more powers to the various state bodies (police, army, secret services, etc.) and using different methods and techniques to involve the widest possible sections of society. However, there is no single position, standard or even consensus on these instruments, and some of them often generate serious controversy in society at large. In my study, I will deal with some of these, and I will look at how and in what way different societies have viewed terrorist attacks in the past and how they have tried to prevent and deal with them. In addition, I will look at Internet communication and how it can be used to reduce terrorist attacks.

Keywords Terrorism · Terror attack · Prevention · Social aspects

4.1 The Concept of Terrorism

Before discussing the prevention of terrorist attacks, it is important to clarify the concept of terrorism. There are several definitions that are similar in certain elements, but there is still no universally accepted definition [1]. The Latin word terror means fright terror. Although often used as an individual term, in a social context the term 'a régime de la Terreur' was first used to describe the rule of the Jacobins during the French Revolution of 1793–1794, which meant a reign of terror over society [2]. As I said, there is no universally accepted definition, but there are several that are more widely accepted, some of which I will share with you. According to the Encyclopedia Britannica dictionary, terrorism is: "The calculated use of violence to create a

J. Besenyő (✉)
Bánki Donát Faculty of Mechanical and Safety Engineering, Óbuda University, Budapest, Hungary
e-mail: besenyo.janos@uni-obuda.hu

T. A. Kovács and I. Fürstner (eds.), *Critical Infrastructure Protection: Advanced Technologies for Crisis Prevention and Response*, NATO Science for Peace and Security Series C: Environmental Security, https://doi.org/10.1007/978-94-024-2308-2_4

general climate of fear in a population and thereby to bring about a particular political objective. Terrorism has been practiced by political organizations with both rightist and leftist objectives, by nationalistic and religious groups, by revolutionaries, and even by state institutions such as armies, intelligence services, and police [3]." The United Nations General Assembly defines terrorism as: "Criminal acts intended or calculated to provoke a state of terror in the public, a group of persons or particular persons for political purposes are in any circumstance unjustifiable, whatever the considerations of a political, philosophical, ideological, racial, ethnic, religious or any other nature that may be invoked to justify them [4]." The United Nations also sees terrorism as an international challenge, not a local or regional issue, and envisages international cooperation as a solution. This approach was even more strongly advocated after the terrorist attack on Israeli athletes at the 1972 Munich Olympics, but a number of countries in the Middle East, Africa and Asia rejected it because of their personal involvement or sympathy with organisations known as terrorist groups [5]. According to the terminology adopted by the European Union, an attack can be considered a terrorist attack if it is intended to achieve the following objectives: "Seriously intimidating a population, or Unduly compelling a government or international organisation to perform or abstain from performing any act, or Seriously destabilising or destroying the fundamental political, constitutional, economic or social structures of a country or an international organisation [6]." NATO's similar, however different approach: "The unlawful use or threatened use of force or violence, instilling fear and terror, against individuals or property in an attempt to coerce or intimidate governments or societies, or to gain control over a population, to achieve political, religious or ideological objectives [7]. "Although the US is not considered an international organisation, its influence and its prominent role in the fight against terrorism have led many to adopt its terminology and use it in their own. I will therefore describe it. US legislators have defined terrorism as follows: "the term "terrorism" means premeditated, politically motivated violence perpetrated against non-combatant targets by subnational groups or clandestine agents [8]." Interestingly, the FBI has modified this concept somewhat as follows: "the unlawful use of force or violence against persons or property to intimidate or coerce a Government, the civilian population, or any segment thereof, in furtherance of political or social objectives [9]." This also indicates, as I will refer to later, that the concept is not used uniformly, almost everyone (even within a given state/government/community) uses their own version.

Of course, not only the above-mentioned organisations, but also academics studying terrorism have developed their own definitions. Perhaps one of the best-known definitions is that of Professor Alex P. Schmid, who has formulated the following: "Terrorism is an anxiety-inspiring method of repeated violent action, employed by (semi-) clandestine individual, group or state actors, for idiosyncratic, criminal or political reasons, whereby—in contrast to assassination—the direct targets of violence are not the main targets. The immediate human victims of violence are generally chosen randomly (targets of opportunity) or selectively (representative or symbolic targets) from a target population, and serve as message generators.

Threat- and violence-based communication processes between terrorist (organization), (imperilled) victims, and main targets are used to manipulate the main target (audience(s)), turning it into a target of terror, a target of demands, or a target of attention, depending on whether intimidation, coercion, or propaganda is primarily sought [10].". According to another renowned terrorism expert Bruce Hoffman, terrorism is "ineluctably political in aims and motives; violent—or, equally important, threatening violence; designed to have far-reaching psychological repercussions beyond the immediate victim or target; conducted either by an organisation with an identifiable chain of command or conspiratorial cell structure (whose members wear no uniform, or identifying insignia) or by individuals or a small collection of individuals influenced, motivated, or inspired by the ideological aims or example of some existent terrorist movement or its leaders, or both; and perpetrated by a subnational group or nonstate entity [11]." Eteri Tsintsadze-Maass and Richard W. Maass in the journal Terrorism and Political Violence have written the following about terrorism: "politically motivated violence perpetrated against non-combatant targets by nonstate actors. Terrorists commit violent acts as tools of coercion, hoping to force a desired political change by threatening further violence [12]." According to Carsten Bockstette, a German military officer and researcher at the George C. Marshall European Center for Security Studies, who has also worked on terrorism in practice "Terrorism is defined as political violence in an asymmetrical conflict that is designed to induce terror and psychic fear (sometimes indiscriminate) through the violent victimization and destruction of non-combatant targets (sometimes iconic symbols). Such acts are meant to send a message from an illicit clandestine organization. The purpose of terrorism is to exploit the media in order to achieve maximum attainable publicity as an amplifying force multiplier in order to influence the targeted audience(s) in order to reach short- and midterm political goals and/or desired long-term end states [13]." The famous law professor Rosalyn C. Higgins, who was also a judge of the International Court of Justice, defined terrorism as follows: "Terrorism is a term without any legal significance. It is merely a convenient way of alluding to activities, whether of States or of individuals widely disapproved of and in which wither the methods used are unlawful, or the targets protected or both [14]." A much shorter version was put forward by Professor Boaz Ganor of the International Institute for Counter Terrorism in Herzliya, Israel: "Terrorism is the intentional use of, or threat to use violence against civilians or against civilian targets in order to attain political ends [15]." Similarly to Ganor, the Croatian philosopher Igor Primoratz formulated a shorter version as follows: "Terrorism is intimidation with a purpose: the terror is meant to cause others to do things they would not otherwise do [16]." As you can see, the definitions put forward by the various experts are all correct, they have much in common, but they are not uniform. Perhaps this is why James M. Poland formulated his opinion as follows: "a consistent classification of terrorism is scarce and unreliable. The several studies explored here confirm the fact that diagnosis remains a matter for the individual ability of the researcher [17]."

Summing up the various opinions, the following statements on terrorism can be considered well-founded:

- It is not an end in itself; it is politically inspired and has political aims;
- It uses actual violence or threats to achieve its goal;
- Primarily targeting civilians (non-combatants) and critical infrastructure;
- Terrorism is a powerful threat to societies and individuals' sense of security;
- Not only different groups and individuals but also states can be actively involved in terrorism.

4.2 Terrorist Attacks Throughout History and Responses Given to Them

Terrorist attacks are most often carried out by groups who, for some reason, are unable to get the majority of society to accept their legitimate or perceived religious, national, ideological or other demands, and therefore try to achieve them by violence. There have been cases where this tactic has worked, but in many cases it has created a spiral of violence that has caused serious damage both to society and to the groups carrying out the terrorist attacks. A good example is the Zealots of Israel (Sicarii) who first made political demands and then, in the first century, sought to expel the Roman Empire which was occupying Israel. Almost from the beginning, they used terrorist methods to achieve their goals. They stalked, attacked and stabbed to death people of their choice (representatives of the Roman Empire or Jews who collaborated with them) at various events, and then fled into the crowd. Later, they took an active part in the uprising against the Roman Empire, where they also massacred their compatriots who wanted to settle relations between the Roman Empire and Israel through negotiations. The Romans eventually brutally suppressed the revolt and Israel lost even a semblance of its former independence [18]. A similar organisation was the Shi'ite "Assassins" sect (Hashashin/Isma'ilis), which was persecuted by the majority Sunni states and which, like the Zealots, sought to eliminate its opponents by assassination. Although the organisation had legitimate demands (an end to the persecution of Shiites), these soon disappeared. The 'mountain elder', who led the sect and also created a separate state formation, transformed the organisation into a terrorist group that could be hired by anyone. If his interests dictated, he carried out political assassinations ordered by Sunni rulers or even crusaders without any problems. The organisation's fate was ultimately sealed when they confronted the Mongols, who launched an attack against them between 1253 and 1256, conquered the territories they ruled and exterminated the group's members [19].

If you look at the background of these groups, you can see that they cover a very broad spectrum. They can be national-ethnic, religious, political-ideological (right-wing, left-wing, etc.), separatist, liberation groups, which are often mixed up and often researchers cannot even clearly decide whether they are terrorist organisations or movements fighting for legitimate demands. These include the Lebanese Hezbollah organisation, whose military wing is often described as a terrorist organisation, while its political wing is seen as a 'reliable, responsible' political actor with considerable influence in Lebanon in almost all areas [20]. The Red Army Faction

(Rote Armee Fraktion or Baader-Meinhof Group) [21], the Italian Red Brigades (Brigatte Rose) [22], the French Action Directe ('Direct Action') [23] or the Peruvian Shining Path [24] were also politically motivated and sought to achieve their goals almost immediately after their formation by means of terror. Perhaps for this reason, they were less controversial for being terrorist organisations rather than liberation organisations. Today, however, these organisations have disappeared. Unlike them, the Irish Republican Army (IRA), which was founded and operated on similar principles, was able to replace terror with the negotiating table and come to an accommodation with the "arch enemy." The organisation has thus become a legal and credible political organisation that has come to terms with its former terrorist past [25].

However, the activities of religiously based terrorist groups are increasingly challenging, especially Islamist, jihadist terrorism, which has now become internationalised, with some of its representatives having taken control of a state (Al-Qaeda/Taliban/Afghanistan) and ISIS having for a time established its own state in the Middle East, which could only be eliminated through international cooperation.

At the end of the nineteenth century, terrorist attacks by lone assassins, known as anarchists, against various targets became increasingly common. The targets ranged widely, including representatives of state power, members of the economic and cultural elite, or anyone identified by anarchists with 'oppression'. The attacks, which were mostly successful, aroused the fear of the leaders of the state power, who used a variety of means—but mostly violent ones—to try to dismantle the various anarchist groups [26]. Christopher Andrew, Professor Emeritus of Modern and Contemporary History at the University of Cambridge, calls the period 1880–1914 the 'golden age of assassins' [27]. In his book, he also points out that it was very difficult to take action against those with a personal motivation and a thirst for revenge, because it is always more difficult to find a person and thwart an attack than to find a group, where the number and diversity of members make it easier to obtain information about the group and perhaps to get the authorities to include one of their informers. Lone wolf attackers, also known as "lonely wolves", remain a serious challenge for societies to this day [28].

Although international law does not recognise the concept of "state terrorism", many experts believe that it exists [29]. In this case, certain governments use terror against their internal and external opponents. According to contemporary accounts, John Calvin, after gaining control of Geneva in the sixteenth century, used a wide range of terror tools to govern the city. As I mentioned earlier, this was typical of the French Revolution, where the Jacobins who ran the state introduced widespread terror not only against their enemies/opponents, but against almost anyone, even their own. European colonial states in Africa (the British in South Africa, the Germans in Namibia), but also in their own territories, such as the Soviet Union, Hitler's Germany, fascist Italy and Francoist Spain, acted in the same way against their enemies/opponents [30]. These states often classify their opponents as terrorists to make it easier to act against them, as Ben Saul put it in his book on the subject: "Despite the shifting and contested meaning of "terrorism" over time, the peculiar semantic power of the term, beyond its literal signification, is its capacity to stigmatize, delegitimize, denigrate, and dehumanize those at whom it is directed, including political opponents. The term is ideologically and politically loaded; pejorative; implies moral, social, and value judgment; and is "slippery and much-abused." In the absence of a definition of terrorism, the struggle over the representation of a violent act is a struggle over its legitimacy. The more confused a concept, the more it lends itself to opportunistic appropriation [31]." As an extremely good example, Syrian President Bashar al Assad has on several occasions treated dissident opposition groups as terrorists, many of whose members have been brutally murdered by the secret services and internal security forces [32].

4.3 Measures Taken to Prevent Terrorist Attacks

It is clear from the above that terrorism has always been/is to some extent a threat to societies. This threat has been prevented and reduced primarily by legal, law enforcement and intelligence means. However, in many cases these have triggered a new spiral of radicalisation and violence [33]. Moreover, the various countries have mostly acted individually and only rarely jointly against the various terrorist groups. This is also evident in the European Union, which only started to cooperate in an organised way in 1975 (TREVI group), and the European Council adopted an terrorism law in 1977. The scale of the problem is illustrated by the fact that several other groups fighting terrorism (the Berne Club, the Kilowatt group and the Vienna group) have been set up by European countries. This is hardly surprising given that between 1950 and 1995, more than 8 000 terrorist attacks were carried out in Western Europe alone, resulting in the deaths of more than 2 500 people [34]. Since the fight against terrorism did not produce the expected results, the attacks of 11 September 2001 were followed by an international coalition led by the USA, with the active participation of the European Union, to combat Al-Qaeda and other terrorist organisations. But these efforts have not been a breakthrough [35]. Although soldiers have been used for internal security purposes in the past, it has become increasingly accepted that military means should be used against terrorists. However, these have not produced the expected results and, according to many experts, are contributing to the escalation of terrorism [36]. Recognising this, the European Union, together with other members of the international community, is increasingly focusing on prevention and looking for ways to reduce radicalisation and prevent potential terrorist acts. Following the terrorist attacks in Paris in 2015 and in Brussels in 2016, a special committee on terrorism was set up in the European Parliament in July 2017 as part of the fight against terrorism, with the primary objective of preventing radicalisation and terrorist attacks [37]. The following are some of the tools used to prevent terrorist attacks.

4.4 Protection of Borders

The first place to stop terrorists entering Europe is the EU's external borders. Control of these is becoming increasingly important, and from time to time European states have also reintroduced/imposed border controls at their internal borders, partly for security reasons and partly because of the COVID epidemic [38]. In December 2016, due to increased terrorist attacks in the EU and foreign fighters with a European background fighting on the side of the Islamic State and returning from Syria, the screening of those entering and leaving EU states and the checking of their documents against the Schengen Information System (SIS) and other relevant EU databases was tightened. This measure amended and tightened the provisions of the Schengen Borders Code (SBC) adopted in 2006 [39]. The new biometric entry and exit registration system, the Entry/Exit System (EES), was introduced from 2022 [40]. Non-EU nationals who do not need a visa are checked using the European Travel Information and Authorisation (Etias) system. To protect the EU's approximately 80,000 km (13,000 km of land borders and nearly 66,000 km at sea) of external borders, the EU Border and Coast Guard Agency (Frontex) will deploy 10,000 permanent staff from 2027. This staff will not only be responsible for border protection but also for managing migration towards the EU [41]. In 2021, the Integrated Border Management Fund (IBMF) was established to support the development of border management in the Member States. At the same time, the Internal Security Fund (ISF) was created to provide resources to fight terrorism, organised crime and cybercrime [42].

4.5 Foreign Fighters

Foreign fighters are people who are actively involved as combatants in a conflict in another country where their own country is not involved. Although the strengthening and tightening of the protection of the EU's external borders has reduced the number of foreign fighters, hundreds of them have remained outside Syria in ISIS/Daesh organisations or joined other terrorist organisations. At least the same number of nationals have some connection to terrorist groups, which is why a special law has been passed to ban European citizens from joining terrorist organisations abroad [43]. This is also important because foreign fighters returning from abroad can easily pass on their experience to those susceptible to radicalise them and then carry out attacks against targets already in Europe [44]. The 2020 Counter-Terrorism Agenda will also address the issue of foreign fighters and the possibilities to counter them [45].

4.6 Information Sharing (Exchange of Information)

In this case, European states share information in various ways to help each other and EU law enforcement/border protection agencies. I will mention three of these forms of cooperation.

4.6.1 Schengen Information System (SIS)

The Schengen Information System (SIS), created in 1995, is an information database that can be used for Schengen (European) border control and cooperation between the pan-European law enforcement (EUROPOL) and the EU countries' own law enforcement agencies (police, border guards, in some cases secret services). The database can be used to search for missing persons, wanted persons and vehicles. The SIS can be used primarily for border protection, migration management and the fight against terrorist organisations or other criminal groups. Although the system has proved its worth, there have been a number of shortcomings in its application and it has been criticised on human rights grounds [46].

4.6.2 Air Passenger Data (Passenger Name Record/PNR) and Advance Passenger Information/API

In 2016, EU lawmakers obliged airlines operating in EU countries to hand over passenger data (name of the passenger, travel dates, itineraries, seats, baggage, contact details and means of payment) to the authorities, who use it to prevent, detect, investigate and prosecute terrorist offences and serious crime [47]. As with the use of the SIS system, human rights issues and criticisms have been raised [48]. In order to complement the PNR database, advance passenger information/API (name, date of birth, nationality, type and number of the travel document, seating information and baggage information) will be sent to the European authorities prior to travel, as adopted in early 2024. The simultaneous use of PNR and API will significantly facilitate and speed up the work of the authorities and the screening of passengers at risk [49].

4.6.3 Europol's Terrorism Situation and Trend Reports

It is not a database, but it provides a thorough overview of the activities of terrorist groups and terrorist threats that the European Community has faced in recent years. It will assist the work of intelligence services, military and law enforcement agencies, as well as academic researchers in the field of terrorism. The reports not only provide information, but also make recommendations to legislators and forecasts [50].

4.7 Maintaining Police and Judicial Cooperation

It is also important that EU and national legislators and law enforcers work closely together to combat terrorism in a way that does not or only to a limited extent violates the fundamental rights of European citizens [51]. In the 1970s, European countries were increasingly targeted by terrorist groups, and legislators and law enforcement forces cooperated with each other in many cases, but not systematically. The 2001 Action Plan on Combating terrorism (SN/140/01) [52] and the 2002 Framework Decision on terrorism (2002) [53] were the first steps in this direction because of the 2001 terrorist attacks in New York. In 2003, within the framework of the mechanism known as the "Prüm decision", lawmakers and police forces cooperated on several occasions. This became increasingly common after the Madrid terrorist attacks in 2004 and the London attacks in 2005, supported by the "2005 EU Counter-terrorism Strategy" and the "Hague Programme" [54]. This process accelerated significantly after the 2015 Paris attacks, but there is still work to be done [55].

4.8 Cutting the Financing of Terrorism

If we want to stop the activities of terrorist organisations, one of the most effective tools is to eliminate the financial and logistical base of terrorists. In this context, in 2018 the EU enacted a Money Laundering and Terrorist Financing Act, which will make the European financial system more transparent and controllable [56]. Despite this, there is evidence that various terrorist groups in Iraq and Syria continue to receive financial support from individuals in the EU [57]. Although the EU is also trying to make the use of cryptocurrencies more transparent, it has so far achieved only limited results [58]. There are also questions about the Hawala system, the use of which is less verifiable and therefore used by terrorist groups [59]. In my opinion, this also includes the use of various sanctions (sanctions against terrorist groups) against terrorist groups, the success of which is also disputed. The EU has already applied sanctions against terrorist groups and those who support terrorism since the early 2000s. As those familiar with the subject know, UN Security Council Resolution 1390 (2002) imposed sanctions against organisations and individuals supporting Osama Bin Laden and Al-Qaeda, which the EU also applied through Common Position 2002/402/CFSP and Council Regulation (EC) No. 881/2002 of 27 May 2002. But the EU has already sought to sanction organisations and individuals supporting various terrorist groups [60]. The EU has also created its own "terrorist list", which only partially corresponded to lists drawn up by the UN or the US [61].

Organisations or individuals on the list were sanctioned, as were their supporters. In 2016, new rules were introduced to sanction terrorist groups [62].

4.9 Reducing Access to Dangerous Weapons

Preventing terrorist organisations and radicalised individuals from gaining access to weapons is a priority for EU countries. This ambition was reaffirmed in the Paris Statement after the bloody attack on the Charlie Hebdo newspaper on 7 January 2015, which stated that "the EU is working to reduce the supply of illegal firearms throughout Europe [63]". This was followed by the adoption of the Internal Security Strategy 2015–2020, also known as the "European Agenda on Security", and later the "Securing Arms, Protecting Citizens", which, among other issues, addressed the issue of access to arms. Not only did they make it more difficult to obtain access to weapons and to obtain weapons licences, but also to obtain copies that could be deactivated (collectors') or made operational. This is also important in the light of the fact that a number of terrorist attacks have been carried out in the EU with previously deactivated firearms which the attackers have rendered operable. The various authorities are also tightening controls on hunters, sport shooters, museums and arms collectors [64]. The purchase of materials and equipment that can be used to make home-made bombs and explosive devices has also been tightened. Since terrorists have used such devices on several occasions, for example in the Paris attacks in 2015, the Brussels attacks in 2016 and the Manchester attacks in 2017, lawmakers considered it important to ensure that no one has access to such materials.

4.10 Preventing Online Radicalisation

Radicalisation and recruitment by terrorist organisations, which increasingly takes place via the internet, is seen as a precursor to physical attacks, as EU law enforcement agencies are increasingly effective in tackling terrorist organisations and extremists. However, the use of the Internet, including social networking sites, gives these organisations a wide margin of manoeuvre, which the authorities are trying to limit by various means [65]. Already in 2013, Europol set up the European Cybercrime Centre (EC3) to fight cybercriminals, terrorists and other radical groups active on the internet. The organisation publishes an annual "Internet Organised Crime Threat Assessment (IOCTA)" report on various cyber-attacks and threats [66].

In 2021, the EU Parliament adopted legislation (EU Regulation 2021/784) that forces internet service providers to remove terrorist or terrorist-supporting content from their sites within one hour of being notified by the authorities (quick and smooth removal of terrorist content online). The monitoring and control of internet sites remains the responsibility of the authorities, not the operators, but a significant number of operators are also taking action on their own initiative against terrorist or radical organisations whose sites and content they delete [67]. The application of the legislation is not yet uniform in all cases, and its application raises a number of questions, but it is certainly an improvement on the previous situation. At the same time, European lawmakers are also increasingly active in tackling criminal groups operating in cyberspace, which in many cases are linked/collaborating with terrorists [68]. Of course, radicalisation can take place not only on the Internet but also in schools, public institutions or even prisons. That is why the European Parliament adopted in December 2020 the EU Security Union Strategy 2020–2025 [69] and the new Counter-Terrorism Agenda [70], which give priority focus on the prevention of terrorist attacks.

However, these tools have limited results in the fight against terrorism. For a greater success, the support and active action of citizens, of those living in the EU, in short, of the community, against terrorism and radicals, is needed. In the next part of my paper, I will propose ways to involve the community in the fight against terrorism on online platforms.

4.11 Opportunities for Community Involvement in the Fight Against Terrorism

Law enforcement agencies, intelligence services and the military are trying to protect us from the physical dangers of terrorism, but thanks to dramatic advances in communications technology over the last few decades, the Internet has spread across the globe, with hardly a corner of the world without internet access. Almost simultaneously with the rise of the Internet, online social platforms, video and image sharing services have emerged, with the main attraction of social networking, content sharing and the use of #hashtags to allow users to search the web by topic. The potential of the Internet and social media has been recognised by jihadist terrorist organisations, who have taken advantage of the above-mentioned opportunities to flood and are flooding social media platforms with their propaganda [71].

Social media stakeholders and law enforcement agencies are working together to combat this kind of activity in the online space, using the power of artificial intelligence. With billions of new pieces of content added to the internet every day, human monitoring is not feasible and AI monitoring has its limitations, so much harmful content is not filtered out. The specific language, slang or abstract references used by terrorist organisations may be unintelligible to AI and therefore require human assistance.

In this situation, the only way forward in the fight against online terrorist content seems to be to involve the public, i.e. communities. Communities of volunteers can actively contribute to the control, restriction and obstruction of law enforcement agencies and social media stakeholders. Practical examples of this already exist, but in my view there is still untapped potential.

In what follows, I would like to present a theoretical sketch of the possibilities for the involvement and applicability of self-organised or state-supported communities, recruited from volunteers.

4.12 Offline (Local Communities)

As you can see in the figure, for completeness I have also outlined some of the possible tasks of offline communities in the real world. These micro-communities can effectively carry out their support activities by knowing the local context. Especially in signalling activities such as propaganda and recruitment, especially among young people. They can contribute to community protection by organising permanent or periodic monitoring of areas at risk. They can report radicalised persons to the authorities, etc. (Fig. 4.1).

4.13 Online Communities

Online communities can basically be divided into two broad categories, online communities of non-IT professionals and organisations composed of IT professionals or at least volunteers with higher user skills.

For both groups we can distinguish.

- state-supported groups and
- fully autonomous, grassroots communities.

State-supported communities are usually based on some kind of state or municipality-supported communities, such as ethnic-based traditional associations or youth branches of political parties, defence-based communities, etc. (Iranian Islamic Revolutionary Guard Corps IT support organisations, opinion leaders, trolls, etc.).

Grassroots online communities, self-sustaining communities based on a common interest or political view, or even online platforms, where members may not know each other personally, or may be geographically distant, but are bound together by a common interest [72]. Such organisations can also be highly effective in supporting Law Enforcement Agencies (LEAs) and social media stakeholders.

The role of communities in these two categories is not fundamentally different. Such communities could help with the following tasks.

- Reporting terrorist-related content in the online space
- Suppression of harmful content, (trolling)
- Creating and sharing enlightening, educational content specifically on platforms favoured by at-risk age groups (TikTok, SnapChat, etc.)
- Point out the contradictions of terrorist propaganda in blogs and posts, and the criminal law consequences

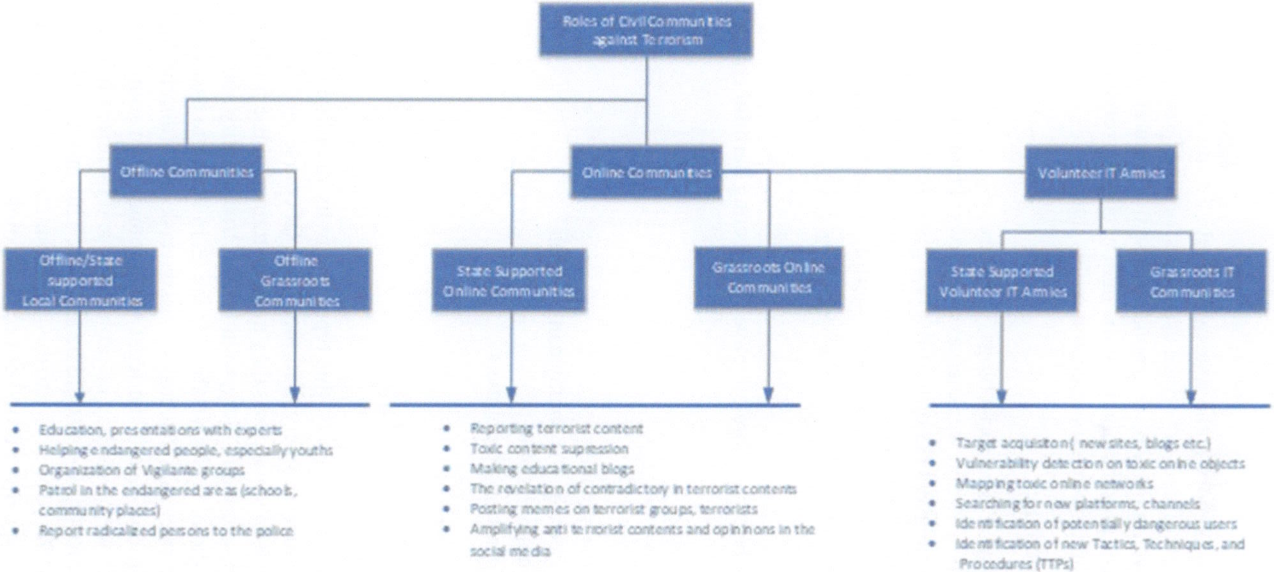

Fig. 4.1 Roles of civil communities against Terrorism

- The dissemination of memes discrediting terrorist content
- To strengthen and disseminate anti-terrorist content in the online space, especially on social media.

The second main category is the state-supported IT army, made up of IT professionals or volunteers with above-average user skills, or the IT volunteer force, organised on a completely voluntary basis around a cause or common interest or purpose. Since the focus of this presentation is on the fight against terrorism, the following is a list of tasks that can be taken into account in this fight.

Such Tasks May Include:

- Target detection such as terrorist blogs, websites
- Identify vulnerabilities and weaknesses and forward them where necessary
- Mapping the online networks of terrorist organisations
- Collecting and possibly identifying users of terrorist content
- Identifying potentially dangerous users, alerting the authorities
- Identify new tactics, techniques and procedures,

Advantages

- Community members are more effective in detecting harmful content
- Flexibility
- Low operating costs

Disadvantages

- No control over the members,
- Unchecked membership, the enemy can infiltrate
- Their activity is unpredictable
- Not uniform preparation of members

4.13.1 Security Projection

Depending on the specific activity, this type of activity in online communities may need to take into account the safety of its members. Of course, in differentiated ways, depending on the task, in some cases they must be trained to use pseudo-profiles and other measures to ensure their safety, to hide their true identity and to make it more difficult to identify them. It is advisable for the team leader to assess the members' knowledge in this respect and, if necessary, assign tasks accordingly. Another important consideration in this case is the legal background, as in some countries direct communication with terrorists may be prohibited, and therefore infiltration into the group for information purposes should be avoided. This task falls to the authorities. Members' attention should be drawn to this. Communication within the group should take place on an encrypted platform between endpoints.

4.13.2 Lesson to Learn, or Take Away

Of course, the above tasks and options are only examples. They may vary depending on the local legal environment, customs, culture and technical possibilities, but I am sure that in every country communities can and should be involved in some form in the fight against terrorism.

4.14 Conclusion

From what I have presented, it is clear that terrorism has been a major challenge almost from the beginning for different societies, which have tried to fight it with different means. Experience so far suggests that a simple law enforcement/military approach alone is not sufficient and often counterproductive, and that there is a need for a joint effort by legislators, law enforcement and the rest of society to combat the various terrorist radical groups. In particular, it is important to involve local communities in tackling online radicalisation, as the authorities alone are unable to address this challenge.

References

1. Frampton M (2021) History and the definition of terrorism. In: English R (ed) The Cambridge history of terrorism. Cambridge University Press, Cambridge, pp 31–57
2. Rappaport M (2015) The French revolution and early European revolutionary terrorism. In: Law RD (ed) The Routledge history of terrorism, vol 63. Routledge, London
3. Terrorism (2024) Encyclopaedia Britannica. https://www.britannica.com/topic/terrorism. Last accessed 16 Feb 2024
4. 1994 United Nations Declaration on Measures to Eliminate International Terrorism annex to UN General Assembly resolution 49/60, Measures to Eliminate International Terrorism, of December 9, 1994
5. Sproat PA (1997) An investigation of the Concept of State terrorism. PhD thesis, Department of Politics, University of Newcastle Upon Tyne, pp 114–119. https://core.ac.uk/download/pdf/153776026.pdf. Last accessed 16 Feb 2024
6. Directive (EU) 2017/541 of the European Parliament and of the Council of 15 March 2017 on combating terrorism and replacing Council Framework Decision 2002/475/JHA and amending Council Decision 2005/671/JHA. https://eur-lex.europa.eu/eli/dir/2017/541/oj
7. North Atlantic Treaty Organization, NATO Standardization Office (NSO): AAP-06 Edition 2019: NATO Glossary of Terms and Definitions (2019). https://www.coemed.org/files/stanags/05_AAP/AAP-06_2019_EF.pdf. Last accessed 16 Feb 2024
8. USC 2656f: Annual country reports on terrorism. Text contains those laws in effect on February 25, 2024. https://uscode.house.gov/view.xhtml?req=granuleid:USC-prelim-title22-section2656f&num=0&edition=prelim. Last accessed 16 Feb 2024
9. Hoffman B (2006) Inside terrorism. Columbia University Press, New York, p 31
10. Schmid AP (ed) (2011) The Routledge handbook of terrorism research (PDF), vol 74. Routledge Handbooks, Routledge, pp xii
11. Hoffman B (2017) Inside terrorism, 3rd edn. Columbia University Press, New York, pp 43–44

12. Tsintsadze-Maass E, Maass RW (2014) Groupthink and terrorist radicalization. Terror Political Violence 26(5):735–758
13. Bockstette C (2008) Jihadist terrorist use of strategic communication management techniques. George C. Marshall Center Occasional 777 Paper Series 20. http://www.marshallcenter.org/mcpublicweb/MCDocs/files/College/F_ResearchProgram/occPapers/occ-paper_20-en.pdf. Last accessed 16 Feb 2024
14. Higgins R (1996) The general international law of terrorism. In: Higgins R, Flory M (eds) International law and terrorism. Routledge, London, New York, p 28
15. Ganor B (1998) Defining terrorism. Is one man's terrorist another man's freedom fighter? Int Policy Inst Count Terror 12, Herzliya
16. Primoratz I (1990) What is terrorism? J Appl Philos 7(2):129–138. http://www.jstor.org/stable/24353583. Last accessed 16 Feb 2024
17. Poland JM (2005) Understanding terrorism. Pearson Education Inc., Upper Saddle River, New Jersey, p 16
18. Coleman JS (1988) Free riders and zealots: the role of social networks. Sociol Theory 6(1):52–57. https://doi.org/10.2307/201913. Last accessed 16 Feb 2024
19. Virani SN (2023) An old man, a garden, and an assembly of assassins: legends and realities of the Nizari Ismaili Muslims. Iran 61(2):272–284
20. Besenyő J, Gulyas A, Trifunovic D (2023) Hezbollah and the internet in the twenty-first century. Int J Intell Count Intell 36(3):669–685; Azani E (2013) The hybrid terrorist organization: Hezbollah as a case study. Stud Confl Terror 36(11):899–916
21. Moghadam A (2012) Failure and disengagement in the red army faction. Stud Confl Terror 35(2):156–181
22. Sundquist VH (2010) Political terrorism: an historical case study of the Italian red brigades. J Strateg Secur 3(3):53–68. http://www.jstor.org/stable/26463145. Last accessed 16 Feb 2024
23. Gregory S (2003) France and the war on terrorism. Terror Polit Violence 15(1):124–147; Dartnell M (1990) France's action directe: terrorists in search of a revolution. Terror Polit Violence 2(4):457–488
24. Escalante EE (2021) The shining path of Peru: an organizational theory for conflict. Free Market Institute Research Paper No. 3941380; McCormick GH (1987) The shining path and Peruvian terrorism. J Strat Stud 10(4):109–126
25. Wawrzyniec K (2020) The Irish republican army, or the nationalistic face of terrorism. A theoretical context and selected examples. Przegląd Zachodni 3(376):143–156. https://www.iz.poznan.pl/plik,pobierz,5026,fd965842acf18708385866fd257526a8/08.%20Konarski.pdf. Last accessed 16 Feb 2024
26. Poland JM (2005) Understanding terrorism. Pearson Education, Inc., Upper Saddle River, New Jersey, pp 34–35
27. Andrew C (2018) The secret world. A history of intelligence. Yale University Press, New Haven, pp 425–448
28. Bates RA (2012) Dancing with wolves: today's lone wolf terrorists. J Pub Prof Sociol 4(1). https://digitalcommons.kennesaw.edu/jpps/vol4/iss1/1. Last accessed 16 Feb 2024
29. Sproat PA (1997) An investigation of the concept of State terrorism. Ph.D. thesis, Department of Politics, University of Newcastle Upon Tyne. https://core.ac.uk/download/pdf/153776026.pdf. Last accessed 16 Feb 2024
30. Hagenloh PM (2015) State terrorism in early Twentieth-Century Europe. In Law RD (ed) The Routledge history of terrorism. Routledge, London, pp 159–176
31. Saul B (2006) Defining terrorism in international law. Oxford University Press, Oxford, p 3
32. Heims J (2022) Putin and Assad, partners in crime: why Russian forces steal wheat from Ukraine. International Centre for Defence and Security. https://icds.ee/content/uploads/dlm_uploads/2022/11/ICDS_Analysis_Putin_and_Assad_Partners_in_Crime_Jonas_Heins_November_2022.pdf. Last accessed 16 Feb 2024
33. Sonderegger L (2014) Torture and the fight against terrorism. Crime Law Soc Change 62:337–353. https://doi.org/10.1007/s10611-013-9469-3. Last accessed 16 Feb 2024

34. Mäkelä J (2003) Combating terrorism in Nordic countries: a comparative study of the Military's Role, vol 19. Department of Security and Strategic Studies, Swedish National Defence College. https://www.diva-portal.org/smash/get/diva2:428775/FULLTEXT01.pdf. Last accessed 16 Feb 2024
35. Tappeiner I (2005) The fight against terrorism. The lists and the gaps. Utrect Law Rev 1(1). https://utrechtlawreview.org/articles/https://doi.org/10.18352/ulr.5. Last accessed 16 Feb 2024
36. Duyvesteyn I (2008) Great expectations: the use of armed force to combat terrorism. Small Wars Insurg 19(3):328–351
37. Piotr B (2023) Understanding EU counter-terrorism policy. EPRS: European Parliamentary Research Service. Belgium. https://policycommons.net/artifacts/3495506/understanding-eu-counter-terrorism-policy/4296067/. Last accessed 16 Feb 2024
38. Gülzau F (2023) A new normal for the Schengen area. When, where and why member states reintroduce temporary border controls? J Borderlands Studies 38(5):785–803. https://doi.org/10.1080/08865655.2021.1996260. Last accessed 16 Feb 2024
39. Regulation (EC) No 562/2006 of the European Parliament and of the Council of 15 March 2006 establishing a Community Code on the rules governing the movement of persons across borders (Schengen Borders Code). https://eur-lex.europa.eu/legal-content/EN/TXT/?uri=celex%3A02006R0562-20131126. Last accessed 16 Feb 2024
40. Regulation (EU) 2017/2226 of the European Parliament and of the Council of 30 November 2017 (EU) No 1077/2011. https://eur-lex.europa.eu/legal-content/EN/TXT/?uri=CELEX%3A02017R2226-20210803. Last accessed 16 Feb 2024
41. Poecke TV, Wouters J (2022) The nexus between organized crime and terrorism: policy responses from the European Union and the Council of Europe. In: Paoli L, Fijnaut C, Wouters J (eds) The nexus between organized crime and terrorism: types and responses. Edward Elgar Publishing, Cheltenham, pp 469–471
42. Ilcus C (2023) European integrated border management: state of play and perspective, pp 21–24. https://doi.org/10.2139/ssrn.4662575. Last accessed 2024/02/16.
43. Preventing terrorism: clampdown on foreign fighters and lone wolves, European Parliament, Press Releases. https://www.europarl.europa.eu/news/en/press-room/20170210IPR61803/preventing-terrorism-clampdown-on-foreign-fighters-and-lone-wolves. Last accessed 16 Feb 2024
44. Pokalova E (2020) The returnee challenge. In: Returning Islamist foreign fighters. Palgrave Macmillan, Cham, p 2. https://doi.org/10.1007/978-3-030-31478-1_1. Last accessed 16 Feb 2024
45. European Commission (2020) Communication from the commission, a counter-terrorism agenda for the EU: anticipate, prevent, protect, respond, COM 795 final, Brussels (9.12.2020). https://eur-lex.europa.eu/legalcontent/EN/TXT/?uri=CELEX%3A52020DC0795&qid=1631885972581. Last accessed 16 Feb 2024
46. Vavoula N (2023) (Covert) Surveillance of foreign terrorism fighters via the Schengen Information System (SIS). New J Eur Crim Law 14(2):206–230. https://doi.org/10.1177/20322844231178927. Last accessed 16 Feb 2024
47. Ergün GO (2021) Police and judicial cooperation in criminal matters in the European Union. Ankara Üniversitesi Hukuk Fakültesi Dergisi 70(1):145–70. https://doi.org/10.33629/auhfd.900044. Last accessed 16 Feb 2024
48. Jones C (2012) Making fundamental rights flexible. Statewatch Anal. https://www.statewatch.org/media/documents/analyses/no-169-eu-pnr-us-aus-comparison.pdf. Last accessed 16 Fb 2024
49. Council of the European Union (2024) Air passenger data: Council and European Parliament reach provisional agreement to increase security and enhance border management. https://www.consilium.europa.eu/en/press/press-releases/2024/03/01/air-passenger-data-council-and-european-parliament-reach-provisional-agreement-to-increase-security-and-enhance-border-management/. Last accessed 11 Mar 2024
50. Poecke TV, Wouters J (2022) The nexus between organized crime and terrorism: policy responses from the European Union and the Council of Europe. In: Paoli L, Fijnaut C,

Wouters J (eds) The nexus between organized crime and terrorism: types and responses. Edward Elgar Publishing, Cheltenham, pp 466–467; Europol (2023) European Union Terrorism Situation and Trend report (2023). https://www.europol.europa.eu/publications-events/main-reports/tesat-report. Last accessed 16 Feb 2024
51. Dabetić A, Matijašević M (2023) European police cooperation in the counter terrorism efforts of the 21st Century. MB Univ Int Rev 1(2):127–133. https://doi.org/10.61837/mbuir010223127d. Last accessed 16 Feb 2024
52. Conclusions and plan of action of the extraordinary European Council meeting on 21 September 2001 (SN/140/01), https://www.consilium.europa.eu/media/20972/140en.pdf. Last accessed 16 Feb 2024
53. Document 32002F0475—council framework decision of 13 June 2002 on combating terrorism. https://eur-lex.europa.eu/legal-content/en/ALL/?uri=celex%3A32002F0475. Last accessed 16 Feb 2024
54. The 2005 EU Counter-terrorism Strategy. https://data.consilium.europa.eu/doc/document/ST%2014469%202005%20REV%204/EN/pdf. Last accessed 16 Feb 2024
55. Andreeva C (2021) EU counter-terrorism's gradual institutionalisation: information-sharing and Cooperation in Law enforcement and intelligence post-2015. PhD thesis of School of Law and Government, Dublin City University, pp 4–11. https://doras.dcu.ie/26183/. Last accessed 16 Feb 2024
56. Koster H (2020) Towards better implementation of the European Union's anti-money laundering and countering the financing of terrorism framework. J Money Laund Control 23(2):379–386. https://doi.org/10.1108/JMLC-09-2019-0073. Last accessed 16 Feb 2024
57. Atanasiu M (2020) Lone Wolfes Islamist terrorists as non-state actors and their impact on security policy of the European Union. Strateg Impact 77(100–125):104. https://www.ceeol.com/search/article-detail?id=937881. Last accessed 16 Feb 2024
58. Haffke L, Fromberger M, Zimmermann P (2020) Cryptocurrencies and anti-money laundering: the shortcomings of the fifth AML directive (EU) and how to address them. J Bank Regul 21:125–138. https://doi.org/10.1057/s41261-019-00101-4. Last accessed 16 Feb 2024
59. Besenyő J, Somogyi T (2024) Hawala and terror financing since 9/11. In: Romaniuk SN et al (eds) Money laundering and terrorism financing through Hawala money transfer operators. Palgrave Macmillan
60. Council Common Position of 27 December 2001 on the application of specific measures to combat terrorism, 2001/931/CFSP of 27 December 2001. https://eur-lex.europa.eu/legal-content/EN/TXT/?uri=celex%3A32001E0931. Last accessed 16 Feb 2024
61. Margulies P (2016) Terrorist sanctions: the clash in US and EU approaches, Lawfare. https://www.lawfareblog.com/terrorist-sanctions-clash-us-and-eu-approaches. Last accessed 16 Feb 2024
62. Council Decision 2016/1693/CFSP of 20 September 2016 concerning restrictive measures against ISIL (Da'esh) and Al-Qaeda and persons, groups, undertakings and entities associated with them, https://eur-lex.europa.eu/legal-content/GA/TXT/?uri=CELEX:32016D1693. Last accessed 16 Feb 2024
63. Poecke TV, Wouters J (2022) The nexus between organized crime and terrorism: policy responses from the European Union and the Council of Europe. In: Paoli L, Fijnaut C, Wouters J (eds) The nexus between organized crime and terrorism: types and responses. Edward Elgar Publishing, Cheltenham, p 459
64. Poecke TV, Wouters J (2022) The nexus between organized crime and terrorism: policy responses from the European Union and the Council of Europe. In: Paoli L, Fijnaut C, Wouters J (eds) The nexus between organized crime and terrorism: types and responses. Edward Elgar Publishing, Cheltenham, pp 460–462
65. Weimann G (2016) The emerging role of social media in the recruitment of foreign fighters. In: de Guttry A, Capone F, Paulussen C (eds) Foreign fighters under international law and beyond. T.M.C. Asser Press, The Hague, pp 78, 80–83. https://doi.org/10.1007/978-94-6265-099-2_6. Last accessed 16 Feb 2024

66. Poecke TVWouters J (2022) The nexus between organized crime and terrorism: policy responses from the European Union and the Council of Europe. In: Paoli L, Fijnaut C, Wouters J (eds) The nexus between organized crime and terrorism: types and responses. Edward Elgar Publishing, Cheltenham, pp 467–469
67. McCarthy Hartman N (2023) Countering terrorist content online: removal = success?: a critical discourse analysis of the EU regulation 2021/784. Student thesis (basic), Swedish Defence University (2023). https://www.diva-portal.org/smash/get/diva2:1840522/FULLTEXT01.pdf. Last accessed 16 Feb 2024
68. Poecke TV, Wouters J (2022) The nexus between organized crime and terrorism: policy responses from the European Union and the Council of Europe. In: Paoli L, Fijnaut C, Wouters J (eds) The nexus between organized crime and terrorism: types and responses. Edward Elgar Publishing, Cheltenham, p 467
69. European Commission (2020) Communication from the Commission to the European Parliament and the Council, First Progress Report on the EU Security Union Strategy, COM(2020) 797 final, Brussels (9.12.2020). https://eur-lex.europa.eu/legal-content/EN/TXT/HTML/?uri=CELEX:52020DC0797. Last accessed 16 Feb 2024
70. European Commission (2020) Communication from the commission, a counter-terrorism agenda for the EU: anticipate, prevent, protect, respond, COM 795 final, Brussels (9.12.2020). https://eur-lex.europa.eu/legal-content/EN/TXT/?uri=CELEX%3A52020DC0795&qid=1631885972581. Last accessed 16 Feb 2024
71. Gulyás A, Demeter M, Besenyő J (2023) The Lernaean hydra on the internet: deplatformization-resistant media ecosystem of the Islamic State. Media War Confl 16(4):1–24. https://journals.sagepub.com/doi/https://doi.org/10.1177/17506352231206306. Last accessed 16 Feb 2024
72. Solon O (2017) Global network of 'hunters' aim to take down terrorists on the internet. The Guardian (Friday 21 July 2017). https://www.theguardian.com/technology/2017/jul/21/global-network-hunters-against-terrorists-on-the-internet. Last accessed 16 Feb 2024

Chapter 5
Fuzzy Logic-Based Risk Assessment Evaluating Physiological Values

Felisberto David Wandi Chivela, Zoltán Papp, and Edit Laufer

Abstract Health monitoring plays a crucial role for those involved in sports and physical activities, as each individual presents unique physiological responses under varying physical loads. Traditional health assessment methods, however, often fall short in taking this difference in individual response into account, emphasizing the need for more personalized approaches. This paper introduces a hierarchical fuzzy model designed to evaluate the risk levels associated with physical activities. Using a personal statistics-based approach to fine-tune membership functions, the model effectively personalizes evaluations to better reflect individual health profiles. The model shows both numerical and linguistic assessments of risk, demonstrating a significant correlation between the improved membership functions and the existing medical recommendations. This study not only showcases the potential of fuzzy logic in improving health monitoring but also lays important groundwork for future research focused on developing innovative mathematical methods to represent patient statistics and refine membership function fitting.

Keywords Health monitoring · Risk evaluation · Fuzzy logic

F. D. W. Chivela
Doctoral School of Applied Informatics and Applied Mathematics, Óbuda University, Budapest, Hungary

Z. Papp (✉)
Subotica Tech College of Applied Sciences, Subotica, Serbia
e-mail: papzoli@vts.su.ac.rs

E. Laufer
Bánki Donát Faculty of Mechanical and Safety Engineering, Óbuda University, Budapest, Hungary

T. A. Kovács and I. Fürstner (eds.), *Critical Infrastructure Protection: Advanced Technologies for Crisis Prevention and Response*, NATO Science for Peace and Security Series C: Environmental Security, https://doi.org/10.1007/978-94-024-2308-2_5

5.1 Introduction

The integration of Internet of Things (IoT) technology in healthcare has brought significant advancements in patient monitoring and preventive care. IoT-based solutions now enable continuous remote monitoring of physiological parameters, transforming how healthcare providers and patients manage health. In recent years, the importance of IoT-based monitoring has become indispensable, especially highlighted during the COVID-19 pandemic, which has shown the critical need for technologies that reduce in-person visits while ensuring comprehensive health tracking for early intervention. These solutions allow for prompt recognition of health issues, improving safety and promoting quick medical responses. Such advances contribute significantly to disease prevention, management of chronic conditions, and overall patient safety by enabling healthcare providers to maintain real-time access to essential health data [1, 2].

IoT-enabled health monitoring systems have a wide range of devices that can track vital health information remotely. These systems range from basic gadgets, such as heart rate and blood pressure monitors, to advanced technologies like wearables and implanted devices. Popular examples include devices such as cardiac pacemakers, fitness-tracking wearables like smart wristbands, and sophisticated hearing aids, all of which utilize sensors to gather vital health information. By forming a network of intelligent devices capable of collecting, processing, and transmitting health data, IoT-based systems support connectivity between home monitoring and hospital-based systems, improving access to patient data across healthcare settings. Such connectivity benefits not only elderly patients or those with chronic illnesses but also individuals recovering from conditions and even healthy individuals tracking their fitness levels. These systems promote continuous health monitoring, encouraging a proactive approach to healthcare management [3, 4].

The functionality of these IoT health monitoring systems depends on various features, such as wireless communication, non-invasive data collection, portability, and ease of use. These characteristics make them particularly suitable for the management of chronic disease and disease prevention, allowing users to access essential health information conveniently. The designs are compact and user-friendly, which further improves accessibility for a wide variety of patients. However, with these technological advances also come significant challenges, particularly regarding the complexity of medical data. Health data does not always fit into precise numerical categories, as the distinction between "standard levels" and "indicative abnormalities" values can often be ambiguous and context dependent. Addressing this ambiguity is indispensable to making these devices as effective as possible in promoting personalized healthcare and accurate diagnosis [5, 6].

To address these challenges, soft computing techniques, especially fuzzy logic, provide flexible solutions for handling the inherent uncertainty and imprecision in medical assessments. Fuzzy logic is highly effective at managing vague and subjective information, aligning well with the qualitative nature of medical descriptions and evaluations. It is well-suited for tasks that require human-like reasoning, making it valuable for translating complex health data into easily interpretable results. Through its ability to accommodate linguistic variables, fuzzy logic offers healthcare professionals and patients a way to understand health indicators on a more intuitive level, promoting informed decision-making. In healthcare, fuzzy logic is particularly advantageous, as it allows for the classification of physiological parameters in a manner that accounts for more detailed variations in individual health conditions [7].

Personalized patient models benefit significantly from the fuzzy logic approach, yet creating these models remains complex. Health outcomes are influenced by numerous factors, some of which may interact in unknown or unpredictable ways. As fuzzy models are customized for individual needs, they can be fine-tuned based on the unique conditions and lifestyle factors of each patient. However, to fully account for every possible aspect is still challenging. Methods for personalizing health assessments include using patient-specific data, such as individual medical recommendations, and establishing criteria based on statistical data related to demographic factors, such as age, sex, and fitness level. These adjustments can be further refined through fuzzy-neural systems, which adjust membership functions using input–output data, allowing for an even greater level of customization to meet the specific needs of each patient [8, 9].

The health benefits of regular physical activity are well-documented, covering a wide range of improvements in both physical and mental well-being. Regular exercise boosts cardiovascular health, reduces the risk of chronic illnesses, and aids weight management by building muscle and burning calories. However, these benefits can only be fully achieved when exercise is conducted within safe limits customized to an individual's physical condition. Exceeding these limits can lead to overexertion or injury, which illustrates the importance of personalized exercise plans based on individual capabilities and health profiles [10, 11].

The purpose of this study is to develop a risk assessment model that customizes health evaluations based on individual patient characteristics, with a focus on the "Current Physical Status" subsystem. This subsystem benefits from patient-specific membership functions that enhance the accuracy and relevance of health assessments. This study emphasizes the importance of patient-specific statistics in the assessment process, thus emphasizing the need to address individualized health monitoring and risk management. Through this approach, the study seeks to improve the effectiveness of health monitoring systems by incorporating customized fuzzy logic models that consider a broader range of influential health factors.

The paper is organized as follows. Section 5.2 describes the hierarchical structure of the model, which serves as the foundation for the proposed risk assessment framework. Section 5.3 examines the 'Current Physical Status' subsystem, elaborating on the customization of membership functions used to improve patient-specific evaluations and the risk assessment process. Section 5.4 presents the statistical approach. Section 5.5 shows a case study as a practical application of the proposed model. Section 5.6 discusses the findings and implications of incorporating patient-specific data, and Sect. 5.7 concludes with a summary of the results and potential future research directions.

5.2 Hierarchical Model Structure Overview

The design of the proposed model includes a hierarchical, clustered structure, allowing for both scalability and ease of evaluation. The evaluation uses fuzzy logic, specifically the Mamdani-type inference system. The organization of the model aligns with the logical flow of the evaluation process, facilitating the classification of input parameters according to their logical relationships. Parameters can be categorized into three primary groups: the patient's medical condition, their athletic behaviors, and the surrounding environmental factors.

To provide detailed understanding of the patient's health, the 'Medical Condition' category represents the most relevant and complex subsystem within the model. Individual health conditions primarily influence the patient's 'Activity Load', and the majority of interactions among input parameters occur within this subsystem. The first input, Disease Condition, includes chronic illnesses such as hypertension, diabetes, and heart disease; although considered quasi-permanent factors, the severity of these conditions can change over time. The second input, Current Physical Status, provides information regarding the patient's immediate health, based on indicators such as heart rate and both systolic and diastolic blood pressures. This assessment can be improved by incorporating additional parameters and factors that impact this data, which are essential in developing personalized membership functions.

Parameters within the primary categories can be further distinguished based on their stability. This includes permanent factors (e.g., sex), quasi-permanent factors that change infrequently (e.g., chronic illnesses and occupation), and dynamic variables that fluctuate in real-time (e.g., blood pressure and heart rate). This structural representation is based on the model described in [12], as illustrated in Fig. 5.1.

The Environmental Condition subsystem combines various subfactors of the surrounding environment. Factors such as temperature, humidity (TH), and wind (TW) are combined due to their impact on thermal sensation, which is especially relevant for outdoor sports. However, temperature and humidity can also influence risk levels in indoor settings [13].

To detail the patient's exercise habits, the Activity Load subsystem is used, which specifies the intensity, duration, and frequency of their physical activities. This subsystem shows how vigorously the activity is performed (Intensity), the length of each session (Duration), and how often the activity occurs each week (Frequency).

The parameters in the Current Physical Status subsystem, which is the primary focus of this paper, experience real-time changes. Continuous monitoring of these parameters allows for consistent control. Therefore, it is crucial to customize this subsystem for each individual patient, reiterating the need of a flexible and adaptable risk assessment framework. While the individual parameters within each subsystem can be assessed in real-time assessment, the other components of the model are evaluated beforehand offline, prior to initiating real-time evaluations.

In the diagram associated with this model, the left section identifies risk factors that impact assessed risk level throughout the evaluation process. The central area categorizes the parameters associated with each risk factor group into three distinct

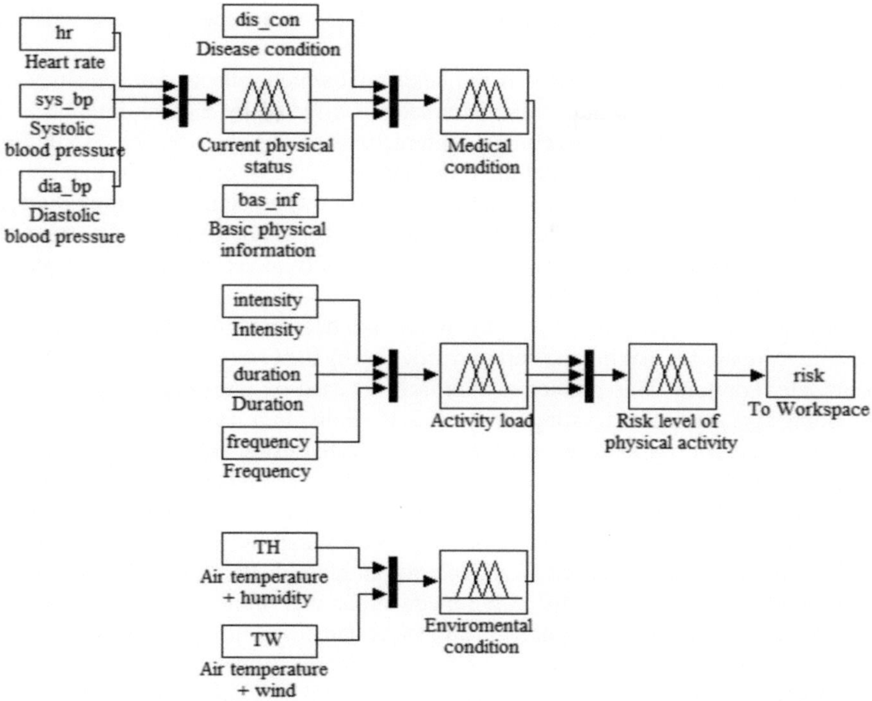

Fig. 5.1 The hierarchical fuzzy model framework

blocks. The highest level of the hierarchy, located on the right side, calculates the actual risk level by integrating inputs from the identified risk factor groups. These primary categories function as the subsystems of the model, with each subsystem's contribution to the overall risk level being calculated separately during processing.

5.3 Evaluation Subsystem for the Current Physical Status

The target zones that will be described in this chapter are based on the Polar heart-rate zones [14], the membership functions of the target heart rate zone (represented as a percentage of the maximum heart rate) can be found in [15]. This categorization corresponds to the original Polar zones: vlow (<50%), light (50–60%), medium (60–70%), mhigh (70–85%), high (85–95%), and vhigh (>95%). This percentage reflects the acceptable intensity level for each individual, varying based on the activity's goals, be it rehabilitation or prevention—as well as the person's athletic experience (e.g., regular athlete versus beginner). As a result, personalized zone limits can be calculated, and membership functions can be tailored to the specific characteristics of each patient.

The initial parameter assessed is the heart rate, which reflects the frequency of heart contractions per minute. This parameter is affected by various internal and external influences, leading to complex interactions among these factors. To define patient-specific membership functions related to heart rate, we consider the maximum heart rate, which denotes the peak value achievable during activity. Approximately 30 distinct factors can impact this limit, such as age, gender, body mass, time of day, environmental variables, physical fitness, intensity of activity, emotional state, medications, and specific health conditions [16].

Instead of exclusively relying on the maximum heart rate, it is common to use a predictive model for maximum heart rate provided by the OwnZone function in Polar heart rate monitors. This predictive approach is based on personal characteristics, including age, gender, and resting heart rate of the individual. The OwnZone function analyzes heart rate variability, which can be evaluated using the Polar device or an oscillometric blood pressure monitor. Determining the predicted maximum heart rate requires only a brief 5 min assessment, in contrast to traditional techniques such as progressive exercise testing or VO2max evaluation, which typically occur in clinical settings and may not be suitable for all individuals, especially those with heart conditions or older adults taking medications that influence cardiac function and circulation. Moreover, this method allows for daily or even multiple assessments, which account for variations in maximum heart rate over time.

Limits can be established according to the patient's medical condition. A patient undergoing medical treatment or rehabilitation for cardiovascular or respiratory diseases would typically have vlow as the appropriate target zone (<50%). In this context, mhigh corresponds to the original light zone (50–60%), while vhigh represents the combination of the remaining zones (medium, mhigh, high, vhigh) (>60%).

In contrast, for the target zones for a beginner-level athlete, the appropriate zones would be light or medium, which leads to the merging of vlow, light, and medium into a target zone of <70%, with mhigh corresponding to the original mhigh zone (70–85%), and vhigh combining the remaining zones (high, vhigh) (>85%). For someone who regularly engages in exercise, the most suitable target zone is mhigh based on the original classification, thus resulting in a target zone that combines vlow, light, medium, and mhigh into one (<85%), while mhigh represents the original high zone (85–95%) and vhigh corresponds to the original vhigh (>95%).

The second parameter we analyze is blood pressure, which is widely acknowledged as the most significant risk factor for cardiovascular issues. Blood pressure indicates the force of blood against the walls of blood vessels, particularly arteries. This force exhibits rhythmic fluctuations corresponding to the heart's contractions. Systolic pressure represents the maximum pressure created by the contracting left ventricle, whereas diastolic pressure signifies the lowest level during the relaxation phase. To conduct a thorough assessment of blood pressure, both systolic and diastolic measurements must be considered, thereby constituting two additional input parameters within the Current Physical Status subsystem.

Typically, systolic pressure increases during exercise, while diastolic pressure remains stable or may experience a slight decrease. The responses at maximal or submaximal exercise levels vary according to factors such as age, gender, and physical fitness. Generally, older individuals display higher blood pressure measurements, though this relationship inversely correlates with fitness levels, as improved fitness tends to yield lower measured values. Moreover, males typically show higher maximum systolic blood pressure than females. The blood pressure thresholds referenced in this paper are derived from maximum values associated with age and gender and are adjusted according to target zones, which further influence optimal blood pressure measurements. The adjustments are as follows, systolic blood pressure is categorized with the following parameters: low (<187), normal (187–204), increased (204–220), and abnormal (>220). Similarly, diastolic blood pressure is classified into the following categories: low (<80), normal (80–84), increased (85–90), and abnormal (>90). Table 5.1 provides a summary of the maximum systolic and diastolic blood pressure values segmented by age and gender, as detailed in [17].

The output membership functions, which indicate risk levels, are defined as follows: vsafe (<0.2), msafe (0.2–0.4), medium (0.4–0.6), mdangerous (0.6–0.8), and vdangerous (0.8–1) [12]. Here, vsafe represents the lowest risk category, while vdangerous indicates the highest risk level.

The structure of the Current Physical Status subsystem is represented in Fig. 5.2. Input factors are positioned on the far left, while influential factors critical for refining membership functions are located in the middle of the figure. The input factors include heart rate (HR), systolic blood pressure (SBP), and diastolic blood pressure (DBP), whereas influential factors comprise disease condition—indicating whether the individual is diseased or a beginner/advanced athlete—basic physical information such as age and sex, and the training target. This subsystem is a detailed version of the "Current Physical Status" input within the overall model, as depicted in Fig. 5.1 [15].

Table 5.1 Normal blood pressure response

Age group	20–29	30–39	40–49	50–59	60–69	70–79
Systolic (Men)	161–203	164–204	167–209	170–216	173–221	169–223
Diastolic (Men)	59–83	64–88	68–92	71–95	72–96	71–97
Systolic (Women)	136–176	138–182	144–190	153–201	162–210	160–210
Diastolic (Women)	58–82	63–85	67–89	69–93	68–84	73–93

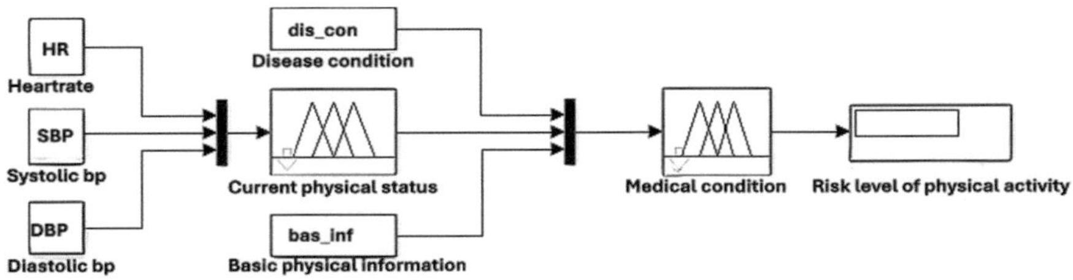

Fig. 5.2 Current physical status subsystem

5.4 Statistical Evaluation and Membership Function Overview

Flexible models that can easily be customized are essential in the generation of membership functions, particularly when dealing with subjective concepts. Without standard evaluation measures, achieving accuracy is challenging, especially in practical problems that require a representation of uncertain or ambiguous data. As a result, various adaptable techniques have been explored for adjusting membership functions to specific data types and application needs.

The absence of definitive guidelines for selecting the appropriate methods adds to the challenge of deriving membership functions from a given data. Dubois and Prade [18] highlight the subjectivity in defining and interpreting membership functions, emphasizing that various interpretations can lead to differing representations. This has led to a range of methods for generating membership values, each aimed at a particular interpretation or desired output.

Multiple methods for constructing membership functions are available, as thoroughly reviewed in [19]. These methods include Heuristic Techniques, where expert knowledge and problem-specific intuition form the basis for function creation. Probability-Possibility Transformations leverage probabilistic data to develop fuzzy membership functions, relying on statistical information to manage uncertainty. Cluster Analysis, such as the fuzzy c-means (FCM) algorithm, classifies data into clusters, with membership degrees assigned to achieve smooth function. Neural Networks are employed to dynamically learn membership functions, which are particularly useful for complex pattern recognition scenarios. Genetic Algorithms draw on evolutionary principles to optimize fuzzy system parameters, improving performance. Additionally, Histograms and Density Estimation represent data distributions visually, aiding in the creation of functions that accurately reflect data patterns. The present study applies the histogram method, which will be further elaborated upon in the following sections.

Medaglia's contribution [20] introduced Bézier curve-based membership functions for convex normal fuzzy sets, presenting an intuitive, efficient alternative to traditional methods. This method simplifies the creation of custom membership functions through straightforward shape adjustments, offering improved adaptability and making it accessible to users with varying levels of technical expertise. Furthermore,

Medaglia's technique is particularly effective in handling diverse data shapes, that can be more challenging for traditional methods that require extensive computations.

Methods for constructing membership functions are well-documented for their application to pattern recognition; however, they are becoming increasingly relevant for medical applications, such as patient monitoring. Common shapes for fuzzy sets include triangular, trapezoidal, and Gaussian functions [21, 22]. In this paper, we make use of trapezoidal functions, which are often represented as follows:

$$f(x; a, b, c, d) = \begin{cases} 0, & x < a \\ \frac{x-a}{b-a}, & a \leq x \leq b \\ 1, & b \leq x \leq c \\ \frac{d-x}{d-c}, & c \leq x \leq d \\ 0, & x > d \end{cases} \tag{5.1}$$

where, $a, b,$ $c$and d represent the trapezoid's parameters.

Devi and Sarma [23] proposed a parametric approach to estimate fuzzy membership functions using histograms and rational function approximation to fit the data. The parameters are derived through least squares and normalization to ensure a maximum function value of one. This approach is particularly useful for classifying unlabeled samples; for each element within the class, histograms are constructed, and the function parameters are determined accordingly.

To effectively customize the model for individual patients, the input membership functions are modified according to personal profiles. For ease of computation and interpretation, trapezoidal functions are used to represent these patient-specific parameters.

5.4.1 Membership Functions Using Histogram Fitting

The interaction between input parameters in a fuzzy logic model often involves a high level of complexity, making it challenging to assess their effects on the resulting values with precision. However, data collection during monitoring sessions can play a vital role in evaluating a patient's current state and establishing a personalized profile for future assessments. By referring to previous measurements recorded under similar conditions, these stored data points can provide context for evaluating the patient's present status. Statistical tools, such as histograms, are then created from these historical values to gain deeper understanding of typical physiological responses. These histograms illustrate the patient's normal reaction under specific conditions.

Once histograms have been constructed, they can form the basis for defining membership functions (MFs). The membership function derived from histogram data is typically piecewise linear, with the highest point aligning with the domain where the histogram reaches its maximum value. The rest of the function is formulated from the remaining histogram values, resulting in a membership function that reflects the patient's standard response patterns under observed conditions. For a more comprehensive discussion of this method, refer to [24, 25].

In this process, histograms are used for the development of membership functions for parameters such as heart rate (HR), systolic blood pressure (SBP), and diastolic blood pressure (DBP). After constructing these histogram-based MFs, adjustments can be made to the original functions to improve their reliability. By incorporating the patient's normal responses along with medical recommendations, these adapted functions—referred to as "improved MFs"—are obtained. This improvement is achieved by averaging the corresponding parameters of the histogram-derived functions with those of the initial membership functions. Through this approach, MFs become more customized to individual patient profiles, enabling more accurate and personalized evaluations.

5.5 Case Study Evaluation

In this section, we analyze the membership functions that represent medical recommendations for an "Advanced Male 20–29" alongside measurement-based statistics for a specific athlete. The Mamdani-type inference system is implemented using MATLAB's Fuzzy Logic Designer, integrating a well-defined rule base for decision-making.

Starting with heart rate, the recommended maximum heart rate and target exercise intensity (as outlined in Fig. 5.2) are key in defining the membership functions. Three primary fuzzy sets are used: "target," which denotes the ideal heart rate zone for the patient; "mhigh," indicating a higher, moderately elevated heart rate; and "vhigh" representing a very high heart rate that is not advisable for this group. These zones, together with the input values, are expressed as percentages of the maximum heart rate. Table 5.2 illustrates the heart rate fuzzy sets for the "Advanced Male 20–29" category, which reflects athletes aged 20–29 who regularly participate in advanced-level sports.

Next, the membership functions for systolic blood pressure (SBP) are presented in Table 5.3. Here, the fuzzy sets are categorized as "low" (lower-than-normal values), "normal" (optimal SBP), "increased" (a slightly higher but still acceptable level), and "abnormal" (values posing a risk to the patient). In a similar manner, Table 5.4 displays the antecedent sets for diastolic blood pressure (DBP), mapping the variations in DBP for this specific category.

With these influential parameters established, we can now adjust the membership functions to align with the unique characteristics of each patient. Equations 5.13–5.15 illustrate the membership functions based on statistics for a 21-year-old elite badminton player. This athlete's personal parameters include a resting heart rate (HRrest) of 60 bpm, a maximum heart rate (HRmax) of 195 bpm, a systolic blood pressure (SBP) of 120 mmHg, a maximum SBP of 235 mmHg, a diastolic blood pressure (DBP) of 55 mmHg, a maximum DBP of 82 mmHg, and a weight of 175 kg. This athlete underwent an incremental treadmill test to measure maximal

Table 5.2 Heart rate target zones for advanced male athletes aged 20–29

Category	Membership function
Target (HR)	$$\mu_{target}(HR_1) = \begin{cases} 0, & if \quad HR_1 < 0 \\ 1, & if \quad 0 \leq HR_1 \leq 0.82 \\ \frac{0.88-HR_1}{0.06}, & if \quad 0.82 < HR_1 < 0.88 \\ 0, & if \quad HR_1 \geq 0.88 \end{cases} \qquad (2)$$
Moderately high (mhigh)	$$\mu_{mhigh}(HR_2) = \begin{cases} 0, & if \quad HR_2 < 0.82 \\ \frac{HR_2-0.82}{0.06}, & if \quad 0.82 \leq HR_2 < 0.88 \\ 1, & if \quad 0.88 \leq HR_2 \leq 0.92 \\ \frac{0.98-HR_2}{0.06}, & if \quad 0.92 < HR_2 < 0.98 \\ 0, & if \quad HR_2 \geq 0.98 \end{cases} \qquad (3)$$
Very high (vhigh)	$$\mu_{vhigh}(HR_3) = \begin{cases} 0, & if \quad HR_3 < 0.92 \\ \frac{HR_3-0.92}{0.06}, & if \quad 0.92 \leq HR_2 < 1.02 \\ 1, & if \quad 1.02 \leq HR_3 \leq 1.08 \\ 0, & if \quad HR_3 > 1.08 \end{cases} \qquad (4)$$

Table 5.3 Systolic blood pressure ranges for advanced male athletes aged 20–29

Category	Membership function
Low (SBP)	$$\mu_{low}(SBP_1) = \begin{cases} 0, & if \quad SBP_1 < 0 \\ 1, & if \quad 0 \leq SBP_1 \leq 130 \\ \frac{130-SBP_1}{10}, & if \quad 130 < SBP_1 < 140 \\ 0, & if \quad SBP_1 \geq 140 \end{cases} \qquad (5)$$
Normal (SBP)	$$\mu_{normal}(SBP_2) = \begin{cases} 0, & if \quad SBP_2 < 130 \\ \frac{SBP_2-130}{10}, & if \quad 130 \leq SBP_2 < 140 \\ 1, & if \quad 140 \leq SBP_2 \leq 188 \\ \frac{188-SBP_2}{10}, & if \quad 188 < SBP_2 < 198 \\ 0, & if \quad SBP_2 \geq 198 \end{cases} \qquad (6)$$
Increased (SBP)	$$\mu_{increased}(SBP_3) = \begin{cases} 0, & if \quad SBP_3 < 188 \\ \frac{SBP_3-188}{10}, & if \quad 188 \leq SBP_2 < 198 \\ 1, & if \quad 198 \leq SBP_3 \leq 204 \\ \frac{214-SBP_3}{10}, & if \quad 204 < SBP_3 < 214 \\ 0, & if \quad SBP_3 \geq 214 \end{cases} \qquad (7)$$
Abnormal (SBP)	$$\mu_{abnormal}(SBP_4) = \begin{cases} 0, & if \quad SBP_4 < 204 \\ \frac{SBP_4-204}{10}, & if \quad 204 \leq SBP_4 < 214 \\ 1, & if \quad 214 \leq SBP_4 \leq 276 \\ 0, & if \quad SBP_4 > 276 \end{cases} \qquad (8)$$

Table 5.4 Diastolic blood pressure ranges for advanced male athletes aged 20–29

Category	Membership function
Low (DBP)	$\mu_{low}(DBP_1) = \begin{cases} 0, & if \quad DBP_1 < 0 \\ 1, & if \quad 0 \le DBP_1 \le 53 \\ \frac{53-DBP_1}{12}, & if \quad 53 < DBP_1 < 65 \\ 0, & if \quad DBP_1 \ge 65 \end{cases} \quad (9)$
Normal (DBP)	$\mu_{normal}(DBP_2) = \begin{cases} 0, & if \quad DBP_2 < 53 \\ \frac{DBP_2-53}{12}, & if \quad 53 \le DBP_2 < 65 \\ 1, & if \quad 65 \le DBP_2 \le 77 \\ \frac{89-SBP_2}{12}, & if \quad 77 < DBP_2 < 89 \\ 0, & if \quad DBP_2 \ge 89 \end{cases} \quad (10)$
Increased (DBP)	$\mu_{increased}(DBP_3) = \begin{cases} 0, & if \quad DBP_3 < 77 \\ \frac{DBP_3-77}{12}, & if \quad 77 \le DBP_3 < 89 \\ 1, & if \quad 89 \le DBP_3 \le 96.5 \\ \frac{108.5-DBP_3}{12}, & if \quad 96.5 < DBP_3 < 108.5 \\ 0, & if \quad DBP_3 \ge 108.5 \end{cases} \quad (11)$
Abnormal (SBP)	$\mu_{abnormal}(SBP_4) = \begin{cases} 0, & if \quad DBP_4 < 96.5 \\ \frac{DBP_4-96.5}{12}, & if \quad 96.5 \le DBP_4 < 108.5 \\ 1, & if \quad 108.5 \le DBP_4 \le 111.5 \\ 0, & if \quad DBP_4 > 111.5 \end{cases} \quad (12)$

oxygen consumption (VO2max), anaerobic threshold, and time to exhaustion. Initial treadmill speed was set at 2.7 km/h with a 10% incline, increasing incrementally every 3 min until voluntary exhaustion. Heart rate variability was also assessed over a 5 min resting period using a Polar monitor in a supine position [26]. Throughout the analysis, the HR, SBP, and DBP were incrementally increased from resting values to near maximum levels.

$$\mu(HR) = \begin{cases} 0, & HR < 0.31 \\ \frac{HR-0.31}{0.15}, & 0.31 \le HR \le 0.46 \\ 1, & 0.46 < HR < 0.62 \\ \frac{0.77-HR}{0.15}, & 0.62 \le HR \le 0.77 \\ 0, & HR > 0.77 \end{cases} \tag{5.13}$$

where HR is the input is the input variable for heart rate.

$$\mu(SB) = \begin{cases} 0, & SB < 130 \\ \frac{SB-130}{10}, & 130 \leq SB \leq 140 \\ 1, & 140 < SB < 180 \\ \frac{190-SB}{10}, & 180 \leq SB \leq 190 \\ 0, & SB > 190 \end{cases} \tag{5.14}$$

where SB is the input is the input variable for the sistolic blood pressure.

$$\mu(DB) = \begin{cases} 0, & DB < 55 \\ \frac{DB-55}{12}, & 55 \leq DB \leq 67 \\ 1, & 67 \leq DB \leq 76 \\ \frac{82-DB}{6}, & 76 \leq DB \leq 82 \\ 0, & DB > 82 \end{cases} \tag{5.15}$$

where DB is the input variable for the diastolic blood pressure.

To improve accuracy, the statistics-based membership functions were compared with medical recommendations to create "improved membership functions" (as shown in Figs. 5.3, 5.4 and 5.5). These improved membership functions (Figs. 5.6, 5.7 and 5.8) were subsequently applied to evaluate the patient's risk level (Table 5.5).

The HR domain for the statistics-based function ranges from 0 to 0.77, while the recommendation function spans 0–0.88. Notably, from 0.62 to 0.88, the athlete-specific model identifies risk levels sooner and offers a more refined risk assessment.

Within the interval [180, 190], the statistics-based function shows SBP values as moderate-to-high risk earlier than the recommendation function, which considers this range at slightly higher values ([188, 198]). This early detection of risk reflects a more sensitive assessment in the personalized model.

For DBP, values in the range [76, 82] trigger elevated risk levels sooner in the athlete-specific function compared to the recommendation function, which does not categorize risk until values approach 89. This increased sensitivity facilitates early risk identification, particularly during high-intensity activities.

Our evaluations of input domains reveal key differences between statistics-based and medical recommendation functions. Specific ranges—[0.62, 0.77] for HR, [180, 190] for SBP, and [76, 82] for DBP—demonstrate the improved risk classification achieved through personalized models. These distinctions reflect the importance of individual health monitoring, which enables fast and precise interventions suited to unique physiological responses, particularly among athletes.

Incorporating the improved membership functions improves the analysis by balancing both the athlete-specific and medically recommended ranges. These functions allow for a gradual alignment with the medical recommendations while maintaining the early warnings occurring through the athlete-specific parameters. By utilizing these "improved" MFs, this approach provides comprehensive monitoring, supporting cardiovascular health management, optimizing performance, and increasing safety for athletes.

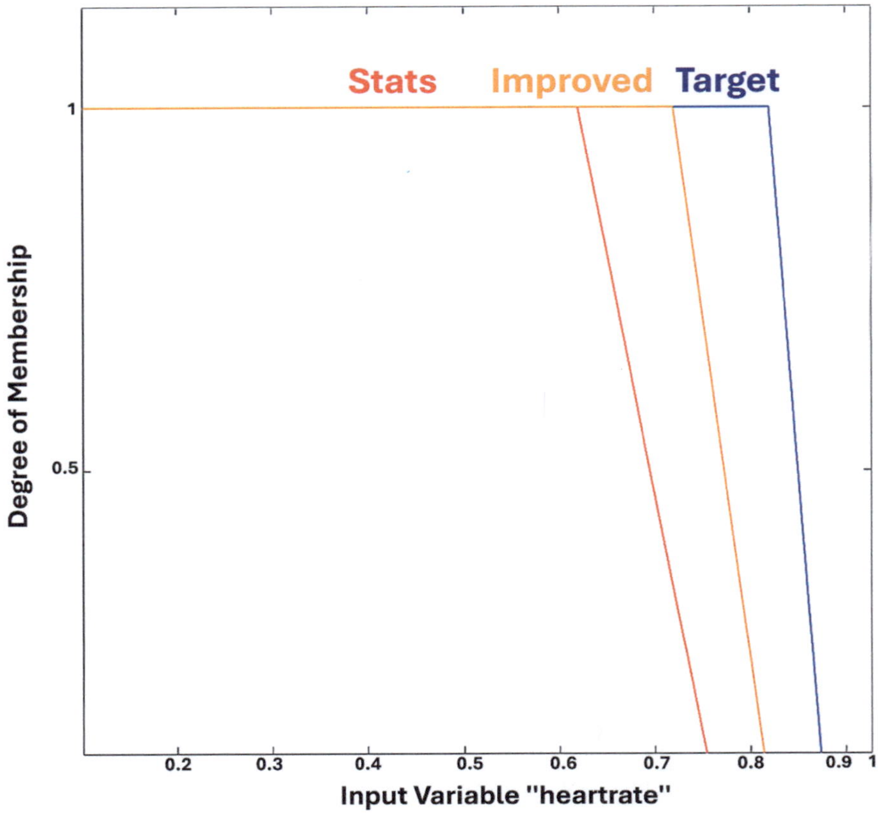

Fig. 5.3 Comparison of heart rate membership functions: statistics-based (input values: [0, 0, 0.62, 0.77]) versus medical recommendations (input values: [0, 0, 0.82, 0.88])

5.6 Results

Table 5.5 provides a detailed risk assessment for a 21-year-old elite male badminton athlete, focusing on critical metrics such as heart rate (HR), systolic blood pressure (SBP), diastolic blood pressure (DBP), and their associated risk levels. This comparison of risk evaluations based on improved membership functions (MFs) alongside established medical recommendations offers valuable insights into the athlete's cardiovascular health across different physiological states.

Analysis of lower heart rate and blood pressure measurements reveals that risk levels predominantly fall into the 'very safe' category. This finding aligns closely with medical recommendations, suggesting that the statistics-based approach effectively reflects low cardiovascular risk in this range. In moderate ranges, both improved MFs and medical recommendations indicate risk levels that fluctuate between 'very safe' and 'moderately safe, ' with a slight variation between the improved MFs

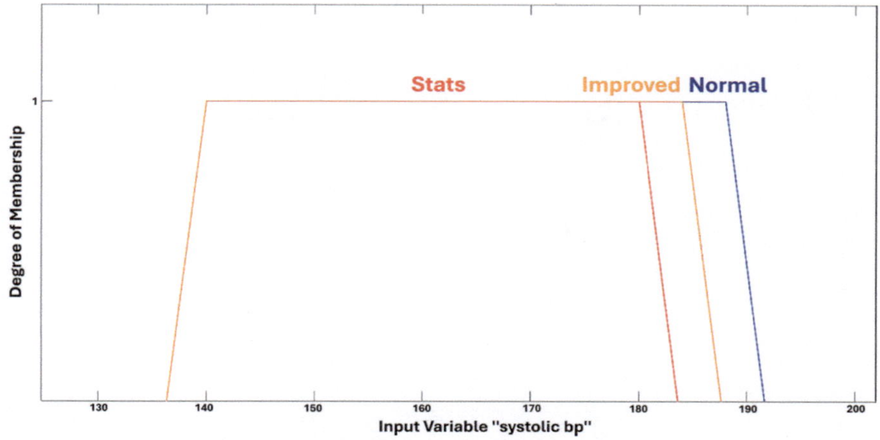

Fig. 5.4 Comparison of systolic blood pressure membership functions: statistics-based (input values: [130, 140, 180, 190]) versus medical recommendations (input values: [130, 140, 188, 198])

and the medical recommendation. As heart rates and blood pressure values rise, both models indicate an escalation in cardiovascular risk, emphasizing the increased danger associated with high levels of physical load.

The alignment of improved MFs with medical recommendations validates the reliability of the model, particularly at higher heart rates where risk levels are more pronounced. Ultimately, this illustrates the critical need for personalized risk assessment frameworks. By customizing evaluations to the unique physiological characteristics of the athlete, we can improve safety measures and optimize performance management strategies, leading to more effective training and health monitoring practices.

Table 5.5 Risk evaluation of a 21-year-old male badminton athlete

Current input values	HR (bpm)	60	70	80	90	100	110	120	130	140	150
	% of HRmax (bpm)	0.31	0.36	0.41	0.46	0.51	0.56	0.62	0.67	0.72	0.77
	SBP (mmHg)	120	130	140	150	160	170	180	190	200	210
	DBP (mmHg)	55	58	61	64	67	70	73	76	79	82
Improved MFs	Risk level (numerical)	0.108	0.226	0.21	0.153	0.106	0.106	0.106	0.361	0.561	0.692
	Categorical description	Very safe	Moderately safe	Moderately safe	Very safe	Very safe	Very safe	Very safe	Moderately safe	Medium	Moderately dangerous
Medical recommendation	Risk level (numerical)	0.111	0.121	0.193	0.131	0.106	0.106	0.106	0.224	0.539	0.698
	Categorical description	Very safe	Very safe	Moderately safe	Very safe	Very safe	Very safe	Very safe	Moderately safe	Medium	Moderately dangerous

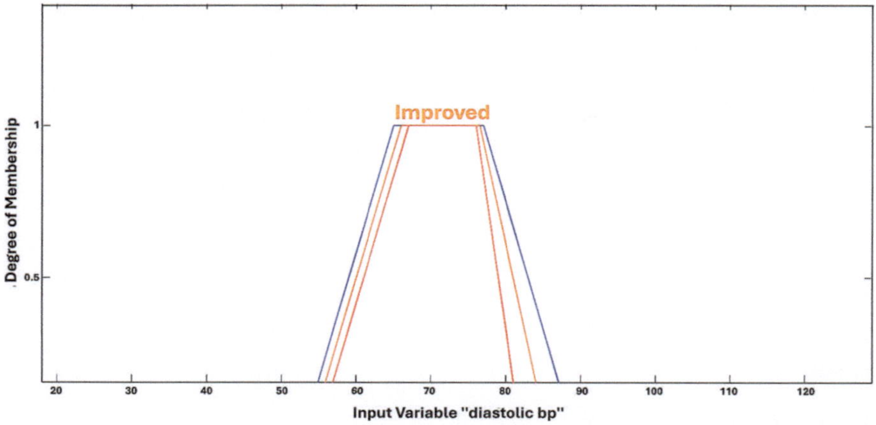

Fig. 5.5 Comparison of diastolic blood pressure membership functions: statistics-based (Input Values: [55, 67, 76, 82]) versus medical recommendations (Input Values: [53, 65, 77, 89])

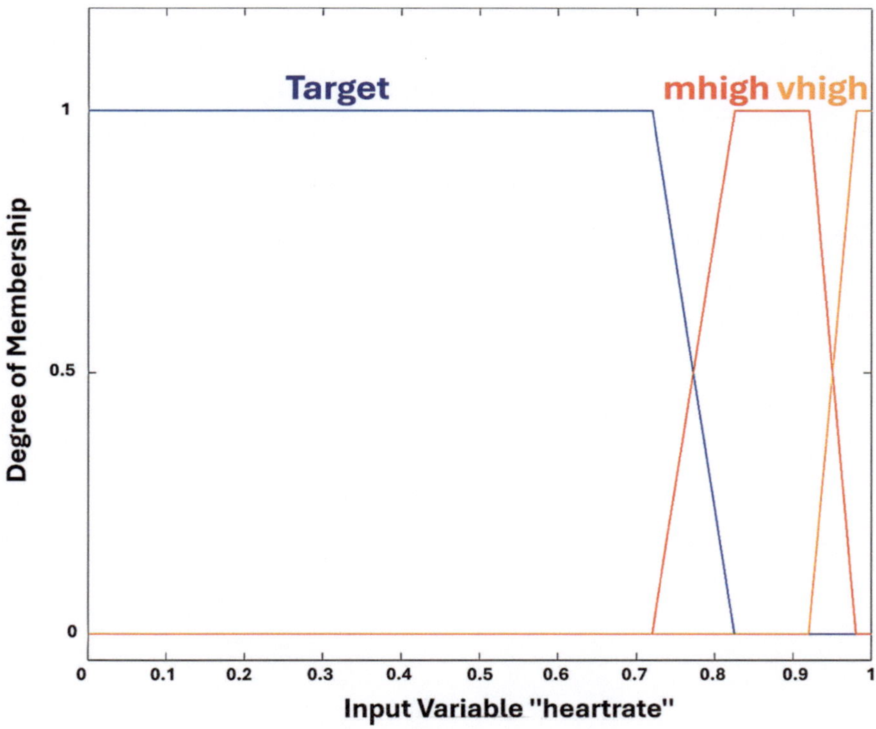

Fig. 5.6 Improved membership functions for Heart Rate (HR)

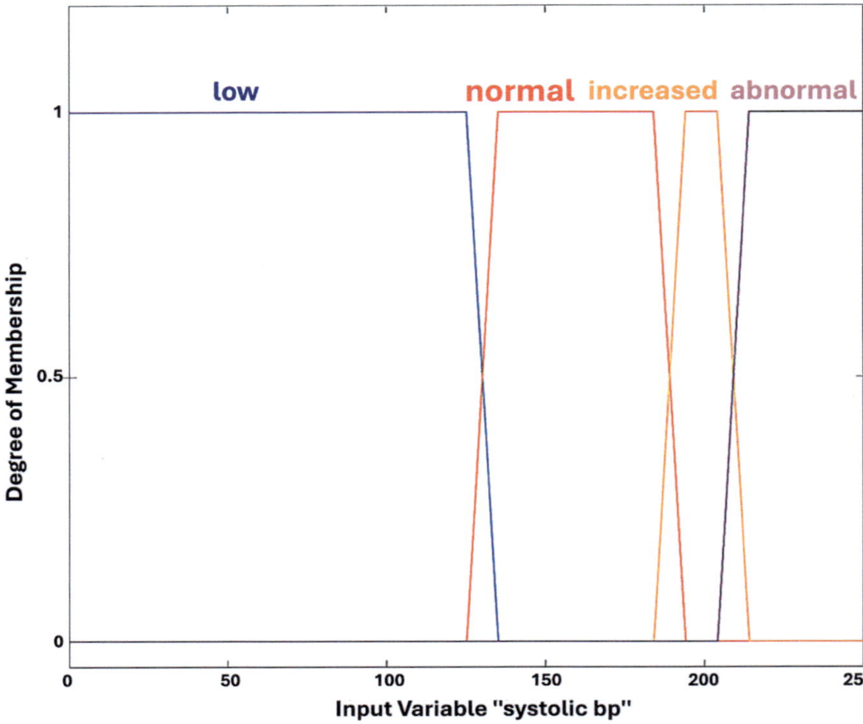

Fig. 5.7 Improved membership functions for Systolic Blood Pressure (SBP)

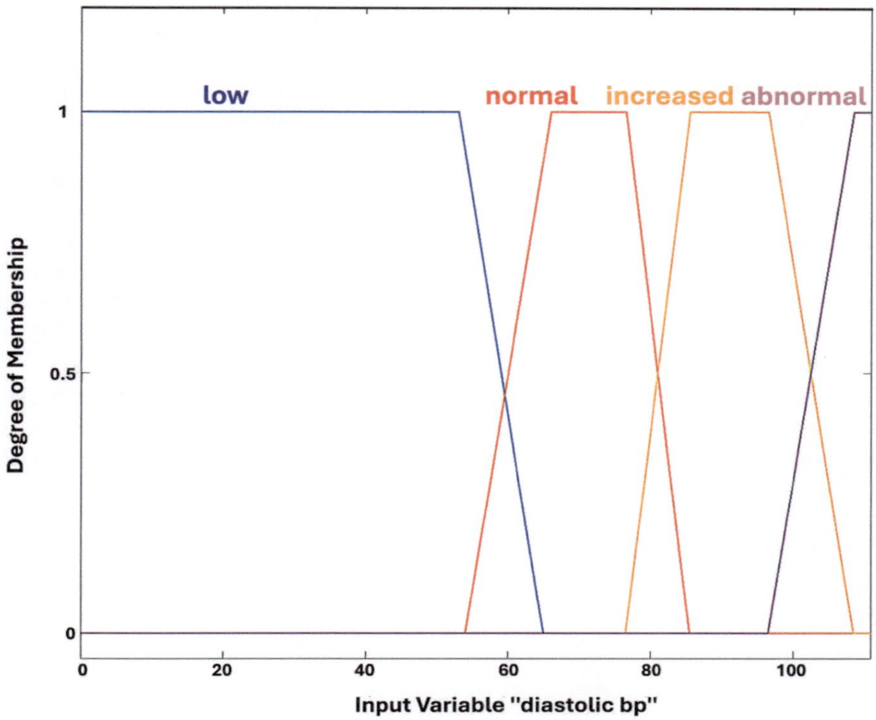

Fig. 5.8 Improved membership functions for Diastolic Blood Pressure (DBP)

5.7 Conclusion

Health monitoring is crucial in sports and physical activity for all levels. Given that individuals exhibit diverse physiological responses and health profiles, the need for personalized assessment methods becomes increasingly relevant. Traditional health monitoring techniques often fail to address these variations, potentially leading to generalized and ineffective evaluations.

Incorporating fuzzy logic into health monitoring systems presents a significant opportunity to deal with the complexities associated with uncertainty, imprecision, and subjectivity inherent in input data. This integration can yield more precise and individualized results, leading to safer practices in physical activity.

Our study analyzed risk levels associated with physical activities using a hierarchical fuzzy model structure, specifically focusing on the current physical status subsystem. During monitoring, various data points—such as heart rate (HR), systolic blood pressure (SBP), and diastolic blood pressure (DBP), alongside sampling frequency, duration, sex, and activity type—are collected and documented. This data can be stored within a personal profile, allowing for personalized evaluations in future assessments. Previous measurements taken under similar conditions can also be factored into this personalized approach.

The research employed a statistics-based approach, generating histograms that illustrate the individual's typical physiological responses under specific circumstances. These histograms facilitate the definition of membership functions (MFs), which are trapezoidal in our case. The peak of each function corresponds to the maximum value observed in the histogram, while the remainder of the function is constructed using the other histogram data points. This process allows for the creation of MFs for inputs like HR, SBP, and DBP. Once the histogram-based functions are established, original MFs can be refined accordingly. This refinement improves reliability by integrating the patient's standard responses with established medical recommendations.

The findings indicate a strong correlation between the improved MFs and the medical recommendations, reinforcing the model's reliability, particularly at elevated heart rates where risk becomes more evident. This study emphasizes the critical importance of personalized risk assessment models in protecting patients' health and optimizing performance management strategies.

Looking ahead, the authors aspire to to develop different mathematical methods that can be used to represent the patient's statistics and for fitting the membership functions.

References

1. Akkaş MA, Sokullu R, Çetin HE (2020) Healthcare and patient monitoring using IoT. Internet Things 11:100173
2. Kotorov R, Chi L, Shen M (2020) Personalized monitoring model for electrocardiogram signals: diagnostic accuracy study. JMIR Biomed Eng 5(1):e24388
3. Ersue M, Romascanu D, Schoenwaelder J, Sehgal A (2015) Management of networks with constrained devices: use cases. IETF Internet Draft
4. Alamri A, Jongeun C, El Saddik A (2010) AR-REHAB: an augmented reality framework for poststroke-patient rehabilitation. IEEE Trans Instrum Meas 59(10):2554–2563
5. Takács M (2009) Extended fuzzy methods in risk management. In: Proceedings of 14th WSEAS international conference applied mathematics, puerto De La Cruz, Spain, pp 300–304
6. Karime A, Eid M, Alja'am JM, El Saddik A, Gueaieb W (2013) A fuzzy-based adaptive rehabilitation framework for home-based wrist training. IEEE Trans Instrum Meas 63(1):135–144
7. Tóth-Laufer E, Takács M, Rudas IJ (2013) Interactions handling between the input factors in risk level calculation. In: 2013 IEEE 11th international symposium on applied machine intelligence and informatics (SAMI), Herl'any, Slovakia, pp 71–76. https://doi.org/10.1109/SAMI.2013.648097
8. Precup RE, David RC, Petriu EM, Radac MB, Preitl S, Fodor J (2013) Evolutionary optimization-based tuning of low-cost fuzzy controllers for servo systems. Knowl Based Syst 38:74–84
9. Tóth-Laufer E, Takács M, Rudas IJ (2012) Conjunction and disjunction operators in neuro-fuzzy risk calculation model simplification. In: 13th IEEE international symposium on computational intelligence and informatics (CINTI 2012), Budapest, Hungary, November 20–22, pp 195–200.

ISBN: 978-1-4673-5204-8. IEEE Catalog Number: CFP1224M-PRT. https://doi.org/10.1109/CINTI.2012.6496759
10. Michelini E (2014) The role of sport in health-related promotion of physical activity: the perspective of the health system. Springer
11. Reiner M, Niermann C, Jekauc D, Woll A (2013) Long-term health benefits of physical activity—a systematic review of longitudinal studies. BMC Pub Health 13:1–9
12. Wu Y, Ding Y, Xu H (2007) Comprehensive fuzzy evaluation model for body physical exercise. In: Risk life system modeling and simulation. Lecture notes in computer science. New York, NY, USA. Springer, pp 227–235
13. Tóth-Laufer E, Takács M, Rudas IJ (2012) Conjunction and disjunction operators in neuro-fuzzy risk calculation model simplification. In: Proceedings of IEEE 13th international symposium on computational intelligence information, Budapest, Hungary, pp 195–200
14. Polar USA (2014) Polar RS800CX user manual. https://support.polar.com/e_manuals/RS800CX/Polar_RS800CX_user_manual_English/manual.pdf
15. Tóth-Laufer E, Takács M, Keresztényi Z (2013) Current physical status evaluation subsystem using user-specific tuned membership functions in sport activity risk calculation. In: 2013 IEEE 9th international conference on computational cybernetics (ICCC). IEEE, pp 185–190
16. Hottenrott K (2007) Training with the heart rate monitor. Quelle & Meyer, Heidelberg, Germany
17. Sieira MC, Ricart AO, Estrani RS (2010) Blood pressure response to exercise testing in Apunts Med Esport. Elsevier, Amsterdam, The Netherlands, pp 191–200
18. Dubois D, Prade H (1994) Fuzzy sets—a convenient fiction for modeling vagueness and possibility. IEEE Trans Fuzzy Syst 2(1):16–21
19. Medasani S, Kim J, Krishnapuram R (1998) An overview of membership function generation techniques for pattern recognition. Int J Approx Reason 19(3–4):391–417
20. Medaglia AL, Fang SC, Nuttle HL, Wilson JR (2002) An efficient and flexible mechanism for constructing membership functions. Eur J Oper Res 139(1):84–95
21. Dombi J (1990) Membership function as an evaluation. Fuzzy Sets Syst 35:1–21
22. Corrente S, Greco S, Słowiński R (2017) Handling imprecise evaluations in multiple criteria decision aiding and robust ordinal regression by n-point intervals. Fuzzy Optim Decis Mak 16:127–157
23. Devi B, Sarma VVS (Year) Estimation of fuzzy memberships from histograms. School of Automation, Indian Institute of Science, Bangalore, India
24. Tóth-Laufer E, Várkonyi-Kóczy AR (2014) A personal profile-based patient-specific anytime risk calculation model. In: 2014 IEEE international symposium on medical measurements and applications (MeMeA). IEEE, pp 1–6
25. Tóth-Laufer E, Várkonyi-Kóczy AR (2014) Personal-statistics-based heart rate evaluation in anytime risk calculation model. IEEE Trans Instrum Meas 64(8):2127–2135
26. Tai CC, Chen YL, Kalfirt L, Masodsai K, Su CT, Yang AL (2022) Differences between elite male and female badminton athletes regarding heart rate variability, arterial stiffness, and aerobic capacity. Int J Environ Res Public Health 19(6):3206

Chapter 6
Application of Heuristic Scanning in Malware Detection

Petar Čisar, Sanja Maravić Čisar, and Attila Pásztor

Abstract A heuristic approach to malware detection is akin to the anomaly detection methods used in intrusion detection systems. It accelerates finding sufficiently effective solutions when thorough research is impractical or too time-consuming. This method relies on general rules, educated guesses, intuition, and common sense. Unlike static signature-based detection, which looks for specific matches, heuristic detection identifies unusual behavior compared to a baseline of normal network activity. It uses rules and algorithms to detect potentially malicious commands without needing a predefined signature. Although signature-based methods may miss new attack types, they produce fewer false positives. In contrast, heuristics can detect new malware but often come with a higher rate of false positives. Therefore, most modern intrusion detection systems software combines both signature and heuristic methods to enhance malware detection and removal. The most effective network security involves using multiple methods simultaneously, often referred to as multi-scanning with various anti-malware engines. This paper provides an overview analysis of heuristic detection mechanisms, as well as examples of malware types for which heuristic detection is particularly suitable.

Keywords Heuristics · Scanning · Malware · Multi-layer detection

P. Čisar (✉)
University of Criminal Investigation and Police Studies, Belgrade, Serbia
e-mail: petar.cisar@kpu.edu.rs; csiszar.peter@nje.hu

P. Čisar · A. Pásztor
GAMF Faculty of Engineering and Computer Science, John von Neumann University, Kecskemét, Hungary
e-mail: pasztor.attila@nje.hu

S. Maravić Čisar
Subotica Tech-College of Applied Sciences, Subotica, Serbia
e-mail: sanjam@vts.su.ac.rs

A. Pásztor
Department of Software Development and Application, Institute of Computer Engineering, University of Dunaujvaros, Dunaújváros, Hungary

T. A. Kovács and I. Fürstner (eds.), *Critical Infrastructure Protection: Advanced Technologies for Crisis Prevention and Response*, NATO Science for Peace and Security Series C: Environmental Security, https://doi.org/10.1007/978-94-024-2308-2_6

6.1 Introduction

Intrusion detection refers to the process of identifying activities that aim to compromise the confidentiality, integrity, or availability of computer resources, which are essential aspects of security. The primary objective of intrusion detection systems (IDS) is to recognize malicious entities attempting to breach these security objectives. Depending on the analytical strategy employed, IDS can either focus on misuse detection—where it identifies known attacks based on pre-defined signatures or rules—or anomaly detection, which monitors network traffic for unusual behaviors that may suggest potential intrusions. While misuse detection is efficient in identifying known threats, anomaly detection excels at identifying new or unknown threats by detecting deviations from standard network behavior [1].

Misuse detection is based on "signatures" (sequences of known attacks) and rules of known attacks. Analysis of static signatures will fail to catch new types of attacks but have usually less false positives. Network anomalies appear in different forms such as [2]:

- protocol anomalies (refers to protocol format and behavior—usually TCP/IP),
- anomalies of the content of application (requires knowledge of application semantics) and
- statistical anomalies (characterization of traffic through statistical measurements—monitoring the behavior of user or system by measuring statistical variables over a certain period).

Applied statistical algorithm must recognize the difference between long-term and short-term changes, to avoid generating false alarms during normal traffic variations. The advantage of statistical anomaly detection is that it can detect attacks that have not been recognized by other detection mechanisms and there is no need for large databases of attack signatures [3].

In the ever-evolving landscape of cybersecurity, the detection and prevention of malware remain paramount. Traditional signature-based detection methods have long served as the bedrock of antivirus solutions. However, the rapid evolution and obfuscation techniques employed by malware creators necessitate more dynamic and adaptive approaches. Heuristic scanning has emerged as a crucial method in this context, providing a robust framework for identifying novel and polymorphic threats that evade conventional signature-based detection [4].

6.2 Heuristic Approach in Malware Detection

Heuristics can be described as an informal decision-making framework that integrates both rational (controlled, deliberate, methodical, and often abstract or rule-based) and intuitive (automatic, rapid, concrete, and associative) components. This hybrid approach is especially crucial in scenarios where decisions must be made swiftly, despite the potential for errors. The balance between these elements allows for quick judgment calls, which are sometimes necessary in time-sensitive situations, even if they come with inherent risks of inaccuracy [3, 4].

In the context of malware detection, heuristic scanning stands in contrast to signature-based scanning, which relies on matching file signatures against a database of known malware . Instead, heuristic scanning employs predefined rules or algorithms to identify potentially harmful behavior or commands that may indicate malicious activity. This approach enables the detection of malware without requiring a pre-existing signature, allowing for the identification of new or unknown threats that might not yet be cataloged in traditional databases [5].

Heuristic scanning involves analyzing code or behavior patterns to identify potential threats based on predefined rules or algorithms. Unlike signature-based methods that rely on exact matches with known malware signatures, heuristic scanning employs a more generalized approach. It evaluates the behavior and characteristics of software, flagging potential threats based on their resemblance to known malicious patterns or anomalous activities [6].

The core principle of heuristic scanning lies in its ability to identify suspicious behaviors and traits that are indicative of malware. This method leverages a set of heuristic rules and algorithms to assess the likelihood that a given file or process is malicious. By examining the execution flow, system calls, and other runtime attributes, heuristic scanning can detect previously unknown or modified malware strains that might not yet have established signatures [7].

Many modern viruses are only slightly altered versions of malware concepts that were developed years ago. Specific detection methods, such as signature-based scanning, have become highly effective for identifying known threats. By identifying a unique signature within the code, antivirus scanners can detect every virus whose signature is stored in their databases [3]. However, the challenge arises when the virus's source code is modified, either by a programmer or through mutation programs. Even minor changes can alter the signature, allowing the virus to evade detection despite behaving in the same manner. The fundamental question is how to identify a virus without prior knowledge of its structure. The solution lies in analyzing its behavior and characteristics rather than relying solely on its code [8].

Heuristic scanning, in its foundational form, involves three core procedures known as metaheuristics. These include [3]:

- Pattern matching: Techniques used to compare and match samples against predefined patterns.
- Automatic learning: A self-adapting system where the control algorithm evolves based on the outcomes of prior operations, improving its performance and effectiveness over time.
- Environment emulation: This process simulates complete hardware environments, enabling software designed for one system to be executed on another.

These metaheuristics collectively enhance the capability of heuristic scanning to detect unknown or polymorphic malware by focusing on behavioral patterns rather than relying on static signatures.

In antivirus software, heuristic scanning is employed to detect potential threats by applying a set of predefined rules. A single suspicious activity is typically insufficient to trigger an alert. However, if a program attempts to remain resident in memory while simultaneously executing instructions to search for executable files, it is likely to be flagged as a virus [9]. Antivirus software classifies these behaviors into sequences, each assigned a flag based on its perceived threat level. Each flag is weighted, and when the combined total for a given program surpasses a specified threshold, the scanner identifies it as a virus. This approach allows for the detection of unknown threats based on behavior rather than just signatures [10].

6.3 Mechanisms of Heuristic Scanning

Heuristic scanning can be broadly categorized into two main mechanisms: static analysis and dynamic analysis. Static analysis involves examining the code or binary file of a program without executing it. This method looks for specific patterns, such as unusual code sequences, suspicious API calls, or anomalies in file structure. Techniques like opcode analysis and control flow analysis are used to identify potential threats. While static analysis is efficient in terms of speed and resource usage, it may face challenges with sophisticated obfuscation techniques that modify code structures to avoid detection [11].

On the other hand, dynamic analysis involves executing the code in a controlled environment, such as a sandbox, to monitor its behavior in real time. This approach allows for the detection of malicious activities that only manifest during execution, such as unauthorized system modifications, network communications, or data exfiltration. Although dynamic analysis provides a more comprehensive view of potential threats, it can be more resource-intensive and slower compared to static methods [12].

Heuristic-based anti-malware software employs various scanning techniques, including [3, 13]:

- File analysis: In this phase, the scanning software conducts a detailed examination of a file to ascertain its purpose, intended destination, and overall intent. For instance, if a file is designed to delete other files, it may be flagged as potentially malicious.
- File emulation: Also known as dynamic scanning or sandbox testing, this method involves executing the file in a controlled virtual environment to observe its behavior. If the file exhibits characteristics typical of a virus, it is likely to be classified as malicious.
- Generic signature detection: This technique aims to identify various forms of malware by leveraging existing virus definitions to detect related threats within the same family.

Considering the previously discussed concepts, the fundamental principles of heuristic scanning are illustrated in Fig. 6.1.

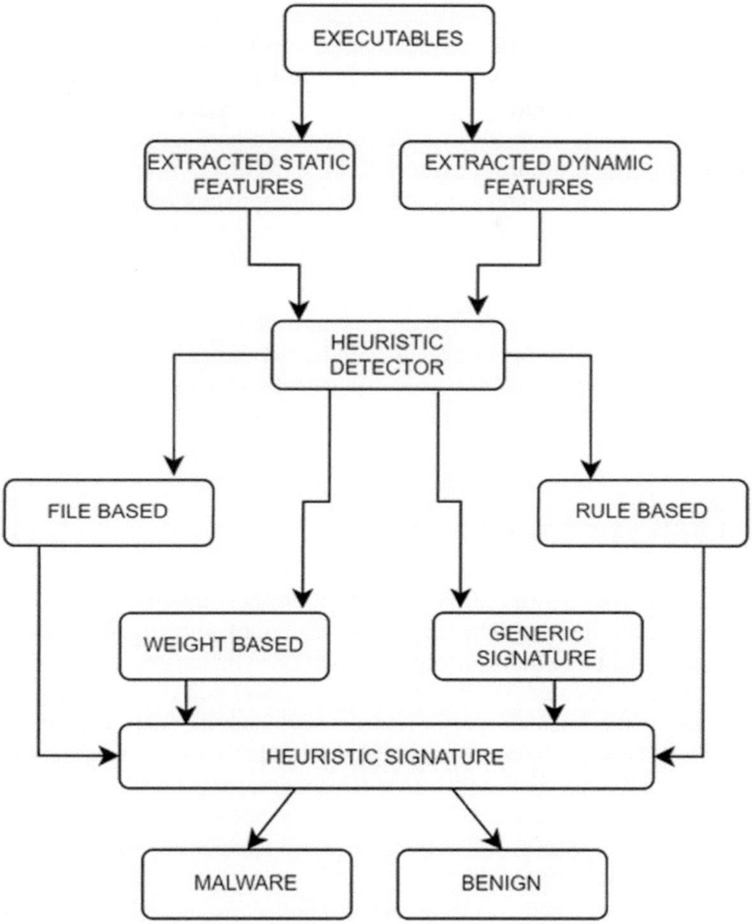

Fig. 6.1 Mechanism of heuristic scanning

6.3.1 Advantages of Heuristic Scanning

Heuristic scanning offers several benefits. It is much faster compared to sandboxing, as it does not require executing the file and observing its behavior, except in some cases where emulation techniques are used. Additionally, the rules in heuristic engines can be updated with the latest threat information without revealing these updates to potential attackers. This method also keeps the specifics of how malware is detected undisclosed, preventing malware creators from adjusting their methods to evade detection. Furthermore, heuristic scanning can identify malware that might bypass sandbox detection [3].

Heuristic scanning provides several significant advantages compared to traditional signature-based detection methods. One of its primary strengths is adaptability, enabling the detection of new and evolving threats. Unlike signature-based methods that depend on predefined identifiers, heuristic scanning identifies malware based on behavioral patterns and characteristics. This makes it particularly effective against novel and polymorphic strains of malware [14].

Heuristic scanning facilitates early detection of suspicious behaviors and potential threats during the initial stages of infection. By identifying these risks proactively, it can help prevent malware from inflicting substantial damage, which is crucial for mitigating the impact of emerging threats [14].

Heuristic scanning reduces the reliance on up-to-date malware signatures, which can be a significant limitation when dealing with newly created or modified malware. This independence from signatures ensures that heuristic scanning remains effective, even when signature databases are outdated or unavailable [14].

6.3.2 Challenges and Limitations

Heuristic scanning has several drawbacks. It often links findings to specific threat names, which can be limiting. The scanning engines look for particular code patterns associated with malicious activity, which presents two main issues: if the malware's creator has not implemented detection for a certain activity, the malware might escape detection, and if the malicious code is disguised, such as being embedded in an encoded file, it can avoid being detected [3]. Additionally, earlier heuristic scanning methods were more prone to false positives because they scanned for a broad range of potentially malicious activities. However, newer techniques, like generic detection, tend to produce fewer false positives by focusing on features or behaviors commonly associated with known threats [7].

Heuristic scanning methods face several challenges in their implementation, notably:

- **False positives**: A key drawback of heuristic scanning is its tendency to generate false positives. Since it identifies malware based on behavioral patterns and heuristics rather than predefined signatures, legitimate programs may sometimes be flagged as malicious. To reduce this, heuristic rules must be carefully calibrated to balance sensitivity with accuracy. While signature-based scanning provides precise detection by comparing software against known virus signatures, it cannot detect new or modified malware. Generic techniques, on the other hand, can catch previously unknown threats but are more prone to triggering false positives [15]. This suggests a trade-off between detection capacity and the likelihood of false positives: as the capacity to detect new malware increases, so does the rate of misidentifications (Fig. 6.2).
- **Performance overheads**: The real-time monitoring required for dynamic heuristic analysis can introduce noticeable performance lags, especially in

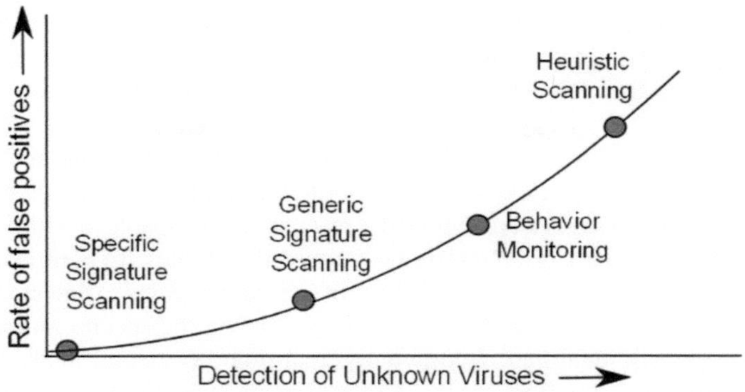

Fig. 6.2 Correlation between new viruses and false positives

resource-constrained environments. Sandboxing and runtime scanning processes consume significant computational resources, potentially degrading user experience [11].

- **Rule complexity**: Heuristic rules need to be continuously updated and refined to account for the rapidly changing landscape of malware. This ongoing process demands substantial resources and expertise to ensure that heuristic methods remain effective in detecting emerging threats without compromising system performance or accuracy [11].

6.3.3 Multi-layered Approaches: Enhancing Malware Detection Through Method Integration

To maximize effectiveness, heuristic scanning is often used in conjunction with other malware detection methods. Combining heuristic scanning with signature-based detection, behavioral analysis, and machine learning techniques can provide a more comprehensive defense strategy. For instance, while heuristic scanning can identify potential threats, signature-based detection can confirm the presence of known malware, and machine learning models can enhance heuristic analysis by identifying complex patterns and correlations [16].

Multi-layered in-depth malware detection is a comprehensive security approach that involves using multiple layers of defense to identify, prevent, and respond to malware threats. This strategy is based on the idea that no single security measure is foolproof. Instead, combining various techniques provides a more robust and resilient defense against malware. In general, the following layers can be identified within this concept [17]:

1. **Prevention layer**: The first layer of defense relies on firewalls, intrusion detection and prevention systems (IDS/IPS), and network access controls to block known

threats and reduce potential vulnerabilities. Additionally, by implementing application authorization, only pre-approved applications are allowed to run, which helps prevent the execution of unknown or unauthorized software, including malware.
2. **Detection layer**: Traditional antivirus software relies on scanning for known malware signatures, such as patterns or hashes of malicious code. Although this method is effective against recognized threats, it is less capable of addressing new or unknown malware, often referred to as zero-day threats. To complement this, heuristic analysis examines the behavior of files or programs to detect suspicious activities, potentially indicating malware even when the specific threat has not been previously identified. Additionally, real-time behavioral analysis continuously monitors programs for signs of malicious activity, such as unusual file access, attempts to contact command-and-control servers, or unauthorized privilege escalation. Another layer of protection involves running suspicious files in a controlled, isolated environment, known as sandboxing, which allows potential malicious behavior to be observed without compromising the main system.
3. **Response layer**: When malware is detected, a pre-established incident response plan is triggered, which may include isolating the infected system, removing the malware, and restoring any affected files from backups. After the incident, a forensic analysis is performed to determine how the malware infiltrated the system, assess its impact, and identify measures to prevent future attacks. Additionally, information about the detected malware and its behavior is shared with other organizations or within a community to enhance collective security defenses.
4. **Recovery layer**: Regular backups are essential to restore systems and data to their original state in the event of a successful malware attack. Additionally, maintaining up-to-date software and systems helps to close vulnerabilities that could be exploited by malware.
5. **Continuous monitoring and improvement**: Regular security audits and assessments are crucial for uncovering weaknesses in the multi-layered defense strategy. Additionally, integrating AI and machine learning can significantly improve detection capabilities by analyzing large datasets, identifying patterns, and predicting new types of malwares based on historical data.

A multi-layered defense strategy provides multiple significant advantages. Each layer of protection offers a safeguard in case another fails, ensuring that if malware bypasses the firewall, it may still be intercepted by antivirus software or identified through behavioral analysis. Such strategies are also highly adaptable, capable of evolving with the constantly changing threat landscape, offering protection against both well-known and emerging malware. Integrating various detection methods minimizes the risk of false positives, ensuring that legitimate software is correctly identified as malware.

An advanced and effective strategy for malware detection involves a multi-layered approach that encompasses several phases: reputation matching, heuristic detection, virtualized execution, and threat analysis. Each layer plays a crucial role in enhancing detection capabilities, ensuring comprehensive protection against various malware threats.

Many anti-malware solutions come equipped with built-in heuristic capabilities. Users can enable these features by adjusting the scanning level within the real-time scanning options. The settings for heuristic scanning are categorized as follows:

- **Off**: This setting disables heuristic scanning, relying solely on the signature database to determine file maliciousness.
- **Low (often recommended)**: This option offers minimal sensitivity in detecting unknown threats, resulting in the fewest false positives. It effectively balances a high level of security with a low rate of false identifications.
- **Medium**: This level enhances sensitivity for detecting unknown threats compared to the 'Low' option but increases the likelihood of false positives.
- **High**: This setting maximizes sensitivity to unknown threats, but it also raises the potential for a higher number of false positives.

6.4 Examples of Malware Best Detected by Heuristic Scanning

Heuristic scanning is a technique used by antivirus software to detect new or unknown types of malwares by analyzing the behavior and properties of files, rather than relying on known signatures. This method can be particularly effective against certain types of malwares that may not yet have established signatures in antivirus databases.

Several malwares types can be identified that might be better detected by heuristic scanning: zero-day exploits, polymorphic malware, metamorphic malware, fileless malware and advanced persistent threats (APT).

6.4.1 Zero-Day Exploits

These are vulnerabilities in software that are exploited by malware before the developer has a chance to patch them. Since there is no signature available for these newly discovered threats, heuristic scanning can identify suspicious behavior indicative of a zero-day exploit. Table 6.1 summarizes notable zero-day vulnerabilities, and the malware associated with their exploitation [18].

6.4.2 Polymorphic Malware

The dynamic nature of polymorphic malware presents significant challenges for detection methods, as these threats are designed to constantly change their code to

Table 6.1 Key zero-day vulnerabilities and their exploitation by malware

Malware	Target	Details
EternalBlue (2017)	Windows operating systems	Zero-day exploit leaked by the Shadow Brokers group, which was allegedly developed by the NSA. It exploited a vulnerability in the Windows SMB protocol. This exploit was used in the WannaCry ransomware attack, which affected hundreds of thousands of computers worldwide, and later in the NotPetya attack
CVE-2021-40,444 (2021)	Microsoft Office	This vulnerability was found in Microsoft Office and was exploited by attackers to deliver malware via malicious Office documents. The exploit involved a flaw in the MSHTML component of Internet Explorer, which is still used by Office for rendering web content
Log4Shell (CVE-2021-44,228) (2021)	Apache Log4j, a popular logging library used in many Java-based applications	This vulnerability allowed remote code execution by simply logging a specially crafted string. It was actively exploited by attackers before a patch was released, making it one of the most severe zero-day vulnerability in recent years due to the widespread use of Log4j
Google Chrome Zero-Day Exploits (2021)	Google Chrome browser	In 2021, Google disclosed several zero-day vulnerability in the Chrome browser that were actively being exploited in the wild. These vulnerability allowed attackers to execute arbitrary code, escalate privileges, and even bypass security features like sandboxing

avoid signature-based detection. Table 6.2 provides an overview of notable polymorphic malware examples, detailing their types and characteristics that contribute to their ability to remain undetected [19].

6.4.3 Metamorphic Malware

Metamorphic malware represents a sophisticated type of malware that changes its code structure with each infection, making it difficult for antivirus software to detect. Unlike polymorphic malware, which only encrypts or obfuscates its code, metamorphic malware completely rewrites its code during each replication. This characteristic allows it to evade traditional signature-based detection methods. Table 6.3 highlights some examples of metamorphic malware, detailing their types, behaviors, and the unique techniques they employ to avoid detection [20].

Table 6.2 Overview of polymorphic malware and its characteristics

Malware	Type	Details
CryptoWall (2014)	Ransomware	A variant of the CryptoLocker ransomware family. It used polymorphic techniques to generate new, unique copies of the malware each time it infected a system. This allowed it to evade detection by security tools and encrypted users' files, demanding ransom payments in Bitcoin for the decryption key
Vobfus (2013)	Worm	A worm that spreads through removable drives and network shares. It is known for its polymorphic nature, which allows it to create new, slightly different versions of itself each time it spreads. This made it difficult for traditional antivirus programs to keep up with the constant changes and detect it effectively
Shiz (2013)	Trojan Horse	A malware strain that uses polymorphic code to evade detection. It primarily spreads through email attachments and is used to steal sensitive information from infected systems. The polymorphic nature of Shiz helps it stay undetected for long periods, allowing attackers to gather data over time
Virlock (2014)	Ransomware	Virlock is a polymorphic ransomware that not only encrypts files but also infects them. It changes its code with each infection, which makes it hard to detect. Additionally, Virlock can replicate itself within the files it encrypts, further complicating removal and recovery efforts

Table 6.3 Examples of metamorphic malware

Malware	Type	Details
W32/Chiton (2009)	Virus	A less common but notable example of metamorphic malware. It infects Windows PE (Portable Executable) files and uses code transposition and obfuscation techniques to change its appearance. Each instance of W32/Chiton is different from the others, even though it retains the same malicious capabilities
Simile (2002)	Virus	Simile (also known as Etap) is one of the most well-known metamorphic viruses. It can infect executable files on both Windows and Linux systems. Simile uses advanced techniques to completely rewrite its code with each infection, making it difficult for antivirus programs to detect and remove. The virus could alter its appearance so effectively that no two copies looked alike, though they all performed the same malicious actions
MetaPHOR (2002)	Virus	MetaPHOR also known as Second Generation Virus Construction Kit (G2) is another example of a metamorphic virus. It is like Zmist in that it disassembles and reassembles its code. MetaPHOR can insert itself into various locations within a host program, significantly altering the original code and making detection extremely challenging

6.4.4 Fileless Malware

Fileless malware presents unique challenges for cybersecurity due to its operation within system memory rather than through files stored on disk. Unlike traditional malware, which leaves detectable traces on a computer's file system, fileless malware hides in plain sight by executing directly in memory, bypassing many conventional detection methods. As it does not rely on files, heuristic scanning, which monitors system behavior, becomes crucial in identifying such threats based on their actions rather than their presence in files [21].

Table 6.4 provides examples of well-known instances of malware that utilize fileless techniques, highlighting their types, primary actions, and specific methods used to evade detection. Each example demonstrates how fileless techniques have been adapted to meet various malicious objectives, from sabotaging infrastructure to stealing credentials.

6.4.5 Advanced Persistent Threats

Advanced Persistent Threats (APTs) represent a class of highly sophisticated, targeted cyber-attacks primarily aimed at espionage, data theft, or long-term infiltration of critical systems. These threats are typically orchestrated by state-sponsored groups or highly skilled cybercriminal organizations that employ complex, persistent techniques to evade detection. Heuristic scanning can be particularly useful in identifying these threats, as it can detect unusual behavior and techniques that may not yet be categorized or recognizable through traditional malware signatures [22].

Table 6.5 outlines examples of APT groups, their primary targets, and specific tactics used to achieve infiltration and persistence within target networks. Each entry demonstrates the unique strategies deployed by APTs to maintain a foothold and collect sensitive information, often for prolonged periods.

6.5 Conclusion

Heuristic scanning represents a critical advancement in the field of malware detection, addressing limitations of traditional signature-based methods by leveraging behavioral analysis to identify previously unknown or modified malware. This approach's capability to detect novel threats based on anomalous activity allows it to respond effectively to the evolving landscape of cyber threats. Despite the challenge of occasional false positives, heuristic scanning significantly improves early threat identification and complements traditional methods within a multi-layered security framework. As cyber threats continue to grow in complexity, the integration of heuristic

Table 6.4 Examples of fileless malware

Malware	Type	Details
Stuxnet (2010)	Worm	While Stuxnet is primarily known as a traditional worm, it used fileless techniques as part of its infection process. Stuxnet exploited zero-day malware and operated within memory to avoid detection. It was designed to sabotage Iran's nuclear program by altering industrial control systems
Lazarus Group Attacks (2017–2019)	Various	The Lazarus Group, a North Korean state-sponsored hacking group, has been known to use fileless techniques in various attacks. These techniques include leveraging PowerShell scripts and WMI (Windows Management Instrumentation) to execute malicious code directly in memory during attacks on financial institutions and cryptocurrency exchanges
SamSam (2018)	Ransomware	Ransomware that has been used in targeted attacks, particularly against healthcare organizations. While it eventually encrypts files on the disk, SamSam initially uses fileless techniques to gain a foothold in the system. It often exploits vulnerabilities or weak passwords to deploy fileless scripts that establish persistence and escalate privileges
Astaroth (2018)	Trojan	Sophisticated piece of fileless malware that uses various legitimate Windows tools like WMIC (Windows Management Instrumentation Command-line) and CertUtil to download and execute its payload directly in memory. Astaroth is primarily used for credential theft and operates without leaving traditional traces on the infected system

techniques with signature-based and behavioral detection strategies offers a comprehensive solution for robust malware defense. Future research should focus on refining heuristic algorithms to balance detection accuracy with efficiency, thereby enhancing their practical applications in real-world cybersecurity systems.

Table 6.5 Advanced persistent threats groups, their targets and attack tactics

Malware	Target	Details
APT28 (Fancy Bear)	Government agencies, military organizations, and media entities, particularly in the United States and Europe	APT28 is a Russian state-sponsored group linked to the Russian military intelligence agency GRU. The group is best known for its involvement in the hacking of the Democratic National Committee during the 2016 U.S. presidential election. APT28 uses a variety of attack vectors, including spear-phishing, zero-day exploits, and malware, to infiltrate and persist within targeted networks
APT1 (Comment Crew)	Primarily U.S. companies across various industries, including aerospace, defense, and technology	APT1 is one of the most well-documented APT groups, believed to be affiliated with the Chinese military. The group conducted extensive cyber-espionage operations, stealing intellectual property, trade secrets, and other sensitive information from numerous companies over several years. The group used spear-phishing emails and other sophisticated techniques to gain access to targeted networks
APT33	Aerospace, defense, and energy sectors, particularly in the United States and Saudi Arabia	APT33 is believed to be an Iranian state-sponsored group. The group has been involved in cyber-espionage and destructive attacks against organizations in the Middle East and elsewhere. APT33 is known for its use of spear-phishing emails, custom malware, and wiper attacks that destroy data on targeted systems
APT34 (OilRig)	Financial, energy, telecommunications, and chemical industries, primarily in the Middle East	APT34 is another Iranian-linked APT group focused on cyber-espionage. The group uses spear-phishing, social engineering, and watering hole attacks to infiltrate networks. Once inside, they maintain a long-term presence to exfiltrate sensitive information. APT34 has been active for several years and continues to pose a significant threat

References

1. Anwar S, Mohamad Zain J, Zolkipli MF, Inayat Z, Khan S, Anthony B, Chang V (2017) From intrusion detection to an intrusion response system: fundamentals, requirements, and future directions. Algorithms 10(2):39. https://doi.org/10.3390/a10020039
2. Mohiuddin A, Abdun M, Jiankun H (2016) A survey of network anomaly detection techniques. J Netw Comput Appl 60:19–31. https://doi.org/10.1016/j.jnca.2015.11.016
3. Cisar P, Joksimovic D (2019) Heuristic scanning and sandbox approach in malware detection. In: Proceedings of international scientific conference "Archibald Reiss days". University of Criminal Investigation and Police Studies, Belgrade pp 299–308
4. Kaspersky: what is heuristic analysis? https://www.kaspersky.com/. Last accessed 7 Oct 2024
5. del Campo C, Pauser S, Steiner E et al (2016) Decision making styles and the use of heuristics in decision making. J Bus Econ 86:389–412. https://doi.org/10.1007/s11573-016-0811-y

6. Al-Asli M, Ghaleb TA (2019) Review of signature-based techniques in antivirus products. In: Proceedings of international conference on computer and information sciences (ICCIS). Sakaka, Saudi Arabia, pp 1–6. https://doi.org/10.1109/ICCISci.2019.8716381
7. Harley D, Lee A (2024) White paper: heuristic analysis-detecting unknown viruses, https://web-assets.esetstatic.com/wls/en/papers/white-papers/Heuristic_Analysis.pdf. Last accessed 7 Oct 2024
8. Azeem M, Khan D, Iftikhar S, Bawazeer S, Alzahrani M (2023) Analyzing and comparing the effectiveness of malware detection: a study of machine learning approaches. Heliyon 10(1):e23574. https://doi.org/10.1016/j.heliyon.2023.e23574
9. Mell P, Kent K, Nusbaum J Guide to malware incident prevention and handling, recommendations of the national institute of standards and technology. https://www.researchgate.net/profile/Karen-Scarfone/publication/329973296_NIST_Special_Publication_800-83_Guide_to_Malware_Incident_Prevention_and_Handling/links/5c26b0c292851c22a34c3ff1/NIST-Special-Publication-800-83-Guide-to-Malware-Incident-Prevention-and-Handling.pdf
10. Trim PRJ, Lee YI (ed) (2023) Managing cybersecurity threats and increasing organizational resilience. MDPI, Basel. https://doi.org/10.3390/books978-3-0365-9645-7
11. Zakeri M, Faraji F, Abbaspour M (2015) A static heuristic approach to detecting malware targets: static malware detection. Secur Commun Netw 8:3015–3027. https://doi.org/10.1002/sec.1228
12. Jones S (2024) What is dynamic malware analysis? https://www.bitdefender.com/en-us/blog/businessinsights/what-is-dynamic-malware-analysis/. Last accessed 17 Oct 2024
13. Bitdefender: what is malware analysis in cyber security and why it is important. https://www.bitdefender.com/en-us/business/infozone/what-is-malware-analysis
14. Aboaoja FA, Zainal A, Ghaleb FA, Al-rimy BAS, Eisa TAE, Elnour AAH (2022) Malware detection issues, challenges, and future directions: a survey. Appl Sci 12:8482. https://doi.org/10.3390/app12178482
15. Mishra U Finding and solving contradictions of false positives in virus scanning. Semant Sch. https://pdfs.semanticscholar.org/6666/30ad1ec7fdf70d1441c1c883264b3bdee20f.pdf
16. Vasani V, Bairwa AK, Joshi S, Pljonkin A, Kaur M, Amoon M (2023) Comprehensive analysis of advanced techniques and vital tools for detecting malware intrusion. Electronics 12(20):4299. https://doi.org/10.3390/electronics12204299
17. Andrews L (2024) Multi-layered security strategies for comprehensive protection, AGS protect. https://www.agsprotect.com/blog/multi-layered-security-strategies-for-comprehensive-protection
18. Zaib R, Zhou KQ (2022) Zero-day vulnerabilities: unveiling the threat landscape in network security. Mesopotamian J Cyber Secur 57–64. https://doi.org/10.58496/MJCS/2022/007
19. Deng X, Mirkovic J Polymorphic malware behavior through network trace analysis. https://www.isi.edu/people-mirkovic/wp-content/uploads/sites/52/2023/10/a25-dengfinal.pdf
20. Irshad M, al-Khateeb H, Mansour A, Ashawa A, Hamisu M (2018) Effective methods to detect metamorphic malware: a systematic review. Int J Electron Secur Digit Forensics 10(2):138–154 (2018). https://doi.org/10.1504/IJESDF.2018.090948
21. Kara I (2023) Fileless malware threats: recent advances, analysis approach through memory forensics and research challenges. Expert Syst Appl 214:119133. https://doi.org/10.1016/j.eswa.2022.119133
22. Rot A, Olszewski B (2017) Advanced persistent threats attacks in cyberspace. In: Proceedings of federated conference on computer science and information systems, vol 12. ACSIS, pp 113–117. https://doi.org/10.15439/2017F488

Chapter 7
Reverse Engineering and Cyber Security

Sanja Maravić Čisar, Robert Pinter, Petar Čisar, and Livia Szedmina

Abstract Reverse engineering is a critical tool in modern cybersecurity, enabling the detection of vulnerabilities, the analysis of malicious code, and the enhancement of defense mechanisms. This paper examines key techniques such as code decompilation, binary analysis, and dynamic behavior monitoring, highlighting their applications in securing IoT devices, traditional software, and mobile applications. A practical example involving the analysis of an Android application demonstrates the steps and tools used in reverse engineering, offering insights into real-world challenges and solutions. Additionally, the paper explores the legal and ethical considerations of reverse engineering, emphasizing the importance of responsible practices. By addressing both technical and ethical aspects, this research underscores the importance of reverse engineering in creating resilient cybersecurity solutions for complex digital systems.

Keywords Reverse engineering · Cybersecurity · IoT · Critical infrastructure

S. Maravić Čisar (✉) · R. Pinter · L. Szedmina
Subotica Tech-College of Applied Sciences, Subotica, Serbia
e-mail: sanjam@vts.su.ac.rs

R. Pinter
e-mail: pinter.robert@vts.su.ac.rs

L. Szedmina
e-mail: slivia@vts.su.ac.rs

P. Čisar
University of Criminal Investigation and Police Studies, Belgrade, Serbia

GAMF Faculty of Engineering and Computer Science, John von Neumann University, Kecskemét, Hungary

P. Čisar
e-mail: petar.cisar@kpu.edu.rs; csiszar.peter@nje.hu

T. A. Kovács and I. Fürstner (eds.), *Critical Infrastructure Protection: Advanced Technologies for Crisis Prevention and Response*, NATO Science for Peace and Security Series C: Environmental Security, https://doi.org/10.1007/978-94-024-2308-2_7

7.1 Introduction

Reverse engineering is the process of analyzing and deconstructing products, systems, or software to understand their functionality and structure. It is widely applied in software development, electronics, and mechanical engineering, but its significance is particularly pronounced in cybersecurity. By dissecting malicious code and identifying vulnerabilities, reverse engineering empowers security professionals to develop effective countermeasures against emerging threats.

One of the most critical areas where reverse engineering demonstrates its value is in protecting critical infrastructure systems. These systems, such as energy grids, transportation networks, and healthcare technologies, are increasingly targeted by sophisticated cyberattacks. Reverse engineering enables the identification of hidden weaknesses within industrial control systems, communication protocols, and embedded devices. This proactive approach is crucial for preventing large-scale disruptions and safeguarding essential services.

The growing complexity of cyber threats requires robust strategies for detecting vulnerabilities and improving defenses. Reverse engineering addresses this need by offering tools and techniques that not only uncover system flaws, but also inform the design of resilient security measures.

This paper is organized as follows: Sect. 7.2 provides insights into key research findings on reverse engineering and its challenges, Sect. 7.3 explains essential methods such as code decompilation, binary analysis, and dynamic behavior monitoring, while Sect. 7.4 focuses on practical applications in mobile security. Finally, Sect. 7.5 highlights the challenges and responsibilities associated with using reverse engineering responsibly.

7.2 Literature Overview

Reverse engineering has established itself as a vital tool in cybersecurity, playing a significant role in identifying vulnerabilities, analyzing malware, and improving defenses. Researchers have extensively explored its applications and challenges in diverse areas such as critical infrastructure, IoT security, and industrial systems.

One line of research has focused on the role of reverse engineering in the detection of vulnerabilities within complex systems. Techniques such as code decompilation and binary analysis have been used to examine software and hardware components for security weaknesses. These methods are particularly important in the protection of critical infrastructure systems, as highlighted in studies that underline their utility to uncover hidden flaws and improve security protocols [1, 2].

In the context of IoT, reverse engineering has been instrumental in addressing emerging security concerns. The layered architecture of IoT systems introduces new vulnerabilities, particularly in connectivity and communication protocols. Researchers have proposed robust security frameworks to mitigate these risks, highlighting the need for interoperable and adaptive solutions [1, 4]. IoT-specific studies have demonstrated the importance of understanding unknown attack patterns and designing secure network architectures to protect connected devices from exploitation [4].

Challenges in reverse engineering, such as obfuscation techniques and anti-debugging mechanisms, have also been a focus of academic inquiry. Techniques such as control flow obfuscation and encryption are frequently used to hinder reverse engineering efforts. However, innovative methods have been developed to counter these defenses, allowing more efficient vulnerability detection and analysis [1, 4].

Research efforts have also been made to explore the cybersecurity implications of industrial control systems, such as SCADA. These systems are increasingly being targeted by sophisticated cyberattacks, making reverse engineering a crucial strategy for securing their components. Studies have highlighted the need for specialized techniques to analyze and protect these systems from physical and digital threats [3].

Advances in artificial intelligence (AI) have further enhanced the effectiveness of reverse engineering in cybersecurity. AI-driven techniques for anomaly detection and malware analysis have improved the ability to identify and respond to cyber threats. These approaches, combined with reverse engineering, contribute significantly to making systems more resilient against evolving attack strategies [3, 4].

International collaboration and knowledge sharing are recognized as essential for addressing global cybersecurity challenges. Studies emphasize the importance of cross-border cooperation in combating threats to critical infrastructure and smart cities. Collaborative initiatives that involve researchers, policymakers, and industry stakeholders are vital to developing effective countermeasures and ensuring robust defenses [3].

7.3 Techniques in Reverse Engineering

Reverse engineering covers a variety of techniques designed to analyze and reconstruct systems, products, or software. These techniques are essential for understanding how components function and interact, providing critical insights into system vulnerabilities, especially in cybersecurity.

One widely used technique in reverse engineering is code decompilation, which involves converting binary or executable code back into a higher-level programming language, such as Java or C++. This method is particularly useful in software security analysis, as it allows researchers to examine the structure of the code and identify potential vulnerabilities [5, 6]. Another common technique is binary analysis, which is used to analyze compiled binaries without needing access to the source code. This technique enables security professionals to reverse engineer malware or inspect suspicious software for weaknesses [7]. Recent research highlighted the challenges in decompiling optimized binaries, where advanced compiler techniques make it difficult to accurately recover the original source code [8]. Tools such as Hex-Rays, integrated with IDA Pro, are often considered industry standards for decompilation in malware analysis [9].

A practical example of binary analysis is SCRUTINIZER, a system that detects code reuse in malware by examining the binary structure. By analyzing these structures, researchers can identify reused code segments, even when the original source code is not available [10]. This technique is especially important in cybersecurity when dealing with obfuscated code, where direct analysis of the binary code is necessary to uncover hidden operations [11]. Binary analysis is often complemented by dynamic analysis, which helps to correlate static binary patterns with runtime behavior, improving the understanding of how the software operates in real time.

Dynamic analysis involves running a program and monitoring its behavior in real time. This technique helps to understand the runtime characteristics of malware, including identifying any obfuscation methods used to avoid detection. It is particularly useful for analyzing malware, where static analysis alone may not reveal the full scope of malicious behavior [7].

A practical example of dynamic analysis is the use of sandboxing techniques, where malware is executed in a controlled environment to observe its behavior without risking the security of the host system. Tools such as Cuckoo Sandbox are commonly used for this purpose, offering a detailed analysis of system changes, including file modification, network activity, and other potentially harmful actions [12]. Dynamic analysis is also valuable for understanding how malware adapts to different environments, allowing researchers to develop more robust detection and mitigation strategies [13].

Hardware disassembly is the physical dismantling of a device to study its components. This technique plays a vital role in securing IoT devices, where attackers frequently target embedded systems. Disassembling a device allows engineers to inspect hardware and firmware, identify potential vulnerabilities, and enhance the security of the system. For example, Chip-Off Forensics is a common hardware disassembly technique used in mobile device forensics. By removing the memory chip from a device, forensic experts can extract and analyze data, even from devices that have been physically damaged or locked [14]. Disassembly is also used in analyzing suspicious hardware, such as microcontrollers or security devices, to reverse engineer embedded software and discover security flaws.

Protocol analysis involves examining the communication protocols used by a system to identify vulnerabilities. This technique is critical in network security, where attackers often exploit weaknesses in protocol implementations. An example is the analysis of the SSL/TLS protocols, which are widely used to secure Internet communications. Researchers use protocol analysis to discover vulnerabilities such as Heartbleed, which allowed attackers to exploit weaknesses in the SSL/TLS implementation to access sensitive information [15]. Advanced protocol analysis techniques use machine learning to automatically detect anomalies in network traffic, making it easier to identify and mitigate protocol-based attacks [16].

The techniques utilized in reverse engineering are continuously evolving, with recent advances focusing on automation and machine learning. For example, artificial intelligence is improving protocol analysis methods to automatically find and take advantage of weaknesses in communication protocols. These advances demonstrate the increasing complexity of reverse engineering tasks and the growing need for more sophisticated tools and methods to address these challenges effectively [17].

Reverse engineering techniques are fundamental to improving cybersecurity, as they allow for a detailed examination of malicious software, the identification of security flaws, and the development of more effective defense mechanisms.

7.4 Reverse Engineering of Android Mobile Applications

Reverse engineering of Android mobile applications is an essential practice in mobile security. It allows developers and security experts to analyze the functionality of applications, identify vulnerabilities, and improve software quality. This process involves deconstructing an Android Package (APK) file to examine its components, behavior, and potential weaknesses. Such an analysis is crucial in ensuring that applications are robust and secure, especially in the field of cybersecurity.

Various tools are used for reverse engineering Android applications, each focusing on specific aspects of the analysis [18]:

- APKTool decompiles APK files into a readable format, providing access to resources such as XML files and the application manifest. It also supports recompilation after modifications, making it a valuable tool for analyzing an application structure.
- JADX is a decompiler that converts APK files to Java source code. This tool is especially helpful for understanding logic and pinpointing vulnerabilities.
- Frida is a dynamic instrumentation toolkit that enables real-time interaction with the behavior of an application. It is used for penetration testing and run-time analysis.
- The Objection, built on Frida, simplifies dynamic testing by providing user-friendly features tailored to mobile application analysis.

Reverse engineering typically employs two main techniques, static analysis, and dynamic analysis [19]:

- Static analysis examines the application without executing it. By analyzing source code, resources, and manifest, researchers can uncover hard-coded credentials, insecure communication protocols, or malicious components. Tools such as APKTool and JADX are commonly used for this purpose.
- Dynamic analysis involves running the application in a controlled environment, such as an emulator or a rooted device. Tools like Frida and Objection allow real-time monitoring and manipulation of application behavior. This approach reveals insights into the interaction of the application with the operating system, the network, and external services.

Reverse engineering of Android applications poses significant technical risks. Through this process, attackers can extract and exploit critical components of an application, including:

1. Back-end server information: Attackers can uncover sensitive details about back-end servers, potentially exposing vulnerabilities.
2. Exploitation of back-end systems: Using extracted data, malicious actors can target the back-end systems, leading to breaches.
3. Access to recent code changes: Hackers can retrieve the latest modifications to the application code, making it easier to exploit new features or patches.
4. Intellectual property theft: Reverse engineering enables unauthorized access to proprietary algorithms, designs, or features.

Reverse engineering not only exposes technical vulnerabilities but also poses several long-term business risks. One major risk is the theft of personal information, where sensitive user data can be compromised, leading to regulatory penalties and loss of customer trust. Additionally, identity theft resulting from reverse-engineered applications can damage an organization's reputation. Another concern is the compromise of back-end systems, as successful attacks may cause system downtime or lead to data breaches. Reverse engineering also allows imitators to create copycat applications, which can reduce the original creator's market share. Additionally, exposed security weaknesses can be exploited for additional attacks, putting the application and its users at further risk.

Although reverse engineering might occasionally reveal areas for performance improvement, it often exposes compliance issues. These non-compliance findings can lead to increased legal risks and challenges for the organization.

Preventing reverse engineering requires a combination of different methods and tools. One common approach is obfuscation, which changes the code by renaming classes, methods, and variables and removing unnecessary parts. This makes the code harder to understand and analyze. Another important technique is code encryption, where sensitive parts of entire code are encrypted to make them unreadable without the correct decryption keys.

Binary protection is also essential, using tools that provide obfuscation, virtualization, and anti-tampering features to protect the binary code of the application. Root detection adds an extra layer of security by preventing the application from being installed on rooted devices, which are more vulnerable to reverse engineering.

Moving critical business logic to server-side systems through server-side processing is another way to reduce the exposure of sensitive components on the user device. Continuous monitoring helps to detect and respond to threats by identifying unusual activities, such as unauthorized access or abnormal network traffic. By combining these strategies, developers can better protect their applications from reverse engineering.

This section has explored the process of reverse engineering Android mobile applications, examining the tools and techniques used in both static and dynamic analysis. These methods allow security professionals to identify vulnerabilities, improve software quality, and understand the functionality of applications. However, a specific reverse engineering example presented below will aid the understanding of how these concepts are applied in practice.

7.4.1 Practical Example of Reverse Engineering

This section will provide a detailed analysis of the CyberGym Android application, available on GitHub [20], to demonstrate the tools and steps involved in reverse engineering a real-world mobile application. This example highlighted the challenges and insights gained during the reverse engineering process [20, 21]. The CyberGym application was provided as an APK file that served as the starting point for the analysis. The challenge was to log into the application by finding the correct 4-digit PIN.

The CyberGym APK was installed and opened, as shown in Fig. 7.1. When the application started, it asked the user to enter the PIN. During testing, it was found that the application allows only three attempts before adding a delay between further guesses.

Since an APK was a ZIP archive, it can be decompiled using tools such as APKTool to extract its resources (see Fig. 7.2).

After decompiling the Android APK file, a folder with the same name as the APK was created. Figure 7.3 provides an example of a decompiled Android application.

For a deeper analysis, the authors used JADX to convert application binary files (DEX) into readable Java code. The following was observed after opening the MainActivity in JADX:

1. Obfuscation techniques, such as the renaming of SQLite database entries to generic names (Fig. 7.4).
2. Two databases were created, including one with a hardcoded encryption key '12,345,678' (Fig. 7.5).

The in-depth analysis of the code started with the 'kkk.db' file. The highlighted code in Fig. 7.6 created a table named 'name' with two columns: 'user' and 'pass'. The values stored in these columns were 'moksh' and 'password'. In the next step the second database was subjected to further analysis. As shown in the code, a table named a was created with two columns: 'z' and 'a'. The values stored in these columns were random, generated by a random number function (Fig. 7.7).

With the databases available locally, their contents can be examined (Fig. 7.8). However, attempting to access them revealed a password screen, requiring further analysis of the source code (see Fig. 7.9). Upon reviewing the code, multiple references to 'sqlcipher' were observed, indicating the use of enhanced encryption for SQLite databases (see Fig. 7.10). Using the advanced search functionality in JADX, references to 'sqlcipher' confirmed its role in the encryption process.

Earlier in the analysis, a hardcoded key was identified in the MainActivity, which was revealed to be the PRAGMA key used for database encryption (Fig. 7.11).

Fig. 7.1 Screenshot of the CyberGym application that asks for a 4-digit PIN

To decrypt the database, a command-line utility called sqlcipher is required. Once sqlcipher is installed and the databases and PRAGMA key are accessible, the contents can be decrypted to complete the challenge and retrieve the PIN code (Fig. 7.12). Figure 7.13 illustrates the application screen after successfully entering the correct PIN code, granting access to the application's main interface.

The analysis of the CyberGym Android application shows how reverse engineering can help to find security weaknesses in mobile applications. Using tools such as APKTool and JADX made it possible to break down the code and resources, revealing important security details such as obfuscation techniques and hard-coded encryption keys. Furthermore, the authors bypassed the application's database encryption using sqlcipher, which in fact underscores the importance of strong encryption and key management.

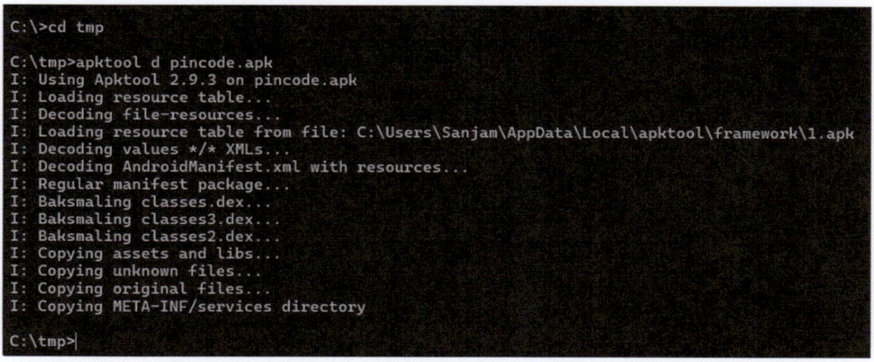

Fig. 7.2 Using APKTool to decompile an APK file

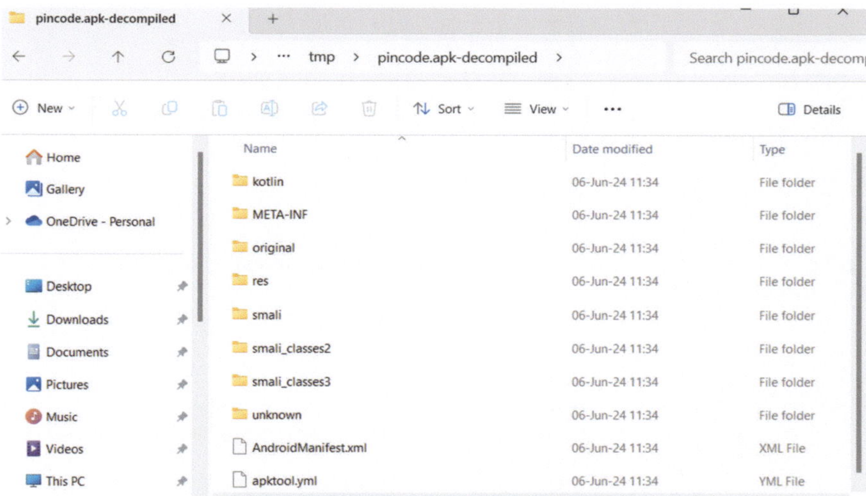

Fig. 7.3 Screenshot of the structure of the Android application folder

This example shows how reverse engineering can uncover both weaknesses and security measures put in place to protect applications. It also serves as a reminder for developers and security professionals to be vigilant and update their applications regularly to defend against new reverse engineering techniques.

```java
import net.sqlcipher.database.SQLiteDatabase;

/* loaded from: classes.dex */
public class MainActivity extends AppCompatActivity {
    Context a;

    /* renamed from: a, reason: collision with other field name */
    Cursor f1003a;

    /* renamed from: a, reason: collision with other field name */
    SQLiteDatabase f1004a;
    int b;
    int c = 2;

    private static String a() {
        return String.valueOf(((int) (Math.random() * 9000.0d)) + 1000);
    }
```

Fig. 7.4 Example of obfuscation in SQLite database entries, showing renamed entries to generic names

```java
}
try {
    String str = new String("123456");
    getDatabasePath("q.db").getParentFile().mkdir();
    deleteDatabase("q.db");
    this.f1004a = SQLiteDatabase.openOrCreateDatabase(getDatabasePath("q.db"), str, (SQLiteDatabase.CursorFactory) null);
    this.f1004a.execSQL("CREATE TABLE IF NOT EXISTS a(z VARCHAR,a VARCHAR);");
    SQLiteDatabase openOrCreateDatabase = SQLiteDatabase.openOrCreateDatabase(getDatabasePath("kkk.db"), "12345678", (SQLiteDatabase.CursorFa
    openOrCreateDatabase.execSQL("CREATE TABLE IF NOT EXISTS name(user VARCHAR, pass VARCHAR)");
    openOrCreateDatabase.execSQL("INSERT INTO name VALUES('moksh','password')");
    openOrCreateDatabase.close();
} catch (Exception e) {
    Toast.makeText(this, "Error: " + e.getMessage(), 0).show();
    Log.e("PinView", e.getMessage());
    e.printStackTrace();
}
```

Fig. 7.5 Snapshot of two databases created during execution, including one with a hardcoded encryption key ('12,345,678')

```java
}
try {
    String str = new String("123456");
    getDatabasePath("q.db").getParentFile().mkdir();
    deleteDatabase("q.db");
    this.f1004a = SQLiteDatabase.openOrCreateDatabase(getDatabasePath("q.db"), str, (SQLiteDatabase.CursorFactory) null);
    this.f1004a.execSQL("CREATE TABLE IF NOT EXISTS a(z VARCHAR,a VARCHAR);");
    SQLiteDatabase openOrCreateDatabase = SQLiteDatabase.openOrCreateDatabase(getDatabasePath("kkk.db"), "12345678", (SQLiteDatabase.CursorFa
    openOrCreateDatabase.execSQL("CREATE TABLE IF NOT EXISTS name(user VARCHAR, pass VARCHAR)");
    openOrCreateDatabase.execSQL("INSERT INTO name VALUES('moksh','password')");
    openOrCreateDatabase.close();
} catch (Exception e) {
    Toast.makeText(this, "Error: " + e.getMessage(), 0).show();
    Log.e("PinView", e.getMessage());
    e.printStackTrace();
}
```

Fig. 7.6 Code snippet showing the creation of the name table with 'user' and 'pass' columns in the kkk.db file

```
    }
    this.f1003a = this.f1004a.rawQuery("SELECT * from a;", null);
    this.f1003a.moveToFirst();
    if (this.f1003a.getCount() <= 0) {
        String a = a();
        String a2 = a();
        this.f1004a.execSQL("INSERT INTO a(z,a) VALUES('" + a + "','" + a2 + "');");
    }
    this.f1003a.close();

private static String a() {
    return String.valueOf(((int) (Math.random() * 9000.0d)) + 1000);
}
```

Fig. 7.7 Code snippet showing the creation of the table with columns z and a and random values generated by a random number function

Fig. 7.8 The databases available locally

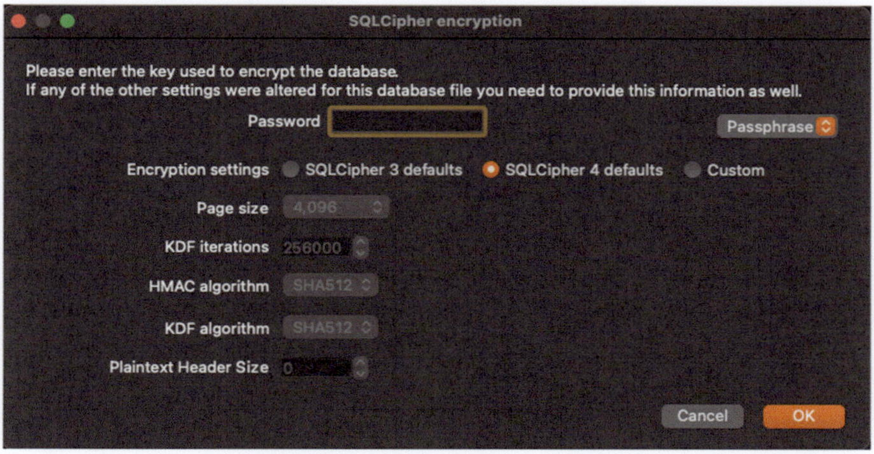

Fig. 7.9 Password screen encountered when trying to access the encrypted database

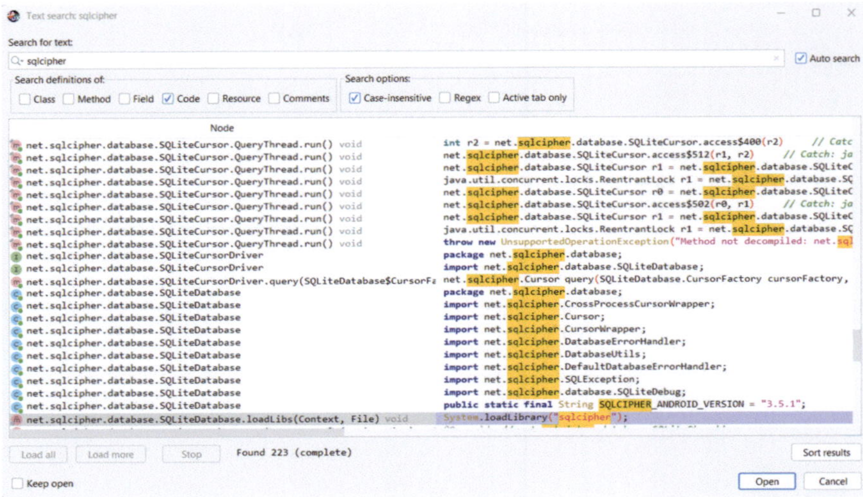

Fig. 7.10 Decompiled code showing multiple references to 'sqlcipher' for SQLite database encryption

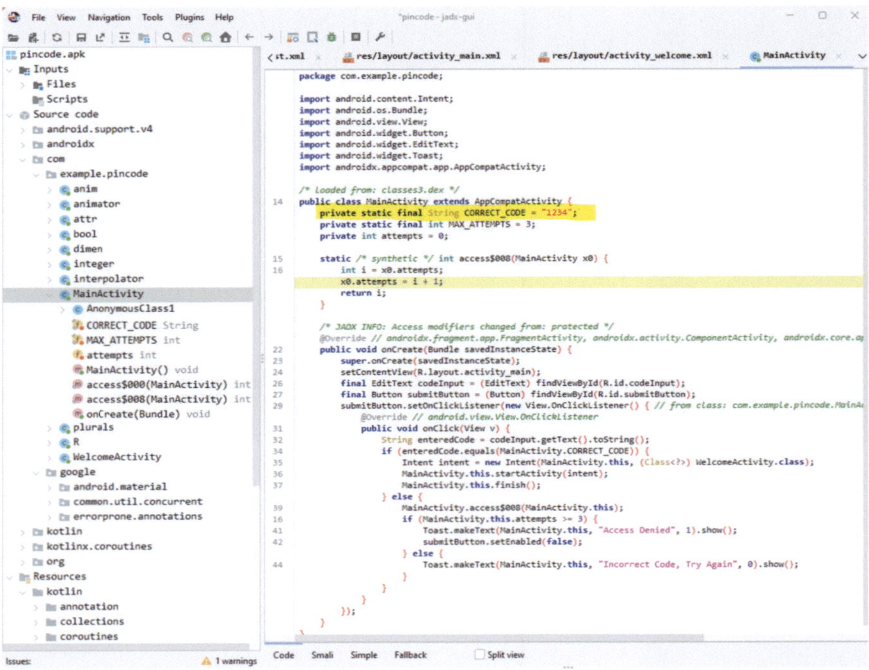

Fig. 7.11 Extracted PRAGMA key extracted from the MainActivity source code used for database decryption

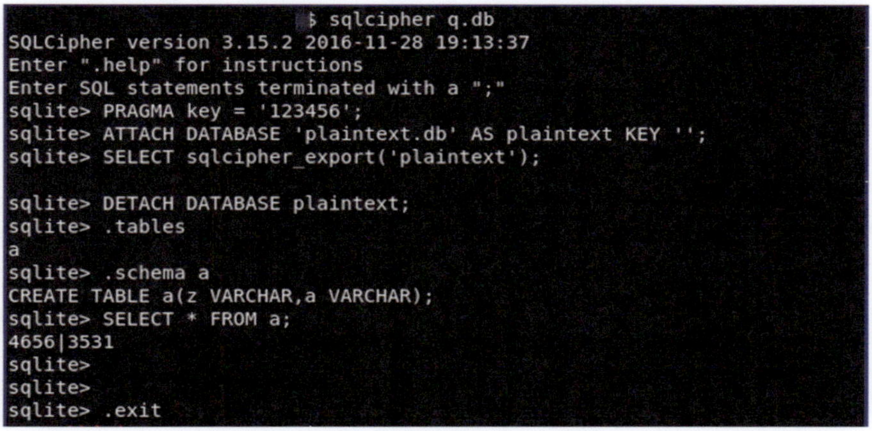

Fig. 7.12 Command-line interface for sqlcipher, demonstrating the process of decrypting the database to retrieve its contents

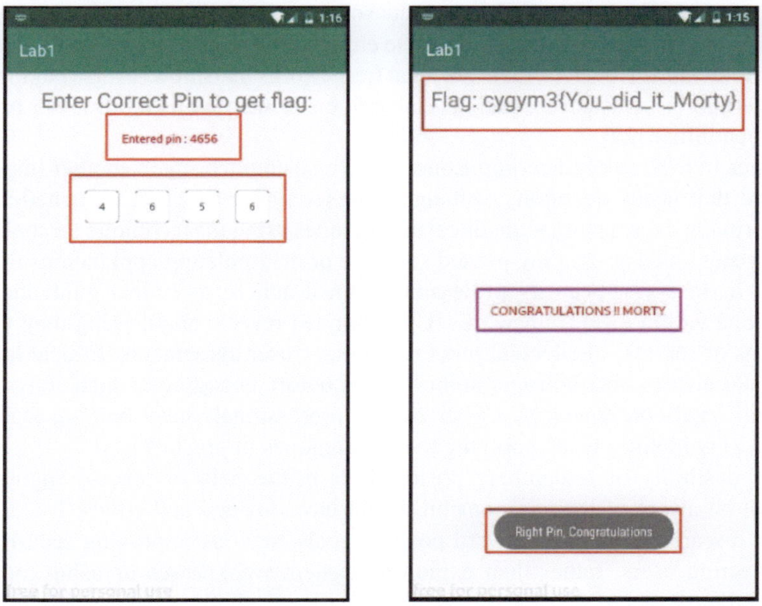

Fig. 7.13 Application screen showing successful log-in after entering the correct PIN code

7.5 Legal and Ethical Considerations

While reverse engineering is an undeniably useful tool in cybersecurity, its implementation also raises various ethical and legal questions. The key aspects related to the legality and ethics of reverse engineering in research and industry applications are the following.

- Legality of reverse engineering: The legal framework that governs reverse engineering varies by country and by specific circumstances. In many cases, reverse engineering is used within the scope of legitimate research, but legal challenges can arise. For example, using reverse engineering on commercial products that are protected by copyright can violate intellectual property and licensing laws. In the United States, the Digital Millennium Copyright Act (DMCA) can restrict the legal use of reverse engineering [22], while the European Union has similar regulations that affect the use of this technique in commercial applications [22].
- Intellectual property rights and copyright: When reverse engineering is applied to software products, it often conflicts with intellectual property and copyright laws. Legislation in many countries protects source code as intellectual property, and disclosing or reusing it through reverse engineering can be considered a violation of these rights. However, there are legal frameworks that allow reverse engineering in certain circumstances, such as research on security vulnerabilities or product compatibility [22].
- Ethics in reverse engineering: Ethics in reverse engineering is another important topic that needs attention. Although reverse engineering can be beneficial for improving cybersecurity, unethical use includes using the technique for malicious purposes, such as creating pirated software or manipulating applications in ways that harm users. Security professionals must adhere to ethical guidelines and operate within legal frameworks. Unauthorized reverse engineering may violate terms of service, intellectual property laws, or user agreements. Ethical hackers should always obtain proper authorization before engaging in such activities to avoid legal consequences. Cybersecurity professionals must balance legal and ethical guidelines when applying these techniques in practice [23].
- Responsibility of researchers: Researchers in the field of reverse engineering must be aware of their responsibilities in terms of law and ethics. It is crucial that research is directed toward positive goals, such as improving security and protecting users, rather than exploiting system weaknesses or using tools for illegal purposes.
- Education and awareness of legal and ethical standards: It is important to educate cybersecurity professionals and students about the legal and ethical standards related to reverse engineering. This reduces the risk of unethical use of tools and techniques and creates a foundation for responsible research and development in the field of cybersecurity.

As shown above, the legality of reverse engineering can be quite complex. In democratic countries, it is important to follow the laws of the specific country and respect intellectual property rights while also considering the terms of license agreements. If these laws and contract terms conflict and there are reasonable doubts about the legality of reverse engineering in a particular case, it is advisable to always consult a lawyer or expert who has sufficient experience in the relevant legal system.

7.6 Conclusions

Reverse engineering plays a vital role in modern cybersecurity by enabling the detection of threats, malware analysis, and strengthening of defenses. Techniques such as code decompilation, binary analysis, and dynamic behavior monitoring allow researchers to uncover vulnerabilities in software and hardware systems. This paper highlighted how these methods are applied to secure Internet of Things (IoT) devices, mobile applications, and traditional software.

One contribution of reverse engineering that is of utmost significance is its impact on securing critical infrastructure. As these systems are increasingly targeted by sophisticated cyberattacks, reverse engineering provides the tools to analyze vulnerabilities in industrial control systems, communication protocols, and embedded devices. This capability is essential for protecting critical sectors such as energy, transportation, and healthcare.

Although reverse engineering reveals potential weaknesses, it also uncovers opportunities for improvement. However, the process is not without challenges, including obfuscation techniques and ethical considerations. To maximize its effectiveness, cybersecurity professionals must balance technical innovations with ethical responsibility and adhere to legal standards.

By continuously advancing reverse engineering techniques, researchers and professionals can contribute to the protection of critical infrastructure and the overall improvement of digital security systems. This highlights the ongoing importance of reverse engineering in creating safer, more resilient technologies.

References

1. Nurgaliyev A, Wang H (2023) Analysis of reverse engineering. In: 15th International conference on advanced computational intelligence (ICACI). Seoul, Korea, pp 1–7. https://doi.org/10.1109/ICACI58115.2023.10146175
2. Bennouk K, Ait Aali N, El Bouzekri El Idrissi Y, Sebai B, Faroukhi AZ, Mahouachi DA (2024) Comprehensive review and assessment of cybersecurity vulnerability detection methodologies. J Cybersecur Priv 4(4):853–908. https://doi.org/10.3390/jcp4040040
3. Tariq U, Ahmed I, Bashir AK, Shaukat K (2023) A critical cybersecurity analysis and future research directions for the Internet of things: a comprehensive review. Sensors 23(8):4117. https://doi.org/10.3390/s23084117
4. Safitra MF, Lubis M, Fakhrurroja H (2023) Counterattacking cyber threats: a framework for the future of cybersecurity. Sustainability 15(18):13369. https://doi.org/10.3390/su151813369

5. Lysne O (2018) Reverse engineering of code. In: The Huawei and Snowden questions. Simula Springer briefs on computing, 4. Springer, Cham, pp 47–55. https://doi.org/10.1007/978-3-319-74950-1_6
6. Manuel D, Islam NT, Khoury J, Nunez A, Bou-Harb E, Najafirad P (2024) Enhancing reverse engineering: investigating and benchmarking large language models for vulnerability analysis in decompiled binaries. arXiv. https://doi.org/10.48550/arXiv.2411.04981
7. Sharma S, Vashisth S, Dhall I (2023) Flexible reverse engineering of desktop and web applications. In: Chatterjee P, Pamucar D, Yazdani M, Panchal D (eds) Computational intelligence for engineering and management applications. Lecture notes in electrical engineering, vol 984. Springer, Singapore, pp 609–624. https://doi.org/10.1007/978-981-19-8493-8_45
8. Katz DS, Ruchti J, Schulte E (2018) Using recurrent neural networks for decompilation. In: IEEE 25th international conference on software analysis, evolution and reengineering (SANER). Campobasso, Italy, pp 346–356. https://doi.org/10.1109/SANER.2018.8330222
9. Mattei J, McLaughlin M, Katcher S, Votipka D (2022) A qualitative evaluation of reverse engineering tool usability. In: Annual computer security applications conference (ACSAC). Austin, TX, USA, pp 1–10. https://doi.org/10.1145/3564625.3567993
10. Mirzaei O, Vasilenko R, Kirda E, Lu L, Kharraz A (2021) SCRUTINIZER: Detecting code reuse in malware via decompilation and machine learning. In: Bilge L, Cavallaro L, Pellegrino G, Neves N (eds) Detection of intrusions and malware, and vulnerability assessment. DIMVA 2021, Lecture notes in computer science, 12756. Springer, Cham, pp 111–126. https://doi.org/10.1007/978-3-030-80825-9_7
11. Jin H, Lee J, Yang S, Kim K, Lee DH (2024) A framework to quantify the quality of source code obfuscation. Appl Sci 14(12):5056. https://doi.org/10.3390/app14125056
12. Jamalpur S, Navya YS, Raja P, Tagore G, Rao GRK (2018) Dynamic malware analysis using cuckoo sandbox. In: Proceeding of the second international conference on inventive communication and computational technologies (ICICCT). Coimbatore, India, pp 1056–1060. https://doi.org/10.1109/ICICCT.2018.8473346
13. Kilgallon S, De La Rosa L, Cavazos J (2017) Improving the effectiveness and efficiency of dynamic malware analysis with machine learning. In: Resilience week (RWS). Wilmington, DE, USA, pp 30–36. https://doi.org/10.1109/RWEEK.2017.8088644
14. Kumar M (2020) Mobile phone forensics-a systematic approach, tools, techniques and challenges. Int J Electron Secur Digit Forensics 13(1):64–87. https://doi.org/10.1504/IJESDF.2021.111725
15. Sirohi P, Agarwal A, Tyagi S (2016) A comprehensive study on security attacks on SSL/TLS protocol. In: 2nd International conference on next generation computing technologies (NGCT). IEEE, Dehradun, pp 893–898. https://doi.org/10.1109/NGCT.2016.7877537
16. Jakkani AK (2024) Real-time network traffic analysis and anomaly detection to enhance network security and performance: machine learning approaches. J Electron Comput Netw Appl Math 4(4):32–44. https://doi.org/10.55529/jecnam.44.32.44
17. Ning B, Zong X, He K, Lian L (2023) PREIUD: an industrial control protocols reverse engineering tool based on unsupervised learning and deep neural network methods. Symmetry 15(3):706. https://doi.org/10.3390/sym15030706
18. Reverse engineering Android apps (CTF challenge) https://medium.com/@prathunan777/reverse-engineering-android-apps-ctf-challenge-baa3a9cbe7d5. Accessed 17 Nov 2024
19. Amin A, Eldessouki A, Magdy MT, Abdeen N, Hindy H, Hegazy I (2019) AndroShield: automated android applications vulnerability detection, a hybrid static and dynamic analysis approach. Information 10(10):326. https://doi.org/10.3390/info10100326
20. GitHub. https://github.com/lucideus-repo/cybergym/tree/master/CyberGym%202. Accessed 17 Nov 2024
21. Corellium. https://www.corellium.com/blog/android-mobile-reverse-engineering. Accessed 17 Nov 2024

22. Ungureanu CT, Tataru SR (2023) The legality of reverse engineering or how to legally decipher trade secrets. In: SHS web conference, 177, pp 1–20. https://doi.org/10.1051/shsconf/202317702001
23. Li N (2023) Legal and ethical issues of reverse engineering the video game. Glob J Polit Law Res 11(1):52–65

Chapter 8
Secure Data Utilization from Web Environment

Zlatko Čović

Abstract This paper addresses key aspects of security in web environments, with a particular focus on API endpoints. The exchange of vast amounts of information within integrated systems requires consistent application of security principles throughout all stages—planning, development, testing, and implementation. The paper provides insights into API endpoints, types of APIs, potential techniques for securing API endpoints, and recommendations for the phases that should be carried out during the development of API endpoints to ensure their security. Special emphasis is placed on raising awareness of security threats among those who develop and use such systems, as well as on the importance of employing appropriate tools for testing API endpoints and other system components. Additionally, the paper explores the challenges of retrieving data from websites without APIs and creating new APIs for exchanging such information, identifying specific security risks and offering recommendations for addressing them. The paper also highlights the importance of introducing new methods into the education of software engineers in the field of information security.

Keywords Web security · API testing · Challenge-based tasks · Vulnerabilities

8.1 Introduction

The increasing reliance on web-based systems for personal, organizational, and societal functions has made secure data utilization a keystone of modern technology. APIs (Application Programming Interfaces) serve as the backbone of integrated web environments, enabling seamless data exchange and interoperability across platforms. However, the same interconnection introduces vulnerabilities that attackers can exploit, leading to breaches, data theft, and compromised systems. This paper explores the critical importance of securing data in web environments, focusing on

---

Z. Čović (✉)
Subotica Tech College of Applied Sciences, Subotica, Serbia
e-mail: chole@vts.su.ac.rs

T. A. Kovács and I. Fürstner (eds.), *Critical Infrastructure Protection: Advanced Technologies for Crisis Prevention and Response*, NATO Science for Peace and Security Series C: Environmental Security, https://doi.org/10.1007/978-94-024-2308-2_8

API endpoints as high-value targets for security threats. By addressing these vulnerabilities, organizations can ensure data integrity, confidentiality, and availability, which are fundamental to trust in digital ecosystems.

In the contemporary digital landscape, a significant proportion of information systems operating within web environments are designed as integrated systems. These systems epitomize the conjunction of multiple components working together to provide seamless functionality and user experiences.

Integrated systems can be broadly defined as complex information ecosystems comprising interconnected modules that perform specific tasks while collectively contributing to the overall system objectives. The most prevalent components of such systems include web applications or websites, mobile applications, services dedicated to data manipulation and exchange, and API endpoints that serve as the communication bridges between these modules.

The development of these components often involves diverse programming languages, frameworks, and platforms, reflecting the specialized requirements and capabilities of each part of the system. This heterogeneity underscores the critical need for secure, efficient, and reliable data exchange mechanisms. Ensuring robust communication between components is essential for maintaining system integrity, protecting sensitive information, and delivering responsive user interactions.

Moreover, integrated systems frequently interact with external resources, such as third-party APIs, to augment functionality or retrieve real-time data. In some cases, these systems rely on data extracted from web pages that lack dedicated public API endpoints, requiring alternative methods of data acquisition. Such scenarios introduce additional challenges, including the need to safeguard against unauthorized access, mitigate data leakage risks, and ensure compliance with regulatory standards.

The authors in [1] investigated independent data extraction from photovoltaic systems (PVS) in households to evaluate cost savings, environmental impact, and sustainability. By analyzing HTTP communication between web panels and PVS units, they retrieved data for a year using an extraction program built with open-source technologies. Plans involve developing API endpoints protected by disposable tokens for integration with mobile apps, enhancing data visualization and optimization, and supporting decision-making in the development of PVS systems for producer-consumers. These advancements would enable secure data use in integrated information systems.

The server can authenticate clients using cookies or sessions over the HTTP protocol with a REST API (Representational State Transfer Application Programming Interface). However, a vulnerability exists that allows hackers to intercept identification information, such as by capturing broadcast packets or utilizing fake proxies or other tools. In [2], a new mechanism is introduced called the disposable token, which enhances token-based authentication for RESTful APIs on the HTTP protocol. This method requires the client to store a pair of public and private tokens generated by the server as part of the authentication process.

APIs play a fundamental role in driving innovation in today's app-centric world. They are crucial to various industries, including banking, retail, transportation, IoT, autonomous vehicles, and smart cities, and are integral to mobile, SaaS, and web applications. APIs are present in customer-facing, partner-facing, and internal systems. However, by exposing application logic and sensitive data like Personally Identifiable Information (PII), APIs have become prime targets for attackers. The rapid pace of innovation would be unachievable without secure APIs in place [3].

This research [4] presents a vulnerable web application designed around OWASP API security risks, aimed at helping students, teachers, web developers, and security testers recognize and practice API security in a safe, legal environment. The study also explores the use of gamification to create an engaging learning experience. Additionally, a scalable container-based service is proposed to ensure optimal performance and availability, especially when the application is tested with resource-intensive web attacks.

The State of API Security Report 2024 released by Salt Labs shows that API attacks are on the rise, with API counts increasing by 167% in the past year. Despite this, 95% of organizations have faced security issues in production APIs, and only 7.5% have implemented dedicated API testing and threat modeling programs. Furthermore, 46% of respondents view API security as a critical C-level discussion [5].

Due to the vast amounts of valuable data, including personally identifiable information (PII), APIs have become prime targets for attackers. Exploiting these vulnerabilities requires minimal skill or resources, as attackers can use a trial-and-error approach to identify and exploit flaws in APIs [5].

In the paper [6], common RESTful API vulnerabilities, such as faulty authentication, excessive data exposure, and injection, have been identified through security testing and mitigation. These vulnerabilities can lead to serious security breaches if not properly addressed. The authors emphasize the need for proper testing and mitigation procedures and suggest future research directions, including the use of machine learning for vulnerability detection and the creation of automated security tools. The paper also highlights the challenges of standardization and the complexity of detecting intricate vulnerabilities, underlining the importance of thorough API security practices.

A significant need has been identified for integrating modern educational methods into software engineering curricula, with an emphasis on simulating real-world scenarios that require the application of appropriate security techniques and methods. Furthermore, it has been observed that some programmers, even after completing formal education and during their professional careers, fail to place sufficient emphasis on security testing, highlighting critical gaps in their understanding of fundamental information security principles [7].

This paper is organized as follows. In the continuation of the introductory part, the terms REST API and integrated web system are explained. Methods of data exchange and authentication within integrated web systems are also listed. In Sect. 8.2, two educational institutions are briefly presented and the importance of introducing safety education into the everyday education of engineers is described. Section 8.3 discusses the security risks of web applications and integrated web systems. Also, the laboratory exercises and topics are described. In Sect. 8.4, tools for testing are presented, as well as the way of performing laboratory exercises and assigning tasks. Several tasks from the laboratory exercises are described in detail.

The paper is organized as follows: Sect. 8.2 introduces the concept and architecture of integrated web systems. Section 8.3 discusses the concept of API endpoints and presents possible implementations of security techniques and methods in API development. It also explores methods for retrieving data from websites that do not provide their own APIs. In Sect. 8.4, the phases of defining, developing, testing, and implementing API endpoints are discussed in detail, with an emphasis on security. Tools for testing API calls are also mentioned. Section 8.5 provides information on the use of new methods in the education of engineers in the field of web security, and Sect. 6 presents the conclusion of the paper.

8.2 Integrated Web Systems

Integrated web systems are complex information systems composed of multiple interconnected components. These systems typically include a web application or website, a mobile application, services for data manipulation and exchange, as well as API endpoints that facilitate communication between different system elements. The integration of these components allows for seamless data flow and functionality, making them essential for modern digital services that require real-time interaction and processing across diverse platforms. In today's interconnected world, where vast amounts of data are shared and exchanged across various platforms, it is crucial to prioritize information security as part of critical infrastructure. Ensuring the protection of sensitive information and safeguarding against potential cyber threats are fundamental to maintaining the integrity, reliability, and trustworthiness of systems that support essential digital interactions and services.

Figure 8.1 shows one of the architectures of integrated web system. One of the key concepts is to implement a division into modular components that can be independently tested and developed, a process known as decoupling. This approach allows for greater flexibility in development, as each module can be worked on separately without affecting the others. By decoupling components, developers can more easily identify and fix issues, streamline testing, and introduce updates or new features without disrupting the entire system. Furthermore, this method promotes better scalability and maintainability, as modular systems are easier to manage and expand over time.

Web Application: The web application interacts with the backend through HTTP/HTTPS or via APIs. It serves as the user interface, allowing users to perform actions and access data.

Mobile Application: Like the web application, the mobile app communicates with the backend using RESTful APIs or GraphQL. It provides a mobile-friendly interface and allows users to interact with the system on the go.

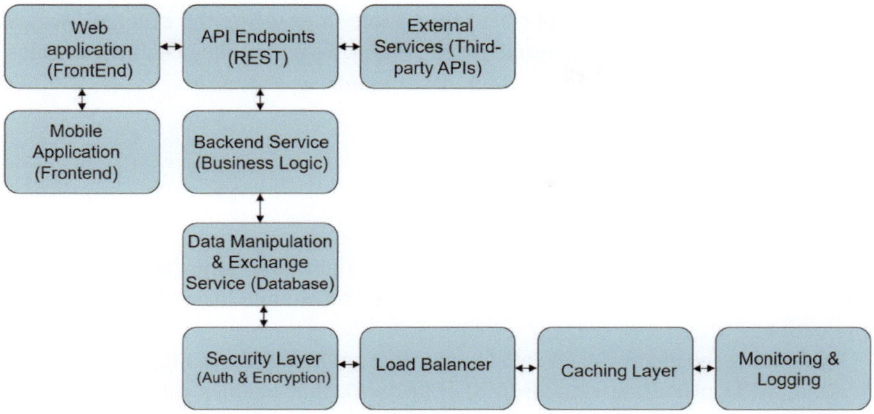

Fig. 8.1 The hierarchical fuzzy model framework

Backend Service: The backend handles all the business logic and data processing. It is responsible for managing user requests, processing data, and interacting with other services.

Data Manipulation and Exchange Service: This component handles CRUD operations (Create, Read, Update, Delete) and often includes database management systems for storing and retrieving data.

API Endpoints: RESTful APIs or GraphQL endpoints facilitate communication between the frontend and backend, as well as with external services. These endpoints implement security methods such as authentication and encryption to protect data and ensure secure access.

External API: These are third-party services integrated into the system, such as payment gateways or social media platforms. They are accessed securely through API calls.

Security Layer: This layer includes authentication and authorization mechanisms (e.g., OAuth, JWT) to ensure that only authorized users can access the system. It also ensures secure data transmission using HTTPS and implements data encryption for sensitive information.

Load Balancer: The load balancer distributes incoming traffic across multiple servers to maintain high availability, optimize performance, and ensure reliability in case of server failure.

Caching Layer: Caching mechanisms, such as Redis or Memcached, are used to improve system performance by storing frequently accessed data and reducing the load on backend services.

Monitoring and Logging: Tools like Prometheus, Grafana, and the ELK stack (Elasticsearch, Logstash, Kibana) are used for monitoring system activities and logging events, which helps track performance, troubleshoot issues, and ensure system health.

Each of these components plays a critical role in ensuring the efficient, secure, and reliable operation of an integrated web system. Together, they enable a seamless user experience while maintaining high performance and security.

8.3 API Endpoints

API, which stands for "Application Programming Interface," refers to a collection of rules, protocols, and tools that facilitate the development and interaction of software applications.

By defining the methods and data formats that applications can utilize for communication, APIs provide a standardized framework that enables different software systems to connect and exchange information seamlessly. This standardization simplifies integration and ensures interoperability between diverse applications.

APIs serve as a crucial foundation for modern software development by providing structured mechanisms for different software components to interact effectively. Here are the key roles APIs play:

Interface: APIs act as a bridge, enabling communication between various software components. This interaction can occur either within the same system or across entirely separate systems.

Abstraction: By abstracting the underlying implementation details, APIs expose only the essential objects or actions developers need. This simplifies the process of leveraging complex functionalities without requiring in-depth technical knowledge.

Data Exchange: APIs facilitate seamless data exchange between systems. For instance, a weather application can retrieve up-to-date weather information from a remote server through an API.

Functionality: APIs provide access to specific functionalities of other software without requiring an understanding of their internal workings. For example, a mobile app can utilize an API to operate the phone's camera.

Standardization: Through APIs, developers can establish consistent interactions between software components. This standardization aids in maintaining application reliability and simplifies scalability.

APIs (Application Programming Interfaces) are categorized based on their accessibility and usage to meet diverse operational and business needs. Below are the main API types:

- Open or Public APIs: Open APIs are available to external developers or the public with minimal restrictions. They foster innovation, encourage third-party integrations, and expand ecosystems. Common examples include APIs from social media platforms or payment gateways. Access is often managed through authentication and rate limits.

- Partner APIs: Partner APIs are shared with specific business partners under controlled conditions and governed by agreements. They enable secure integration between organizations, such as payment processors collaborating with e-commerce platforms.
- Internal APIs: Internal or private APIs are used within an organization to enable communication between systems or teams. They improve efficiency, streamline workflows, and support modular development, such as integrating HR and payroll systems.
- Composite APIs: Composite APIs execute multiple API calls in a single request, ideal for scenarios requiring data from various sources. For example, a travel booking system may retrieve flight, hotel, and car rental data simultaneously, reducing latency and improving user experience.

A RESTful API (Representational State Transfer API) is a web service architecture designed to enable seamless interaction between distributed systems over the web. It adheres to REST principles, which emphasize statelessness, scalability, and the use of standard HTTP methods such as GET, POST, PUT, DELETE, and PATCH. These methods facilitate operations like retrieving, creating, updating, deleting, and partially updating resources.

RESTful APIs use structured resources, typically represented in formats like JSON or XML, to exchange data efficiently. They also support additional HTTP methods, such as HEAD for metadata retrieval, OPTIONS for identifying supported actions, and TRACE for debugging purposes, depending on the implementation.

By leveraging RESTful APIs, systems can maintain loose coupling, ensure scalability, and promote interoperability, making them a cornerstone of modern web-based software architecture.

8.3.1 Security of API Endpoints

In the modern era, characterized by intense information exchange and the development of various systems that rely on API services, one of the key challenges is ensuring security during the creation, testing, and implementation of these services. API services are increasingly becoming targets for attackers who may exploit potential security vulnerabilities in their implementation.

API endpoints can be secured through various methods, often employing multiple techniques within the same API system to achieve a higher level of security. The following approaches are critical for ensuring protection:

- Tokens: Include JWT (JSON Web Token), Bearer tokens, access tokens, and disposable tokens, which enable secure authentication and authorization.
- API Keys: Used to identify and control access for users or applications interacting with API resources.
- Authentication with Username and Passwords: A traditional method for verifying user identity.

- Rate Limiting: Restricts the number of requests from a single API source, reducing the risk of DDoS attacks and abuse.
- Access Restrictions Using API Keys or Tokens: Configurable access levels for different users or applications, including IP-based restrictions.
- Handling Invalid Requests and Server Issues: Ensuring proper responses to invalid requests and quickly addressing potential server errors.
- Monitoring and Logging: Continuous monitoring and logging of activities on API endpoints enable the detection and prevention of security threats.

These methods, when combined, significantly enhance the resilience of API services against various types of attacks and misuse. MOZDA NAVESTI RAD testing API?

8.3.2 Secure Data Utilization

In the context of various project developments, data is frequently required from websites that do not offer API endpoints for direct access. In such scenarios, data retrieval can be achieved either manually or automatically, depending on the specific project needs and the frequency of data updates. Although manual data collection is sometimes utilized, it is typically employed in cases where the data changes infrequently, making it feasible to retrieve the information manually. However, this approach is not scalable and is generally reserved for one-time or occasional data retrieval.

For dynamic or frequently updated data, automatic data retrieval becomes essential. In such cases, web crawlers or spiders are employed to automate the process. The first step in this process is conducting a thorough analysis of network communication and the client-side source code of the web page. This involves identifying any potential restrictions or limitations that may govern or restrict access to the data. Web crawlers then visit the web pages, navigate through the site's structure, and systematically extract the necessary data. This method is particularly advantageous when dealing with websites that lack APIs but still contain valuable, structured data that can be utilized in a project.

Once the data has been retrieved, several key steps are required before it can be effectively used. This includes processing and transforming the data into the appropriate format for the intended application. Raw data, for instance, may need to be cleaned, normalized, or structured to meet the system's requirements. Following the processing stage, the data is generally stored in a secure storage solution, such as a database or cloud storage system, where it can be accessed for future use.

Throughout this process, security plays a crucial role, particularly regarding ensuring that data is collected in an ethical and legal manner. It is imperative to comply with privacy policies, terms of service, and data protection regulations when extracting data from websites. Furthermore, the storage and transmission of the data must be secured to prevent unauthorized access, ensuring that sensitive information remains protected at all stages of its utilization.

8.4 Creating Secure API Endpoints

Creating secure API endpoints and their safe usage should be carried out in several interconnected phases. First, it is essential to clearly define:

- how many API endpoints are needed,
- the methods for invoking these APIs,
- the input parameters,
- feedback (such as messages, status codes, data, possibly tokens or other critical information),
- the changes caused by invoking these endpoints within a system,
- the required authorization mechanisms,
- as well as monitoring and logging of API service usage.

After clearly defining these aspects, the development of API endpoints should proceed using the programming language specified by the project or server configurations.

Since these are programmatic codes, it is advisable to conduct UNIT testing for all possible API usage scenarios. API calls that retrieve information should be unit tested in the programming language in which the API was created. For endpoints that accept data or perform client-side functions based on responses (such as "code on demand"), UNIT testing should be done in different programming languages and platforms, as the API will be available in various environments. This is particularly true for public APIs, with or without authentication. In the case of internal API development, UNIT tests are performed in predefined technologies used within the organization.

API calls should be tested in a testing environment with test data that mimics the structure of a real system but differs in content. In addition to UNIT tests, both manual and automated testing should be performed. Manual testing requires manually creating all calls, tracking responses, and identifying errors. Although manual tests allow for thorough control, they are prone to human errors.

Automated testing, unlike manual testing, uses scripts and specialized tools for this type of testing. Some of the most popular tools and frameworks for automated testing are Postman, Insomnia, Selenium, JUnit, REST Assured, Katalon Studio, and others. Automated testing is faster, more consistent, and scalable, making it ideal for large projects, regression testing, and frequent testing.

Within testing, security tests should be conducted, which often also cover functional aspects of the API. It is also important to monitor server status based on logs and request monitoring. During security testing of APIs, it is recommended to follow the guidelines defined in the OWASP API Security Top 10.

Fig. 8.2 SwaggerUI (https://swagger.io/tools/swagger-ui/)

After executing the tests and correcting any errors and issues, it is necessary to document the API calls. One of the most popular tools for documenting APIs is Swagger/OpenAPI Generator. While primarily used for documenting APIs, Swagger (or OpenAPI) also offers tools for testing and mocking APIs. The Swagger UI allows users to interact directly with APIs and test endpoints. Additionally, the OpenAPI Generator can generate client libraries and server stubs, facilitating automated testing in different programming languages [8]. Figure 8.2 shows the Swagger UI.

8.5 Security in Education and Challenge Tasks

Core challenges in the secure development and usage of web applications stem from insufficient knowledge and a lack of awareness regarding potential threats. Developers often fail to implement appropriate security techniques and methods, and neglect to conduct comprehensive security assessments of their applications.

This issue also persists in the education of software engineers. To bridge the gap between theoretical and practical knowledge, students must engage in tasks focused on addressing security challenges.

At Subotica Tech, students are trained both theoretically and practically in various aspects of web security. These include input data filtering, preventing the acceptance of semantically incorrect data, secure file uploads, and secure user management systems such as registration and login processes. Additionally, students learn to handle sessions and cookies securely, protect web servers and directories, and implement cryptographic algorithms. They are also taught to guard against SQL injection, secure mobile applications, and call restricted functions when using dynamic function calls. The curriculum covers creating and securing API endpoints, using tools for

API testing, and implementing token-based authentication for secure communication and data exchange. Furthermore, students are educated on avoiding deprecated methods and restricting access to sensitive system components, as well as analyzing EXIF data in media files and ensuring comprehensive API security. They also learn about security of IoT and network systems [7].

After acquiring both theoretical and practical knowledge in web programming, students are given individual tasks to solve during laboratory exercises.

These tasks are designed to simulate security challenges and are directly related to web application security. Each task is allotted a specific timeframe. Once the time has elapsed, students present their solutions and discuss any issues encountered while working on the assignment. This is followed by a discussion where various approaches to solving the tasks are explored.

As standard frameworks for recommendations on web application and API security, students are introduced to the OWASP Top 10 and OWASP API Security Top 10. To make the learning process more interactive and provide students with a platform to test their knowledge, an online quiz was developed using OpenAI and the GPT-4 mini model. This quiz covers key areas of OWASP standards related to web application security. The initial testing was conducted with third-year computer science students, and after successful results, the system is planned to be expanded to include areas related to OWASP API Security. Figure 8.3 shows the start page of OWASP Top Ten AI Quiz Generator.

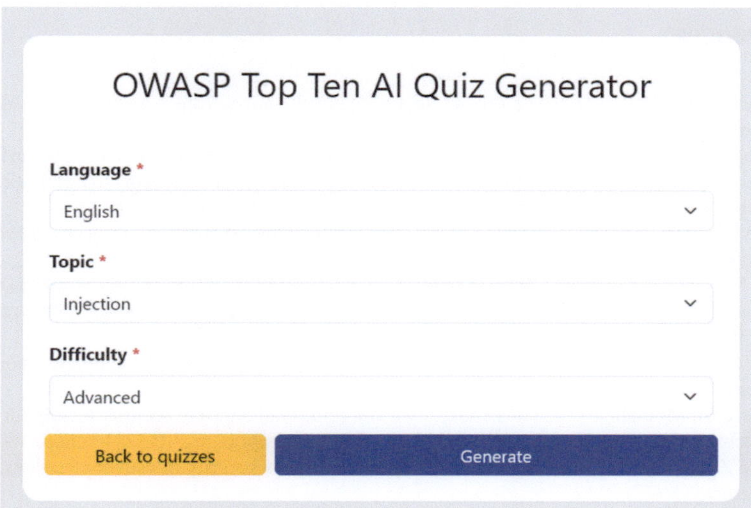

Fig. 8.3 Start page of OWASP Top Ten AI Quiz Generator

8.6 Conclusion

The increasing reliance on web-based systems for personal, organizational, and societal functions has made secure data utilization a keystone of modern technology. APIs (Application Programming Interfaces) serve as the backbone of integrated web environments, enabling seamless data exchange and interoperability across platforms. However, the same interconnection introduces vulnerabilities that attackers can exploit, leading to breaches, data theft, and compromised systems.

Section 8.2 has introduced the concept and architecture of integrated web systems, providing a foundation for understanding their structure. Section 8.3 has discussed the concept of API endpoints, along with the implementation of security techniques and methods in their development. It has also explored the retrieval of data from websites without APIs, addressing the associated security risks. Section 8.4 has provided a detailed analysis of the phases involved in defining, developing, testing, and implementing API endpoints, with an emphasis on security. Additionally, tools for testing API calls have been highlighted. In Sect. 8.5, the importance of incorporating new methods into the education of engineers in the field of web security has been emphasized. The findings presented throughout this paper offer a comprehensive perspective on securing API endpoints and underscore the necessity of enhancing both education and awareness in the field of web security.

References

1. Čović Z, Rajnai Z, Fürstner I (2024) Secure data utilization from photovoltaic systems for optimization purposes. Biztonságtudományi Szemle 6(1):65–77
2. Huang X-W, Hsieh C-Y, Cheng CH (2015) A token-based user authentication mechanism for data exchange in restful API. In: Proc. 18th international conference on network-based information systems. Taipei, Taiwan, pp 601–606. https://doi.org/10.1109/nbis.2015.89
3. OWASP API Security Project Homepage. https://owasp.org/www-project-api-security/. Accessed 27 Dec 2024
4. Idris M, Syarif I, Winarno I (2021) Development of vulnerable web application based on OWASP API security risks. In: 2021 International electronics symposium (IES). Surabaya, Indonesia, pp 190–194. https://doi.org/10.1109/IES53407.2021.9593934
5. State of API Security Report 2024 Homepage. https://content.salt.security/state-api-report.html. Accessed 04 Jan 2025
6. Alharbi SJ, Moulahi T (2023) API security testing: the challenges of security testing for restful APIs. Int J Innov Res Sci Eng Technol 8:1485–1499. https://doi.org/10.5281/zenodo.7988410
7. Čović Z (2024) Using challenge-based tasks for testing the security of web applications. Bánki Közlemények 6(2):7–12
8. Swagger UI Homepage. https://swagger.io/tools/swagger-ui/. Accessed 03 Jan 2025

Chapter 9
Critical Infrastructure Risks from Sabotage

Norbert Daruka

Abstract Recent years have provided clear evidence that critical infrastructures are increasingly at risk of being targeted by malicious attacks. As infrastructures provide essential services that are vital to the functioning of society, such as energy supply, transport, water supply, healthcare and communication systems, their protection can be a particularly important factor. Some infrastructures perform such a specific function for a country or region that the temporary or permanent loss of their services could force fundamental changes in the area. These are critical elements whose protection must be designed with multiple safeguards to ensure that their operation is uninterrupted. The organization of external physical protection is not a simple task for such a facility, but the prevention of internal sabotage is even more difficult. Blast attacks are a particular threat to these systems, as they can cause serious damage to infrastructure, human lives and the economy in a short time. The perpetrators of attacks are often external actors, but increasing attention must also be paid to the threats posed by internal perpetrators, as advances in technology open up new opportunities not only in the security domain but also, unfortunately, in the case of malicious attacks. My aim, even with this publication, is to raise awareness of insider offenders as a source of danger and the risks of their presence.

Keywords Sabotage · Insider offender · Explosive attack

9.1 The Role of Insiders in Attacks Against Critical Infrastructure

Critical infrastructure security is one of the most important issues of our time. These infrastructures provide essential services that are vital for the functioning of society, such as energy, transport, water, healthcare and communication systems.

N. Daruka (✉)
Banki Donat Faculty of Mechanical and Safety Engineering, Obuda University, Budapest, Hungary
e-mail: daruka.norbert@bgk.uni-obuda.hu

T. A. Kovács and I. Fürstner (eds.), *Critical Infrastructure Protection: Advanced Technologies for Crisis Prevention and Response*, NATO Science for Peace and Security Series C: Environmental Security, https://doi.org/10.1007/978-94-024-2308-2_9

Bomb attacks pose a particular threat to these systems, as they can cause serious damage to infrastructure, human lives and the economy in a short time. Attacks are often perpetrated by external actors, but increasing attention must also be paid to the threats posed by internal perpetrators. A specific feature of insider actors is that they have an advantage in the planning and execution of attacks that makes it difficult or impossible to prevent their activities. This includes direct access to infrastructure, a thorough knowledge of systems and procedures, and the ability to bypass security measures. The following section details how insider offenders can influence critical infrastructures, what their motivations are and what measures can be taken to reduce the threat they pose.

Today, critical infrastructure security is a high priority. The challenges of securing these facilities are becoming more and more acute. Infrastructures are particularly at risk, especially those that are essential for the performance of vital societal functions—health, safety, security, economic and social well-being of people—and whose disruption or destruction would have significant consequences if they were not to continue to perform these functions. Today's technological advances provide us with the means to prevent attacks, prevent external attacks, reduce their likelihood of occurrence, and minimize the harmful effects of successful attacks. As individuals and groups interested in successfully carrying out attacks have quickly recognized this defence mechanism and fortunately have not found a workaround, attempts at external attacks have been significantly reduced. However, the method of carrying out internal attacks (sabotage) has become more sophisticated, which not only makes it more difficult to counter the defence sector, but also makes it very difficult to identify the perpetrators and the methods used. The insider perpetrator (saboteur) has a huge advantage because of his knowledge of the infrastructure, the vulnerable units and where he can cause the most damage.

Let's take a step back from the saboteur working to destroy infrastructure at all costs. There are situations where the main goal is not destruction but deliberate disruption, sabotage, even the threat of sabotage in an industrial facility. This could be the triggering of a partial or total shutdown that could cause millions of euros of damage to the facility. I will illustrate this damage with a real example. In an automotive manufacturing plant, a bomb threat was ordered at least twice a month. The company had to close down and evacuate the affected part of the factory in accordance with the rules and regulations in force in the country. Workers were only allowed to return to the production lines and restart the production process after being checked by the bomb squad. This usually took half a shift, so the day's production was almost completely wasted. For each of these stoppages, the company had to account for damages of around €750 000. This example shows that damage can be caused by threats, and it is a fortunate situation, as there is no question of anyone being injured or killed. These situations can easily be prevented by proper internal regulations and by regulating movements in the various sectors. This example also led to the rules being changed and people who had knowledge of the internal processes and key elements of production being more thoroughly vetted [1]. In the case of our example, it was the undercover person from the competitor company who got to know the internal processes, assessed the weaknesses and started to weaken the company from the inside. "Insider threats are perhaps the biggest and most difficult part of the security challenge" [2], as only by maintaining targeted and systematic controls at all times can we have any chance of preventing this method of attack.

The insider threat is already considered a serious risk in many areas, and quite specific terminology has been introduced in different areas. For example, in cyber-security and the energy sector, the term insider threat actor can be used [3]. The term insider threat (insider offender) refers to a person with direct physical access who has the means, opportunity, motive and expertise to plan, organize and carry out an explosive attack. Such insiders may be current or former employees, contractors, suppliers or business partners, all of whom have—or have had—authorized access to the premises or facilities of the organization or facility. This may include those who are able to import explosives or explosive devices indirectly rather than directly. However, we should also not forget those persons with non-physical access who are able to control, program or remotely operate the technological systems of a technology, manufacturing process or facility in such a way that an explosive atmosphere is created under appropriate conditions. In the latter case, a hacker, a skilled programmer, is sufficient to carry out the process. It is not so simple that anyone who is computer literate or has ever worked in a factory or on a production line can carry out the criminal process. This would require professional knowledge, such as explosives or explosion protection engineering, which is relevant in very few cases [1].

All in all, an insider offender is simply an adversary with inside access who has the capability and opportunity to carry out malicious acts, hence the insider security threat.

9.2 The Threat of Explosive Attacks

If we assume that the saboteur wants to cause the most damage to the infrastructure, then the most vulnerable points of the infrastructure should be investigated and controlled. The only way to increase the amount of damage caused is for the perpetrator himself to use a material to carry out the attack that is capable of causing the greatest destruction, i.e. explosives. As I mentioned, with advances in technology, it is very easy to prevent large quantities of materials (e.g. explosives) or structures (e.g. explosive devices) from entering a protected facility, elements of critical infrastructure [1]. However, it is difficult to prevent a minimal amount (a few grams) of explosive material or various precursors from entering the facility. In the case of precursors, these are materials that can be used as components of explosive devices if they are poured, mixed or otherwise manipulated. The fact that these components (precursors) can also be found inside the facilities may also be an important factor, as various cleaning agents, disinfectants, acids, alkalis can all fall into this category [4].

This being the case, the primary concern is not necessarily the introduction of explosives into or the production of explosives within a facility, as combustible gases, vapours or mists or combustible dusts can also cause an explosion within critical infrastructure [5]. Of course, it is not as easy to cause deliberate damage as with a specially designed explosive, but under the right conditions these combustible materials can also be destructive. Before turning to causality, I think it is important to identify the causes.

Now let's turn back to explosives, as not all saboteurs have the knowledge and experience to turn precursors into explosive compounds. Explosives can be easily introduced in extremely small quantities, even into tightly secured areas. In this case we are talking about small quantities that do not even attract the attention of security personnel. Since the insider offender has free access to the infrastructure, he can use the principle of "a lot for a little" to introduce the quantity of explosives needed for an attack into the facility continuously over several days or months. The only condition is that the material must be stable and that the separation and recombination of 1–2 g of the material is not a problem. Under appropriate conditions, a significant proportion of military explosives may be suitable for such tasks. I am not, of course, trying to give ideas to the perpetrators, but rather to draw the attention of those involved in defence to what to look out for when operating a screening device or scanner at access control systems.

The question may arise: If these few grams of explosives are successfully brought into the facility and the necessary amount of explosives are collected, what can be done with them? Several kilos of explosives would be needed to cause serious damage. Well, those who think so are wrong! Given today's technological possibilities—and I include here the possibilities of additive manufacturing—it is possible to design components and elements which, alone or as part of a system, can form bodies which, when loaded with explosives, can, if properly initiated, cause the necessary destruction [6]. This, of course, requires the use of what has been a well-established technique in the field of explosives engineering for decades. Examples include the cumulative effect or the principle of using Explosively Formed Projectiles—EFP [7].

Malicious acts—acts or attempts of unauthorized removal or sabotage—that can cause damage or injury may vary from establishment to establishment. It depends on the local legal background, the profile of the company and many other factors. For example, the main purpose of physical protection systems [8] in different facilities is to protect finished products, components, other materials essential for manufacturing from blast attacks, intentional destruction and to minimise the damage and consequences of an explosion [9].

If we want to define the key classifications or positions who could be potential insider offenders, we see that it could be anyone who has access and/or privileges to the facility's systems, so in particular, but not exclusively: facility management officials; employees of the facility; security personnel, guards; IT system administrators; external contractors, partners; maintenance staff; officials; utility providers (electricity, gas, water, sewerage, internet, and so one); suppliers, couriers; visitors (Fig. 9.1).

For those at risk, the most dangerous is that employees are often unaware that confidential information is being leaked, facilitating successful attacks. In sectors

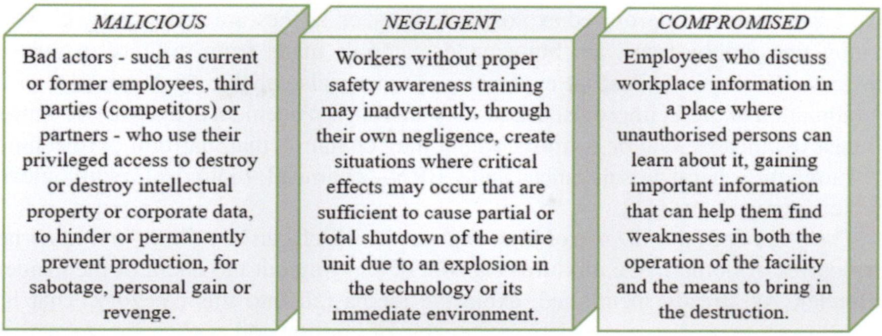

Fig. 9.1 Insider threat categories (*Source* edited by the author)

where leaked (critical) information could have a significant impact on the organization of attacks (e.g. critical elements of the energy sector), it is particularly important to exclude the slightest technical or human error factor.

9.3 The Nature of Blast Attacks

Bomb attacks are some of the most serious and destructive acts of violence in modern times. These attacks can be motivated by a variety of motives and objectives, which determine the methods, locations and effects of their perpetration. If we want to understand the motivations of the individuals who carry out the bombings, or the organizations and interest groups behind them, we need to look at the motivations behind the bombings. These motivational factors may be political or ideological, economic interests or personal motives (individual grievances). The targets in this respect are not necessarily infrastructure, although its elimination may indirectly affect the achievement of the objective. In the case of politically or ideologically motivated attacks, the aim is not to render the targeted infrastructure inoperable, but to create a situation where its removal creates the opportunity to undermine the stability of the ruling power and challenge its ideological views [10]. Attacks of this kind usually occur in the run-up to elections (see the election periods in Afghanistan). I have already given a good example of attacks with an economic motive. Personal motives and individual grievances often play a role in the background of conflicts and attacks. In such situations, the behavior of the actors is usually driven by emotional factors, past experiences and self-interest. These motives may be revenge, jealousy, power or dominance, or dissatisfaction. In this case, it is also important to look at the causes of grievances, which may be experiences of failure, unresolved conflicts, hypersensitivity or rejection.

Since the means of attack—and here we should distinguish from threats—are mostly explosive devices, explosive media, it is also necessary to know them if we want to formulate appropriate countermeasures. Attackers may use improvised explosive devices. Improvised explosive devices, or, as they are colloquially known, booby traps/booby-traps, are homemade materials made from military explosive devices, military or industrial explosives, or materials capable of detonation, or a combination of these, improvised assembly and/or emplacement of a device designed to destroy, incapacitate or confuse, which may contain lethal, harmful pyrotechnic or flammable chemical substances and CBRN—chemical, biological, radiological, nuclear components [11].

The other option is the use of high explosives, which can be military or industrial explosives or home-made mixtures capable of carrying out the intent of the insider offender. As already mentioned, explosive media fall into this category. That is, gases, vapours, mists, or powders are capable of ignition and explosive combustion when exposed to an appropriate ignition source. There is a huge difference between a space explosion and an explosive explosion, but both are capable of wreaking havoc in a given area in a way that can affect the operation of a plant. The only difference is that explosives can cause damage anywhere, in any infrastructure, whereas a space explosion can only occur in an area where an explosive atmosphere can be created and the appropriate ignition source is present [12].

9.4 The Need for Information on the Feasibility of Attacks

To achieve their goals, insiders or saboteurs need to have a wealth of information, which can come indirectly from external, independent sources or directly from personal experience. The goal of intentional harm is not sufficient to carry out an assassination or attack; a serious knowledge base is needed to carry out the operation independently and with as few people as possible. The question may arise: why is a small number of people important? The answer is very simple: the more people involved in the operation, the more people who know the details, the lower the chance of error and the higher the chance that the security organization will spot certain critical elements and individuals.

When saboteurs act independently, they are able to become invisible, so to speak, because the relative scarcity of information means that they may not be noticed by the security services. Consequently, insiders, whether autonomous or a few strong, are free to plan attacks undetected, have time to assess their capabilities and test their security skills without consequences. Insider attackers acting alone can also be effective because they can decide autonomously when it is best to act, which targets are most vulnerable, and whether to partner with other insiders or casual outsiders to achieve their goals [1].

For an explosive attack to be carried out on a critical infrastructure, it must have the following elements: permanent or temporary access to the area in question; knowledge of how the technology works, its vulnerabilities and the consequences of the attack; and the expertise to produce, place and launch the required quantities of explosives in the appropriate place and at the appropriate time.

In the event that the saboteur does not have at least one of these elements at his disposal, other persons, either outsiders or insiders, must be involved in the attack process. Those performing protection duties should monitor these areas, keep them under review and ensure that all elements are not secured at the same time.

In order to be able to carry out explosive attacks, insiders must have physical access to information, systems and equipment related to the technology or technological process. Following this chain of logic, the following elements require particular attention from security professionals: the storage, access and use of databases; access to information and communication systems; access to policies, internal operational documentation; access to protocols and procedures; access to plans, security plan and emergency plan; access to premises, whether office, warehouse, storage, server room; access to equipment, devices, systems, even the security system; access to transport and transport equipment.

It is particularly difficult to organise protection if the insider is entitled to access the items listed above [13]. I believe that if the insider has access to everything that is actually used by the company, the only thing that can prevent him from carrying out his plan is chance. Therefore, every effort should be made to ensure that the items listed above are not collectively accessible to any employee.

In my opinion, it is particularly important to pay attention to those who have access to the most sensitive areas. These people can be maintenance staff, inspectors, operators of the technology. They are people who know the location of the sites, access routes, building layouts, utility networks and, of course, the information needed for operations. This information includes:

- organisational level knowledge: knowledge of the management structure of the organisation, the organisational structure, the location and availability of employees, their qualifications, the relationship between subordinates and supervisors, personal information about employees, and rules, procedures, protocols and policies;
- professional skills: an offender can only be successful if he is fully aware of the weaknesses of the technology and the process;
- security-level skills: to be successful, the saboteur must be familiar with the types of alarm and signalling systems, their installation locations, numbers, guard locations and patrol routes, or elements of security protocol (Fig. 9.2).

If the saboteur has time to plan the attack, he will certainly take into account external partners and suppliers, those who provide raw materials for the infrastructure. Why is this important? Well, at its simplest, there is no risk of introducing explosives or precursors into the target area. The easiest way to do this is for the saboteur or his assistant to know the details of contracts with external partners and suppliers. They should know information such as when deliveries and shipments are expected, and even waste collection agreements. Special access for repair and maintenance should also be considered, as this is one of the easiest ways to get the necessary materials and equipment inside the protected area.

As can be seen from the above, insiders have a wide range of opportunities and conditions favour them, but with the right measures, well-defined responsibilities

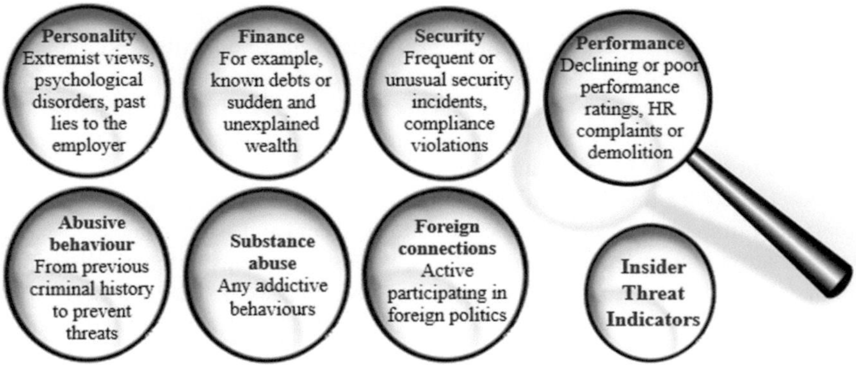

Fig. 9.2 Insider threat indicators (Source: edited by the author)

and effective internal segregation, not only movement but also permeability can be controlled, thus reducing the risk of successful sabotage.

For insiders, we can distinguish between active and passive factors, which can help us understand and categorise insider activity itself. An active insider is always actively involved in the conspiracy. There is therefore a constant risk of being caught. A very good example is that "if an insider gives his own key to an adversary, he is a passive insider, but if he steals or copies his colleague's key, he becomes an active insider" [5]. The passive, non-violent insider confronts security services by stealth and deception, not violence, while the active, violent insider is prepared to use violence against anyone who interferes in any way.

9.5 Options to Defend Against Saboteurs

Physical security and access control is essential to prevent sabotage, theft, data loss or other threats. Organising the protection of your infrastructure should start with enhancing physical security. Access control mechanisms include organising different access control systems, setting access levels, and logging and monitoring who (organisations) enters.

Locks, gates, fences and barriers, as well as the protection of internal spaces, are essential and defining elements of physical barriers. In order to be able to perform their intended function effectively, they must fit in with the range of activities represented by the infrastructure itself. The fact that a gate has the best indicators in the catalogues does not necessarily mean that it can prevent tampering.

Nor should we forget the basic security systems that all infrastructures need to have in place, regardless of the level of vulnerability. On the technology side, these are physical security technologies, camera systems, alarm systems and fire protection systems. If technology is not the most important factor, then staff training and awareness raising should be the main factor. Particular attention should also be paid to security training, communication of policies and their feedback, and reporting of suspicious activities.

If the basic systems are in place, one of the main tasks is to organise regular safety checks. Continuous audits and tests should be carried out alongside vulnerability assessments. Of course, maintenance and upgrades following technological developments should also be a priority.

We also need to recognise that, despite the best organised measures, there are still factors that preclude the possibility of complete security. There will always be factors, developments and technologies that cannot be discovered, put in place and applied in time. In this case, emergency plans (incident management and evacuation plans) are needed to ensure the maximum possible damage recovery, regardless of human technological considerations.

Internal audits and psychological screening can play a critical role in preventing sabotage by identifying and addressing potential insider threats. Internal audits and psychological screening can play a critical role in preventing sabotage by identifying and addressing potential insider threats. The objectives of internal controls for critical infrastructures include monitoring compliance with organizational policies and procedures, timely identification and detection of violations, internal threats and anomalies, and the correct use of resources.

Psychological screening to identify internal threats refers to pre-employment screening (psychological tests and background checks) and ongoing psychological assessments. In some cases, saboteurs may communicate unusual behaviour such as excessive stress, dissatisfaction, disregard for rules to the outside world. These direct threats or outbursts of temper can also be signs that the employee (saboteur) is up to something to get back at his colleague, boss or company. In such cases, psychologists or HR experts have a particularly important role to play in creating a climate of trust in the workplace. Employees can turn to them with confidence and they can be the first to identify the problem.

Nowadays, an integrated approach, i.e. the coordination of controls and screening, can also be a solution to detect problematic cases. A good solution could be to carry out targeted psychological screening based on the results of internal audits, for example when the behaviour of an employee makes this necessary. In such a case, access rights to a risk area can be temporarily restricted by feeding back the results of the screening.

Training and awareness-raising are critical to preventing sabotage, as ill-informed workers are unable to recognise and avert threats. For this training element, it is of paramount importance that employees are prepared to recognise and avoid sabotage and are able to pass on this knowledge to their colleagues. To ensure that employees are optimally prepared to deal effectively with suspicious activities, a thorough knowledge of organisational security protocols should be the first step. This should include the basics of physical security, cyber security and incident management, the psychology of sabotage itself and stress management. This can be achieved through practical training, e-learning modules and various professional workshops.

Communication campaigns are the most effective way to raise awareness in the fight against sabotage. Posters, leaflets, information graphics in offices and monthly newsletters can be the most helpful, but motivational programmes (prizes, competitions to raise awareness) can also have the right impact. What is most often forgotten is leadership, for example, the active participation of managers in safety programmes and open communication about safety policies and objectives. It is also important that workers are able to practice following protocols through real-life exercises and sabotage simulations. Of course, presentations of cases where awareness has helped prevent a threat, presented in workplace training sessions, can also be a useful way forward. Such measures can contribute to a more responsible behaviour of employees and a more effective detection of sabotage threats, thus strengthening the overall security of the organisation.

Information security plays a key role in protecting against sabotage, as data protection is critical for modern organisations. The right information security measures can minimise the risk of sabotage, protect sensitive data and ensure business continuity. This includes the following factors:

- Access control and privilege management: Strict authentication procedures; Role-Based Access Control; Least Privilege Principle; Regular access audits.
- Data protection and encryption: Encryption; Data Classification; Management of data media; Data destruction.
- Monitoring and event management: System log monitoring; Security Information and Event Management (SIEM) systems; Incident Management Plan.
- User Awareness: Regular training; Security protocols; Tests and simulations.
- Technological protection: Firewalls and intrusion detection systems; Antivirus and antimalware software; Data backup and recovery; Network traffic monitoring.
- Compliance with laws and regulations; Compliance with data protection legislation; Establish internal policies.

Protecting information security against sabotage requires a multi-layered approach that includes technological, organisational and human factors. To properly protect information, controlling access, encrypting data, preparing for incidents and raising employee awareness are all vital.

References

1. Daruka N (2024) The risks of insider attacks on critical infrastructure. Russian Aggression and European Security, pp 142–157. ISBN: 9788382703375
2. Bunn M, Sagan SD (2017) Insider threats. Cornell University Press, Ithaca, p 216. ISBN: 9781501705946. https://doi.org/10.7591/9781501705946
3. Nyikes Z, Kovács TA (2019) The innovation of special coating for defence of information leak and radio activated blasting in case of critical infrastructure buildings. Critical infrastructure protection. IOS Press, pp 21–26
4. Horváth T, Ember I (2021) Characteristics of homemade explosive materials and the possibilities of their identification. Revista XXVI(2):100–107

5. Bunyitai Á (2023) Insider threat mitigation in high security facilities. Nemzetbiztonsági Szemle XI–1/2023, pp 49–61
6. Ember I, Ádám B (2022) Kumulatív töltetházak 3D nyomtatása. Hadmérnök 17(3):35–44
7. Lukács L (2023) Explosive technology in domestic military journals from the late 1800s to the present. Budapest, Ludovika, ISBN 978–963–531–983–1. (ePub) Subchapter 3.3, The explosively formed projectile-the Misnay-Schardin effect, pp 210–221
8. Kovács Z (2016) Barriers for physical protection of military facilities against vehicle born improvised explosive devices. Vojenské Reflexie 11(1):86–99
9. Kovács Z (2016) Explosion of improvised explosive device on structures. Hadmérnök 11(1):56–63
10. Ackerman G, Abhayaratne P, Bale J, Bhattacharjee A, Blair C, Hansell L, Jayne A, Kosal M, Lucas S, Moran K, Seroki L, Vadlamudi S (2007) Assessing terrorist motivations for attacking critical infrastructure. Center for Nonproliferation Studies, Monterey Institute of International Studies, UCRL-TR-227068, p 273
11. Daruka N, Szalkai L (2024) Detection of explosives and explosive devices current issues. Németh Kornél. VIII. Turizmus és Biztonság Nemzetközi Tudományos Konferencia tanulmánykötet. ISBN: 9789633962862, pp 436–448
12. Zsarnovszki A, Elek B (2024) Robbanásveszélyes ipari technológiák biztonsági követelményeinek érvényesülése a gyakorlatban. Biztonságtudományi Szemle 6(4):127–142
13. Figuli L, Kovács TA, Nyikes Z, Maňas P (2022) Principles of physical protection of strategic assets against blast effects. Security-related advanced, chnologies in critical infrastructure protec-tion: theoretical and practical approach. Springer Netherlands, Heidelberg, Germany, pp 341–352

Chapter 10
Analysis of Challenges and Opportunities of Using Photovoltaic Systems in the Case of Households in Republic of Serbia by Using Publicly Available Data

Igor Fürstner

Abstract The presented research investigates the challenges and opportunities associated with photovoltaic (PV) systems in Serbian households, focusing on cost efficiency, environmental benefits, and energy independence. Leveraging publicly available data and a case study, the research provides a comprehensive analysis of PV viability in the Republic of Serbia, especially within the framework of the prosumer model and net measuring. The study explores the impact of Serbia's legislative frameworks, including incentives and net metering policies, on household PV adoption. Additionally, it examines the economic implications of Serbia's dual tariff structure, highlighting how PV systems can reduce electricity costs and contribute to grid stability. A key finding reveals that optimizing PV systems capacity and considering storage options significantly affect return on investment and long-term energy savings. While properly dimensioned PV system installations prove cost-effective, additional investment in storage, despite benefits in energy independence, extends the return on investment period considerably. The findings suggest that, although PV systems present substantial benefits, improvements in policy support, storage technology, and grid infrastructure are crucial for maximizing PV potential. Future research should focus on optimizing storage solutions, integrating dynamic tariff systems, and developing advanced grid management techniques to enhance the overall impact of PV systems in Serbia's energy landscape.

Keywords Photovoltaic system · Tariff system · Cost analysis

I. Fürstner (✉)
Subotica Tech—College of Applied Sciences, Subotica, Serbia
e-mail: ifurst@vts.su.ac.rs

T. A. Kovács and I. Fürstner (eds.), *Critical Infrastructure Protection: Advanced Technologies for Crisis Prevention and Response*, NATO Science for Peace and Security Series C: Environmental Security, https://doi.org/10.1007/978-94-024-2308-2_10

10.1 Introduction

The use of PV systems in households has grown rapidly in recent years, fuelled by a combination of technological advancements, cost reductions, and increased awareness of environmental issues. PV systems convert sunlight directly into electric energy using solar panels, which are typically installed on the roofs of residential buildings.

One of the primary advantages of household PV systems is the significant reduction in electricity costs. By generating their own power, homeowners can decrease their reliance on grid electricity, which can be particularly beneficial in regions where energy prices are high or volatile. In many cases, PV systems generate more electricity than a household consumes, allowing homeowners to sell excess power back to the grid. This not only offsets the initial investment in solar panels but also provides a financial incentive for more households to adopt renewable energy.

The environmental benefits of PV systems are substantial. Traditional electricity generation methods, such as coal and natural gas plants, produce greenhouse gases and other pollutants that contribute to climate change and poor air quality. By contrast, solar power is a clean, renewable energy source that produces no emissions during operation. Widespread adoption of PV systems in households can significantly reduce carbon footprints and lessen the demand for environmentally damaging energy sources.

In addition to financial and environmental benefits, PV systems increase energy independence. By producing their own electricity, homeowners become less vulnerable to grid disruptions and power outages. This energy independence can be particularly advantageous in remote or rural areas, where access to reliable electricity may be limited. When combined with battery storage systems, PV installations allow households to store excess energy for use during periods of low sunlight, further enhancing reliability and resilience.

Government policies and incentives play a crucial role in promoting residential PV system adoption. Many countries offer different financial incentives to reduce the upfront cost of PV installations. Additionally, advancements in financing options, such as solar leases and power purchase agreements, have made PV systems more accessible to a broader range of homeowners.

Despite the benefits, there are challenges associated with PV systems, including the initial cost, the need for adequate sunlight, and space requirements for installation. However, with ongoing research and development, these obstacles are gradually being overcome. Energy storage solutions are becoming more affordable, and new, high-efficiency solar technologies continue to emerge, making it feasible for more households to harness solar energy effectively.

The presented research deals with the challenges and opportunities of using PV systems in the case of households in Republic of Serbia by using publicly available data. The research is restricted to the prosumer category of electric energy producers-consumers, which contracts with the supplier of electric energy in the electricity market with net measuring. Through a case study, the current possibilities and limitations regarding the researched topic are discussed.

The reminder of the paper is structured in the following manner: First, a literature review is presented. After that, the legislative background, and the consumption cost calculation is presented. Following that, a case study is presented and the results are discussed. After the discussion, conclusions are presented.

10.2 Literature Review

The deployment of PV systems in Serbia and the wider Southeast Europe (SEE) region has gained increasing attention in recent years, with many countries in the region working to align their energy strategies with European Union (EU) directives on renewable energy and carbon reduction. Serbia, in particular, has made significant progress towards increasing its share of renewable energy sources in its energy mix, with solar PV systems recognized as a pivotal contributor to this transition.

A fundamental driver of PV system deployment in Serbia and the region is the political and policy alignment with EU energy and environmental objectives. Serbia, for instance, has adopted national targets to increase renewable energy capacity as part of its EU accession goals. The EU's Green Deal and associated policies, which emphasize a transition to renewable energy, have significantly influenced policy reforms in SEE countries, including the establishment of incentive mechanisms and subsidies to stimulate investments in this field. Some studies highlight that national frameworks, such as Serbia's Energy Strategy, prioritize renewable energy projects, particularly PV installations, as a means to reduce dependency on coal and achieve a lower carbon footprint [1]. Further analysis suggests that these policy interventions, combined with the EU's financial and advisory support, have fostered a more favourable investment climate for PV systems in SEE [2]. However, the studies emphasize that continuous policy improvement is necessary to overcome bureaucratic barriers that still exist in the region. For instance, the permitting processes can be lengthy and complex, deterring some investors from engaging in PV projects. Serbia and other SEE countries must simplify these procedures to sustain and expand PV capacity in alignment with EU benchmarks [3].

Moreover, the International Renewable Energy Agency reports that PV installation costs have dropped nearly tenfold from 2010 to 2020, highlighting a global trend that benefits the SEE region as well. Economic incentives provided by local governments further enhance PV systems attractiveness, with mechanisms such as feed-in tariffs and tax reductions supporting early adoption [4]. However, despite the lower costs, Serbia and neighbouring countries still require more robust financial mechanisms and public–private partnerships to encourage widespread PV investment, especially in rural and underserved areas [3].

Innovations in PV technologies have further expanded the potential applications and efficiency of PV systems in SEE. Many researchers suggested different possibilities of novel solution for maximizing the utilization of PV systems [5–7]. Such innovations can further enhance PV performance in Serbia, where geographical and climatic conditions are favourable for solar energy production. Furthermore, recent advancements in energy storage systems have been crucial in addressing the intermittency of solar power, allowing for a more stable and reliable energy supply [8].

The geographical position of the SEE region offers ideal conditions for PV system deployment, with high levels of solar irradiance, particularly in southern Serbia and surrounding areas. Serbia's annual solar irradiation ranges from 1,200 to 1,600 kWh/m^2, providing a strong basis for PV applications. Southeast Serbia is especially suited for large-scale PV installations due to the combination of high solar potential and available land. These favourable environmental conditions have led to significant interest in scaling up PV installations in both urban and rural areas. Furthermore, the environmental benefits of PV systems align with the EU's climate objectives by reducing carbon emissions and lowering the ecological footprint of energy production. The use of PV systems contributes to decreased reliance on fossil fuels, which remain a predominant energy source in Serbia [9].

Despite technical and economic feasibility, the deployment of PV systems in SEE faces challenges related to social acceptance, regulatory consistency, and institutional support. Public awareness campaigns and educational initiatives are essential to foster community acceptance and engagement [10].

Additionally, Serbia's institutional frameworks require improvement to support streamlined PV development. Bureaucratic hurdles, such as complex permitting processes and inconsistent regulatory policies, remain significant barriers to investment and implementation. Institutions must adopt a more transparent and efficient approach to regulatory processes to encourage both domestic and foreign investments in PV systems [11].

Grid integration of PV systems poses another challenge due to the intermittent nature of solar energy. Current grid infrastructure in Serbia and the SEE region requires upgrades to accommodate high levels of PV-generated electricity [7]. Energy storage solutions, such as battery storage and pumped hydro storage, are being explored to address these intermittency issues, though these technologies come with additional costs. Successful PV integration will require comprehensive grid modernization, which includes developing flexible and resilient energy systems [8]. As more PV systems are integrated, the grid must adapt to ensure reliability and avoid disruptions. Developing robust grid infrastructure is particularly vital for achieving energy security and reliability across SEE.

Localized case studies provide valuable insights into the potentials and challenges of PV adoption in specific contexts. These case studies show that PV systems can effectively reduce electricity costs and increase energy independence in various settings [12]. Furthermore, studies on mini PV power plants in rural parts of South Serbia, demonstrate how PV can provide sustainable energy solutions in remote and underserved areas [9].

Future projections for Serbia and SEE suggest a substantial increase in PV capacity, driven by both national policies and international commitments. The Vienna Workshop Report outlines scenarios for achieving zero-carbon energy systems in

SEE by 2050, with PV systems identified as a cornerstone of this transition [11]. Serbia's government aims to achieve a significant increase in PV capacity by 2030, in line with EU renewable targets. These ambitions underscore the potential for PV systems to play a leading role in SEE's energy future [6]. The region's future energy landscape is expected to be shaped by continuous advancements in PV technology, supportive policy frameworks, and a strong commitment to sustainable development [10].

In summary, the literature on PV systems in Serbia and SEE shows the promising potential of PV systems to contribute to energy transition goals. The combination of favourable environmental conditions, falling PV costs, supportive policy frameworks, and technological advancements creates a solid foundation for scaling up PV installations across the region. However, challenges related to regulatory frameworks, social acceptance, and grid integration must be addressed to fully realize this potential. Future efforts should focus on streamlining regulatory processes, modernizing grid infrastructure, and enhancing public awareness and support for renewable energy. The SEE region's commitment to the EU's environmental goals, coupled with its strategic investments in PV systems, positions it well to become a leader in renewable energy adoption in Europe.

10.3 Legislative Background

The legislative framework governing the use of PV systems in Serbian households is grounded in the Law on Renewable Energy Sources, enacted in 2021, which provides the primary regulatory foundation for renewable energy generation and integration in Serbia. This law, together with complementary regulations, has established the legal requirements, incentives, and obligations related to PV systems and their application within residential settings.

The Law on Renewable Energy Sources [13] was a significant legislative step in harmonizing Serbia's energy sector with EU regulations, specifically targeting an increase in the share of renewable energy. For households, the law primarily introduces the concept of prosumers (producers-consumers)—households who generate their own electricity through PV systems, either for self-consumption or to feed into the national grid. According to the law, households can become prosumers, enabling them to install PV systems on their properties and engage in self-consumption with the potential to export surplus energy to the grid.

One of the law's critical provisions for household PV system users is the establishment of a net measuring mechanism, allowing prosumers to offset their electricity consumption with their generated energy. This means that any excess electricity produced by household PV system that is fed into the grid is credited against their electricity bill, enhancing affordability and incentivizing PV adoption.

The Law on Efficient Use of Energy [14] complements the Renewable Energy Law by imposing energy efficiency requirements on buildings, which indirectly impacts PV adoption. Households opting to install PV systems must meet certain energy efficiency standards, ensuring that PV-generated electricity is used in a manner that minimizes overall energy consumption and aligns with national efficiency targets.

The Serbian Regulation on Technical Standards and Safety of Photovoltaic Installations [15] outlines the technical and safety requirements for PV systems, including standards for installation, grid connection, and maintenance. This regulation ensures that household PV systems meet minimum safety standards to prevent hazards and ensures system longevity. Moreover, it stipulates that installations must be performed by certified professionals, which adds a layer of quality assurance and protects consumers from substandard practices.

The Law on Renewable Energy Sources establishes that all prosumers have the right to connect their PV systems to the grid, obligating utility companies to facilitate connections for eligible households. To connect a PV system, households must submit an application to their utility provider, accompanied by relevant documentation such as technical specifications, installation plans, and energy efficiency certifications. The Energy Law [16] reinforces the obligation of distribution system operators to support the integration of renewable energy sources into the grid.

In Serbia, household PV installations under a certain capacity threshold (typically up to 10 kW for residential properties) are not required to undergo a full environmental impact assessment, as specified in the Law on Environmental Protection [17]. According to the Law on Renewable Energy Sources, small-scale PV systems benefit from a simplified licensing process, exempting household systems below certain thresholds from burdensome licensing requirements. This exemption significantly reduces administrative barriers, making it easier for households to adopt PV technology without the need for lengthy regulatory procedures.

Serbia's legislative framework for renewable energy, particularly the Law on Renewable Energy Sources, has set a strong foundation for the growth of PV systems in households. The introduction of prosumer rights, financial incentives, and simplified regulatory processes makes PV adoption increasingly feasible for Serbian households. However, ongoing adjustments to regulations and incentives are anticipated to enhance system compatibility with EU energy standards and accommodate the growing interest in household PV solutions. Serbia's commitment to renewable energy, as seen in recent national energy strategies, suggests that further supportive measures for PV systems will be introduced in the coming years to maintain the momentum toward sustainable energy.

10.4 Consumption Cost Calculation

In this chapter, the electric energy consumption cost calculation is presented in the case of households.

10.4.1 Tariff Structure and Consumption Levels

The Serbian household electricity tariff system operates on a dual tariff structure that considers both energy consumption levels and the time of day when electricity is used. This approach incentivizes efficient energy usage by offering different rates based on monthly consumption thresholds and peak/off-peak hours, helping to balance the national grid's demand and encouraging energy conservation.

The tariff system is divided into three primary consumption tiers, each with distinct price implications:

1. Green Tariff: This tier covers consumption up to 350 kWh per month, based on a 30-day billing cycle. The green tariff is the most economical and aims to encourage households to stay within this lower consumption limit, rewarding efficient usage with lower costs.
2. Blue Tariff: This level applies to consumption between 351 and 1600 kWh per month. Households consuming within this range pay a moderate rate, higher than the green tariff but still less than the red tariff. This tier typically applies to households with moderate energy use.
3. Red Tariff: The highest tariff rate applies to households that exceed 1600 kWh per month. The red tariff penalizes high electricity usage, making it the most costly option. This tier is designed to discourage excessive consumption and encourage energy saving behaviours.

In addition to the consumption-based tiers, Serbian households are charged different rates depending on the time of day. This time-based structure includes:

- High Tariff Period: From 7:00 AM to 11:00 PM, when electricity demand is highest, the high tariff rate applies. This period coincides with daytime and early evening hours, when households and businesses typically consume more power.
- Low Tariff Period: From 11:00 PM to 7:00 AM, a lower tariff rate is available to encourage off-peak electricity usage. By offering a discount during night-time hours, the system promotes shifting high-consumption activities—such as washing clothes or running dishwashers—to these off-peak hours.

Households' monthly electricity bills are calculated by applying these two factors (consumption level and time of day) proportionately. For instance, if a household consumes electricity across multiple tariff zones (green, blue, and red) and during both high and low tariff periods, the charges are distributed accordingly. This proportionate distribution ensures that high-consumption households incur higher costs, especially if they use electricity during peak hours.

Overall, Serbia's dual tariff system effectively combines incentives for reducing overall consumption with rewards for shifting usage to non-peak hours, helping to maintain grid stability and encouraging households to adopt energy-saving habits. However, to be able to accurately calculate the costs, other parameters have to be taken into consideration as well.

10.4.2 Input Parameters

The inputs for cost calculation regarding one month are presented in Table 10.1.

Based on the inputs for cost calculation, the inputs for the total cost (TC) calculation can be defined, which are presented in Table 10.2.

Having in mind the inputs from the tables, the total cost can be calculated by using Eq. (10.1).

$$\begin{aligned}\mathrm{TC} = & \mathrm{CCP} + \mathrm{GSF} + \mathrm{CUE} + \mathrm{CIPEP} + \mathrm{IEEF} \\ & + \mathrm{FADS} + \mathrm{CET} + \mathrm{CVAT} + \mathrm{FPMS}\end{aligned} \tag{10.1}$$

10.5 Case Study with Discussion

The case study involves a household generating electricity through a PV system with a total installed capacity of 12 kWp, consisting of 32 monocrystalline modules (model: CHSM60M-HC (BF) 166). The system is equipped with a 10 kW inverter (model: SUN2000-10KTL-M1) and operates without any energy storage.

The data used is for 1 year (from Apr 1, 2023 till March 31, 2024). The following figure shows the PV production (see Fig. 10.1).

It can be seen, that the cumulative production for 1 year is approximately 13500 kWh, while the monthly production ranges from about 300 kWh to almost 2000 kWh.

The following figure compares the PV production and the household consumption (see Fig. 10.2). From the figure one can conclude that the cumulative PV production is slightly lower than the household's cumulative consumption, which implicates that the installed capacity of the PV system is satisfactory. Figure 10.3 presents the calculated costs with the use of PV system, and the costs if there is no PV system.

From the previous figure a conclusion can be made, that the costs are reduced to about one third of the costs of electric energy without the use of PV system. Figure 10.4 shows the exact cost reduction for each month.

If one wants to calculate the return of investment of the PV, than the cost reduction and the initial investment into the system should be taken into consideration. Regarding the presented case study, the initial investment is 1.700.000, 00 [RSD]. Having these data in mind, the yearly return of investment is about 8%, which points towards a conclusion that the total return of investment is more than 12 years (see Fig. 10.5). This calculation was made without taking into consideration that the initial investment took place at the beginning of the process of using the PV, and that the savings would take place during the use of the PV. If this was also taken into consideration, the return of investment would have been longer.

Before further discussion about the previously presented results, it has to be pointed out, that despite the almost ideal ratio between the produced electric energy

Table 10.1 Inputs for cost calculation

Input	Abbreviation	Amount	Unit	Amount validity [Date]
Cost per unit of contracted power	CPUCP	54.258	[RSD/kW]	–
Contracted power (for the household presented in the case study)	CP	11.04	[kW]	–
Guaranteed supplier fee	GSF	145.22	[RSD]	2023/01–2023/11
		160.67		2023/12
Fee for public media service	FPMS	299	[RSD]	–
Compensation for the incentive of privileged electricity producers—per unit	CIPEPU	0.801	[RSD/kWh]	–
Improvement of energy efficiency—per unit	IEEFU	0.015	[RSD/kWh]	–
Fee for the access to the distributions system for the difference between taken and used electricity—per unit	FADSU	3.88	[RSD/kWh]	–
Excise tax	ET	7.5	[%]	–
Value added tax	VAT	20	[%]	–
High tariff—green	T_1	8.336	[RSD/kWh]	2023/04–2023/10
		9.1092		2023/11-
Low tariff—green	T_2	2.084	[RSD/kWh]	2023/04–2023/10
		2.2773		2023/11-
High tariff—blue	T_3	12.504	[RSD/kWh]	2023/04–2023/10
		13.6638		2023/11-
Low tariff—blue	T_4	3.126	[RSD/kWh]	2023/04–2023/10
		3.416		2023/11
High tariff—red	T_5	25.008	[RSD/kWh]	2023/04–2023/10
		27.3276		2023/11
Low tariff—red	T_6	6.252	[RSD/kWh]	2023/04–2023/10
		6.8319		2023/11
Used electricity high tariff-green	UE_1	Variable	[kWh]	–
Used electricity low tariff-green	UE_2	Variable	[kWh]	–
Used electricity high tariff-blue	UE_3	Variable	[kWh]	–
Used electricity low tariff-blue	UE_4	Variable	[kWh]	–

(continued)

Table 10.1 (continued)

Input	Abbreviation	Amount	Unit	Amount validity [Date]
Used electricity high tariff-red	UE_5	Variable	[kWh]	–
Used electricity low tariff-red	UE_6	Variable	[kWh]	–
Taken electricity high tariff-green	TE_1	Variable	[kWh]	–
Taken electricity low tariff-green	TE_2	Variable	[kWh]	–
Taken electricity high tariff-blue	TE_3	Variable	[kWh]	–
Taken electricity low tariff-blue	TE_4	Variable	[kWh]	–
Taken electricity high tariff-red	TE_5	Variable	[kWh]	–
Taken electricity low tariff-red	TE_6	Variable	[kWh]	–

Table 10.2 Inputs for cost calculation [RSD]

Input	Abbreviation	Equation
Cost of contracted power	CCP	$CPUP * CP$
Cost of used electricity	CUE	$\sum_{i=1}^{6} (T_i * UE_i)$
Cost of taken electricity	CTE	$\sum_{i=1}^{6} (T_i * TE_i)$
Compensation for the incentive of privileged electricity producers	CIPEP	$CIPEPU * \sum_{i=1}^{6} TE_i$
Improvement of energy efficiency fee	IEEF	$IEEFU * \sum_{i=1}^{6} TE_i$
Fee for the access to the distribution system	FADS	$FADSU * \sum_{i=1}^{6} (TE_i * UE_i)$
Cost of excise tax (until Aug 2023)	CET	$\begin{pmatrix} CCP + GSF + CTE \\ +CIPEP + IEEF \end{pmatrix} * ET$
Cost of excise tax (from Aug 2023)	CET	$\begin{pmatrix} CCP + GSF + FADS \\ +CIPEP + IEEF + CUE \end{pmatrix} * ET$
Cost of value added tax	CVAT	$\begin{pmatrix} CCP + GSF + CUE \\ +CIPEP + IEEF \\ +FADS + CET \end{pmatrix} * VAT$

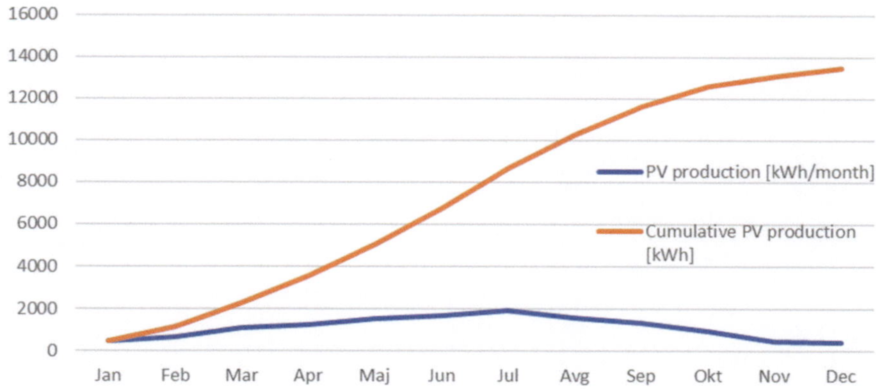

Fig. 10.1 PV production per month and cumulative PV production

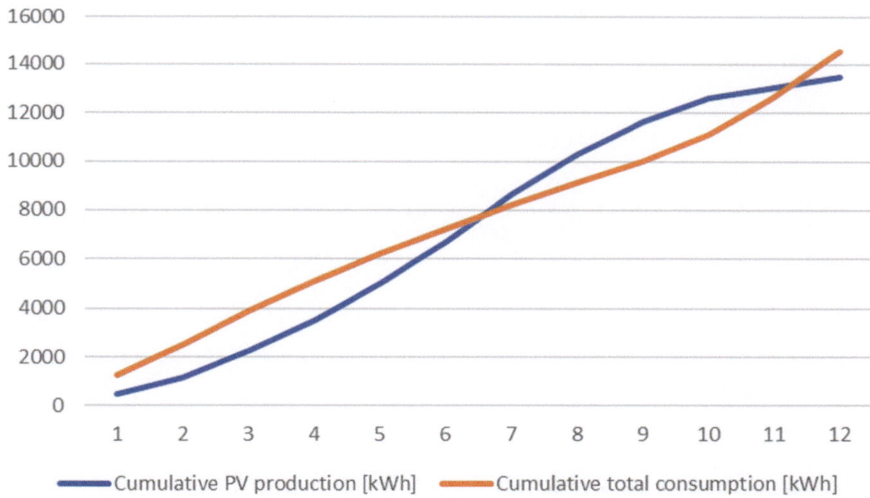

Fig. 10.2 PV production and total consumption

by the PV system and the energy consumption of the household (see Fig. 10.2), there is a difference between the amount of excess electric energy distributed from the PV system to the grid i.e. the supplier, and the energy taken from the grid deduced from the excess energy distributed to the system. Figure 10.6 shows that the yearly excess energy is more than 3500 kWh. This energy is lost for the household. The reason for this lies in the fact that the supplier supplies the previously delivered energy from the PV system only during the tariff in which the energy is produced and delivered, i.e. if the energy is produced during the high or low tariff, it can be used just during high or low tariff.

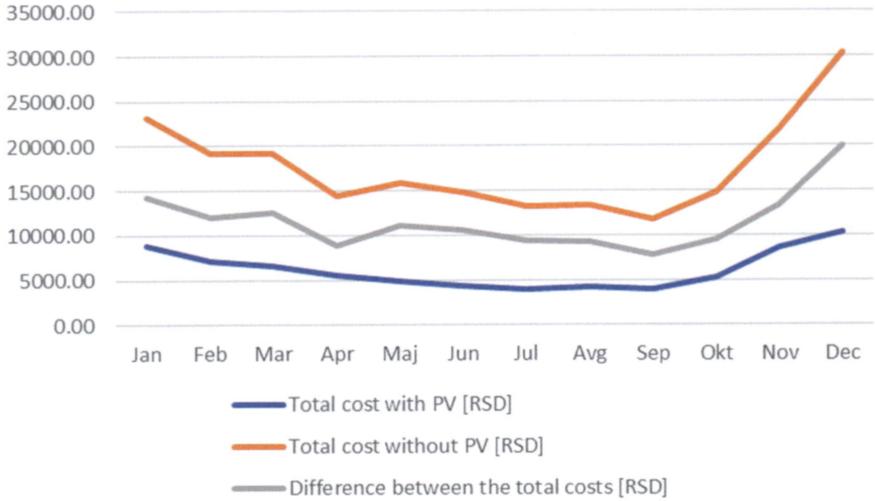

Fig. 10.3 Calculated costs

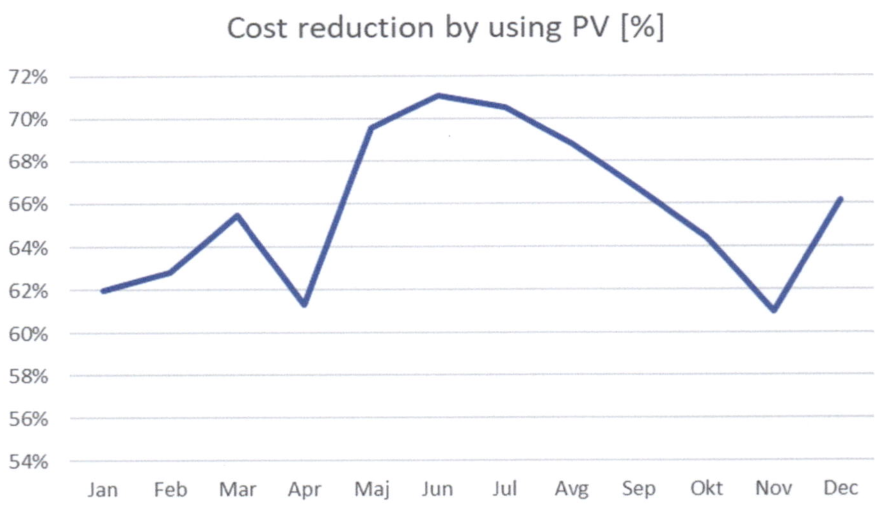

Fig. 10.4 Cost reduction

This means that any energy consumption during the low tariff, which is during the night is taken directly from the supplier. From Fig. 10.7 it can be seen that in this case the total consumption during low tariff could be compensated with previously produced excess energy.

Based on the results, there is evidence that despite the almost equal amount of produced electric energy from PV system and used energy by the household,

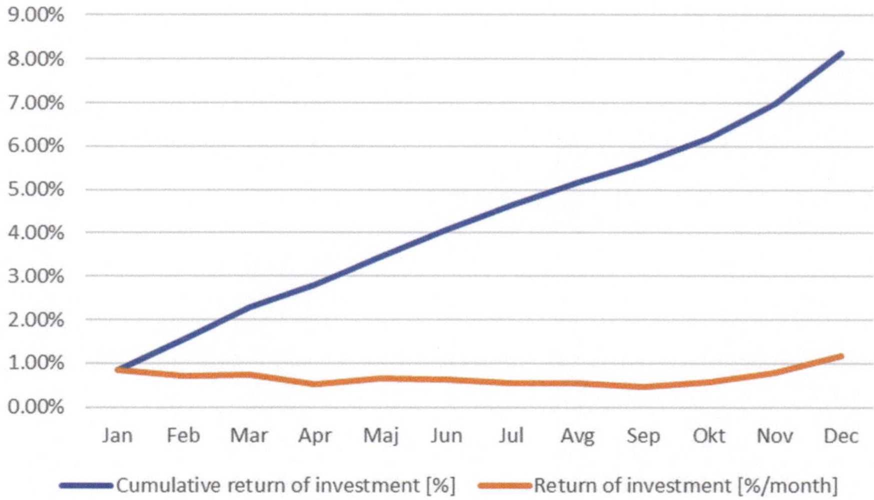

Fig. 10.5 Return of investment

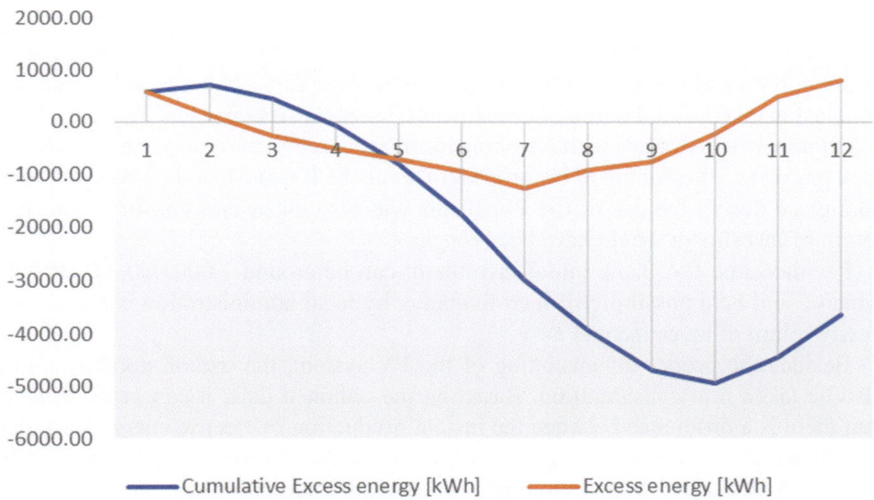

Fig. 10.6 Excess energy distributed from PV to the supplier

due to methodology of electricity consumption calculation, the PV system is over-dimensioned by approximately 27%. Therefore a correction of the system dimensioning is introduced. Instead of 32 modules, 24 modules are used and an 8kW inverter. In this case the household produces electric energy by using a PV system with a total string capacity of 9kWp (24 modules—Monocrystalline—CHSM60M-HC (BF) 166), with a 8KW inverter (SUN2000-8 KTL-M1) and also with no storage.

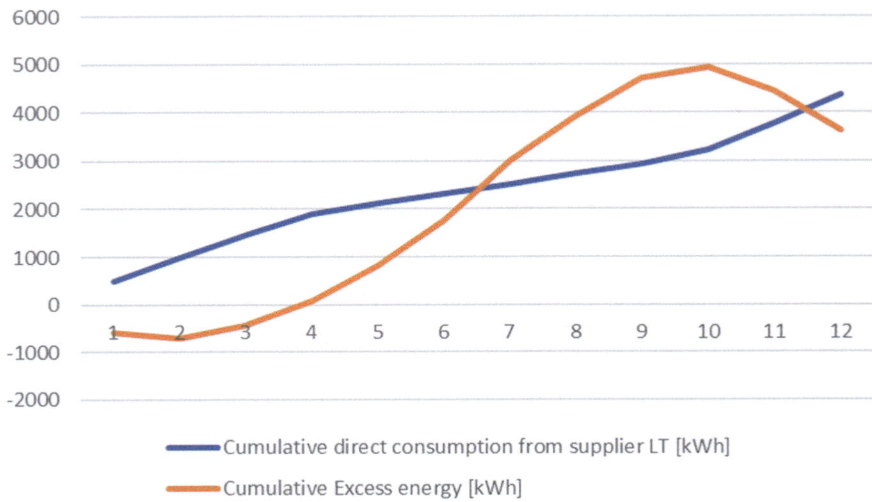

Fig. 10.7 Direct consumption from supplier during low tariff and excess energy supplied to the supplier

In this case, the initial investment is 1.350.000, 00 [RSD]. Having these data in mind, the yearly return of investment is more than 10%, which points towards a conclusion that the total return of investment is less than 10 years (see Fig. 10.8). This calculation was also made without taking into consideration that the initial investment took place at the beginning of the process of using the PV, and that the savings would take place during the use of the PV. If this was also taken into consideration, the return of investment would have been longer.

For the same PV, if the initial investment can be around 1.000.000, 00 [RSD], which could be a possibility if a co-financing by local administration is active, the yearly return of investment is 14%.

Besides the proper dimensioning of the PV system, the use of storage should also be taken into consideration. Based on the obtained data, it can be concluded that there is a difference between the instant production of electric energy from the PV system and instant consumption by the household. If there is an excess electric energy produced by PV it is supplied to the supplier and used later, but it has to be taken into account, that there is a certain cost associated with the use of previously supplied excess energy to the supplier.

Figure 10.9 shows the difference between PV production and direct consumption from PV. It can be noticed that the average difference per day ranges from 4.9 kWh in December to 48.1 kWh in July. If a 10 kWh storage system is installed additionally to the PV system, the costs for electric energy fall, especially during the summer months, while during wintertime there is no significant difference between the costs (see Fig. 10.10). However, the initial investment in this case would be around 2.900.000, 00 [RSD]. Having these data in mind, the yearly return of investment would be just

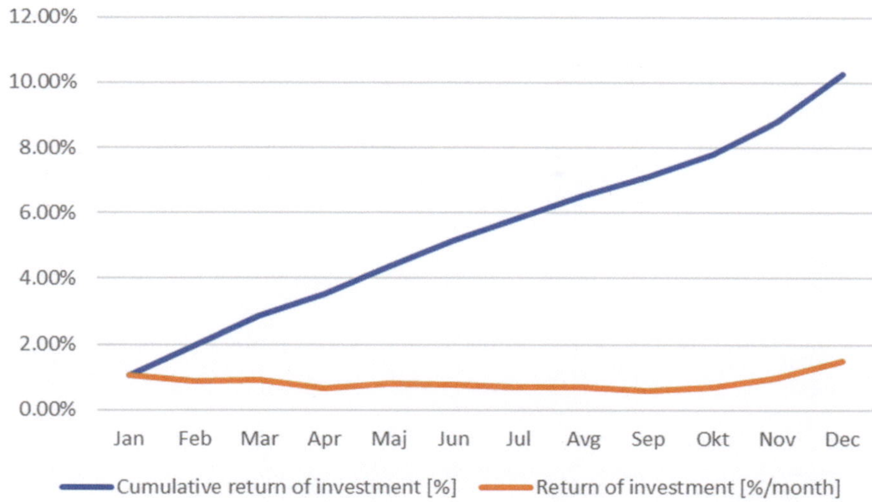

Fig. 10.8 Return of investment for the properly dimensioned PV

about 5.5%, which points towards a conclusion that the total return of investment would be more than 18 years (see Fig. 10.11). This calculation was also made without taking into consideration that the initial investment took place at the beginning of the process of using the PV, and that the savings would take place during the use of the PV. If this was also taken into consideration, the return of investment would have been longer.

Considering the results obtained by the case study, it can be concluded, that for the proper design of a PV system the specifics of the tariff system should be taken into consideration. Also, having in mind the current electric energy prices in Republic of Serbia in the case of households, the return of investment rate is rather low regardless of whether there is co-financing or not, and the usage of storage system is not economical at this moment, because the supplier accepts all excess energy produced by the PV.

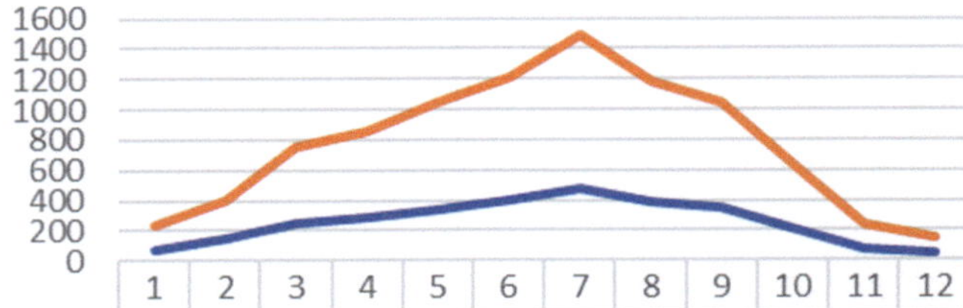

	1	2	3	4	5	6	7	8	9	10	11	12
Difference between PV production and direct consumption from PV [avg. kWh*10/day]	79	143	245	286	339	403	481	384	347	208	78	49
Difference between PV production and direct consumption from PV [kWh]	244	401	759	858	1050	1210	1490	1190	1040	645	234	153

Difference between PV production and direct consumption from PV [avg. kWh*10/day]

Difference between PV production and direct consumption from PV [kWh]

Fig. 10.9 Difference between PV production and direct consumption from PV

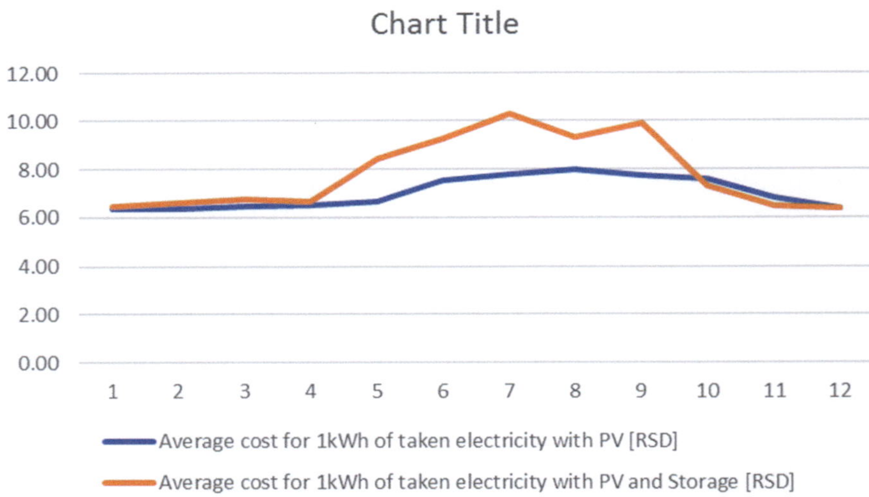

Fig. 10.10 Average costs of electric energy with and without storage

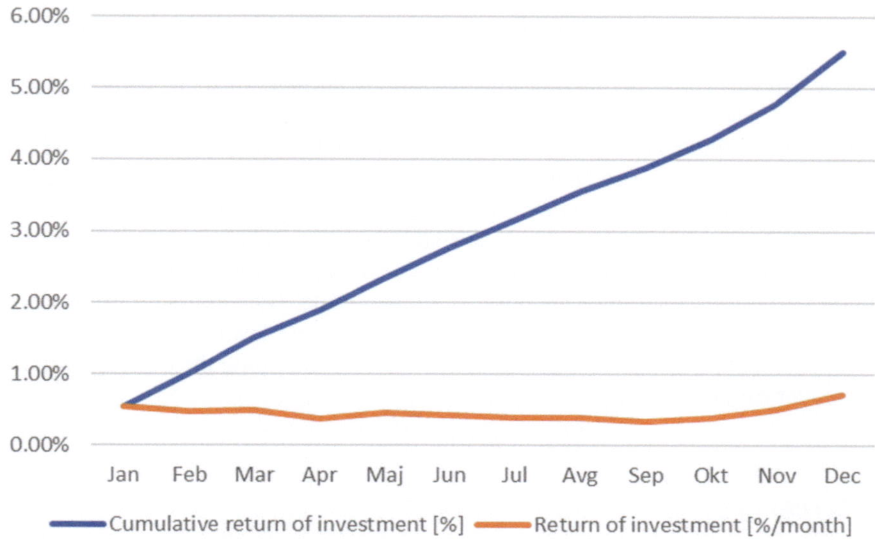

Fig. 10.11 Return of investment for PV with storage

10.6 Conclusion

The presented research provides an insightful examination of PV systems in Serbian households, emphasizing financial, environmental, and energy independence benefits. Through case studies and cost analyses, it demonstrates the substantial potential for PV installations to lower energy costs and support Serbia's transition toward renewable energy. However, challenges remain, particularly with respect to tariff systems, energy costs, and the need for efficient energy storage solutions. These challenges affect both cost-effectiveness and energy stability, particularly in balancing PV energy production with household consumption patterns.

While the initial investment costs remain high, the study shows that strategic dimensioning of PV systems and careful consideration of storage options can enhance investment returns, though additional costs for storage often extend the return-on-investment timeline. This points to the necessity of ongoing legislative support and financial incentives, especially for energy storage solutions, to enable wider adoption and optimize the use of renewable energy resources.

To advance PV effectiveness, future research should focus on optimizing storage solutions. Since grid reliability is impacted by PV intermittency, research into cost-effective battery storage systems or hybrid storage solutions (e.g., battery plus thermal storage) could improve household energy independence and grid stability. Investigations into scalable storage options could benefit both residential and larger-scale applications. Exploring dynamic tariff structures that adapt based on PV production levels and grid demand could incentivize households to optimize their PV usage. Additionally, demand response mechanisms could encourage households to adjust their energy consumption patterns in real time to align with PV production, enhancing both economic and grid benefits. As more PV systems are integrated into Serbia's energy infrastructure, optimizing grid operations through smart technologies and distributed energy resources could ensure stability. Research into smart grid systems and predictive models for PV output and consumption could aid in managing the supply–demand balance, reducing the need for costly infrastructure upgrades. Further investigation into the cost dynamics of storage technologies and the impact of potential government incentives could aid in making storage solutions more accessible. Studies could provide evidence-based recommendations for policy frameworks that support both PV and storage adoption at a larger scale. By addressing these areas, Serbia can continue to refine its renewable energy strategy, leveraging PV systems as a cornerstone of a sustainable energy future while overcoming the existing barriers to broader adoption and optimization.

References

1. Borozan S, Mateska AK, Krstevski P (2021) Progress of the electricity sectors in south east Europe: challenges and opportunities in achieving compliance with EU energy policy Energy Rep 7
2. Capuder T, Pandžić H, Delimar M (2017) Integration of renewable energy sources in southeast Europe: a review of incentive mechanisms and feasibility of investments. Renew Sustain Energy Rev 75
3. Politis S et al (2016) Energy scenarios for south eastern Europe: a close look into the western Balkans. In: Vienna workshop
4. Gvero P (2023) Renewable energy sources and their role in the future of energy system of Republic of Srpska. Sustain Dev Nat Resour Manag
5. Bajc T, Kostadinović D (2023) Potential of usage of the floating photovoltaic systems on natural and artificial lakes in the Republic of Serbia. J Clean Prod 367
6. Fejzic E, Niet T, Wade C, Usher W (2023) Aligning the western Balkans power sectors with the European green deal. Environ Res Commun
7. Franki V, Višković A (2021) Multi-criteria decision support: a case study of Southeast Europe power systems. Util Policy 65
8. Dominković DF et al (2016) Zero carbon energy system of south east Europe in 2050. Appl Energy 162
9. Mitković M, Đekić J, Igić MZ, Mitković PB (2018) Analysis of electric power production results in South Serbia: recommendations for improvement of operation of first mini photovoltaic power plants. Therm Sci 22(1A)
10. Szabó L, Mezősi A, Pató Z (2016) South East Europe electricity roadmap: regional report
11. Politis S, Zucker A, Stambolis C, Vujasinović Z, Szabó L (2016) Energy scenarios for south eastern Europe: a close look into the western Balkans
12. Milosavljević D et al (2014) Assessment of the possibilities of building integrated PV systems in some spa resorts in Serbia. Sylwan
13. Law on Renewable Energy Sources, Official Gazette of the Republic of Serbia, no. 40/2021 (2021)

14. Law on Efficient Use of Energy, Official Gazette of the Republic of Serbia, no. 25/2013 (2013)
15. Regulation on Technical Standards and Safety of Photovoltaic Installations, Official Gazette of the Republic of Serbia (2021)
16. Energy Law, Official Gazette of the Republic of Serbia, no. 145/2014 (2014)
17. Law on Environmental Protection, Official gazette of the Republic of Serbia, no. 135/2004 (2004)

Chapter 11
Robot Navigation Based on Wireless Sensor Networks and Its Security Challenges in Industrial Environments

László Gogolák

Abstract The rapidly increasing development in automation demands the use of wireless devices. Localization and navigation play a crucial role in the application of these devices. Navigation has seen numerous methods developed in today's advanced world. In my work, I investigated the application of widely used WiFi technology worldwide. The chosen method was RSSI-based mapping, localization, and navigation. The research aims to test and implement this method, collecting and evaluating data through laboratory measurements. Ultimately, the results obtained demonstrate the accuracy of a wireless sensor network-based localization and navigation using a single technology. In this work, we present a wireless sensor network (WSN) navigation solution implemented in a real industrial robot environment, allowing the robot to determine the location of an objects for gripping. The mapping process was achieved through practical experiments, and the results were validated with precise measurements. The use of wireless technology raises several security concerns, which we analyze to assess the application's robustness from this perspective. This study thus addresses both the operational and security dimensions of WSN-based robotic navigation.

Keywords Wireless sensor network (WSN) · RSSI-based localization · Indoor positioning system · Security in wireless networks · Signal interference mitigation

11.1 Introduction

The rapid advancement in technology has created a significant demand for precise and swift data acquisition within the field of wireless sensor networks (WSNs). Localization has become a critical requirement across various WSN applications, including robotic networks, where highly accurate and reliable positional information

L. Gogolák (✉)
Department of Mechatronics and Automation, Faculty of Engineering, University of Szeged, Szeged, Hungary
e-mail: gogolak@mk.u-szeged.hu

T. A. Kovács and I. Fürstner (eds.), *Critical Infrastructure Protection: Advanced Technologies for Crisis Prevention and Response*, NATO Science for Peace and Security Series C: Environmental Security, https://doi.org/10.1007/978-94-024-2308-2_11

is essential. This need for localization is not only relevant in robotics but also critical to the operation of everyday electronic devices within the framework of the Internet of Things (IoT). For indoor positioning, multiple radio communication technologies are employed, such as Ultra-Wideband (UWB), Bluetooth Low Energy (BLE), and WiFi [1].

GPS, the Global Positioning System, operates as a network of satellites and receiving devices designed to determine positions on Earth with impressive accuracy, in some cases down to the centimeter level. GPS-compatible devices pinpoint locations through latitude, longitude, and altitude coordinates, offering not only precise location data but also exact time information. The GPS infrastructure includes a constellation of twenty-four satellites in precise orbital paths around the Earth, each completing an orbit in twelve hours. This satellite array continuously broadcasts radio signals, which form the foundation of GPS technology. GPS receivers calculate their position by measuring the travel time of signals from the satellites, relying on the constant speed of radio waves to determine distance and, hence, location. Accurate localization requires signals from at least four satellites, with more satellites improving precision. In cases where only one satellite distance is known, the receiver's possible location spans the surface of a sphere defined by that distance. The intersection of multiple such spheres narrows down the receiver's precise location a method known as trilateration. GPS is widely used in aircraft, ships, trains, submarines, and space exploration, as well as in everyday applications like automotive navigation, where the device projects calculated coordinates onto electronic maps for intuitive route guidance. However, GPS signals can be disrupted by obstacles or atmospheric factors, making it unsuitable for covered or indoor environments where a line of sight (LoS) is obstructed [2].

Ultra-Wideband (UWB) is a short-range communication protocol that operates at high frequencies across a wide bandwidth, capable of transmitting signals in nanosecond pulses to achieve real-time positioning. It is particularly useful within a 10-m range due to low power consumption and high data transfer capabilities (500 MHz). However, UWB signals do not penetrate solid obstacles like walls, limiting their application in complex environments [3]. Bluetooth Low Energy (BLE), developed by the Bluetooth Special Interest Group, is a wireless communication technology originally intended for healthcare, fitness, and security applications. Unlike classic Bluetooth, BLE operates at reduced power while covering a comparable communication range, making it an economical solution. BLE uses the same 2.4 GHz radio frequency as traditional Bluetooth, though with advantages in terms of cost-effectiveness and energy efficiency [4]. Radio Frequency Identification (RFID) is another wireless technology, employing electromagnetic or electrostatic coupling in the radio frequency portion of the electromagnetic spectrum for unique identification of objects, animals, or people. RFID systems consist of three main components: the scanning antenna, transceiver, and transponder. Transponders, or "tags", activate upon receiving signals from an RFID reader, subsequently sending identification data back to the reader's antenna. RFID readers may be stationary or mobile, with reading ranges influenced by tag type, reader type, frequency, and environmental factors [5].

WiFi technology, another viable option for localization, is indispensable today, providing a robust and accessible platform for geolocation. WiFi enables relatively high accuracy for local positioning and exceeds BLE or UWB in range, although UWB offers better precision at closer ranges. The advantage of WiFi over GPS lies in its applicability within indoor or obstructed environments, as it relies on signal strength rather than satellite connectivity for localization. Using the Received Signal Strength Indicator (RSSI) as a measure of signal strength allows for effective positioning within a WLAN system [6]. This study focuses on RSSI-based localization. By leveraging the spatial limits of radio signals, the RSSI metric can provide highly accurate positional data relative to known reference points. This technology enables effective monitoring and movement tracking within designated areas, facilitated through a Wireless Local Area Network (WLAN) configuration where Access Points (APs) allow Station (STA) devices to connect and exchange RSSI-based location data [7, 8] This research employs ESP32 microcontrollers, WiFi-enabled devices capable of supporting necessary data transfer rates, to implement the proposed system.

In recent studies on the integration of wireless sensor networks (WSNs) within robotics, a strong emphasis has been placed on cybersecurity to address vulnerabilities in automated environments. Ismail et al. [9] developed a lightweight multilayer detection system specifically designed to counter cyber-attacks in WSNs, leveraging mobile robots to manage internal security threats effectively. Their approach addresses the unique vulnerabilities posed by WSNs in dynamic settings where rapid response to internal breaches is essential. Additionally in work [10] autors reviewed the vulnerabilities and protective measures relevant to robotic systems dependent on WSN support, stressing the importance of WSN security to prevent potential disruptions in robotic functions that could compromise critical industrial processes. This work provides insights into safeguarding robotic frameworks against cyber-attacks.

Together, these studies present comprehensive insights into securing WSNs within robotics and IoT, offering protocols and methodologies to combat a wide range of threats, from internal attacks to large-scale cyber intrusions. These frameworks contribute to a safer implementation of WSNs in modern automated systems, ensuring both functionality and resilience in the face of cybersecurity challenges.

11.2 Wireless Localization Technologies

11.2.1 Received Signal Strength Indicator

The Received Signal Strength Indicator (RSSI) is a metric that shows the strength of a WiFi signal [11]. Its unit of measurement is decibel milliwatt [dBm], representing a power level relative to 1 milliwatt, and it produces values in the set of negative integers. The RSSI value depends on the distance from the router or access point and the spatial positioning of the device. By leveraging these properties, it's possible to map out both planar and three-dimensional spaces. In my work, I focused on a planar area, or "workspace". The most common mapping methods are Triangulation and Fingerprinting. Our work focuses on a planar area, I opted for the Fingerprinting method.

Typically, wireless signals are sent from a WiFi router and received by WiFi-enabled devices, such as smartphones, laptops, wireless security cameras, and other devices. The term "received" indicates that the RSSI is a metric based on an actual received signal at a given moment. Signal Strength of the RSSI measurements show the signal strength at a specific point in time. In other words, RSSI numerically represents WiFi signal strength at a particular moment. Indicator simply means that RSSI is an indicative measure or metric that shows signal strength. RSSI levels exist because WiFi signals lose strength as they propagate to devices. Consequently, RSSI values are negative from the first measurements, reflecting this signal degradation. In practice, it's impossible to achieve an RSSI value of zero (which would mean perfect signal strength), but the goal is to get as close to zero as possible. To give a clearer understanding of signal proximity, RSSI values are categorized into levels, each representing a specific strength.

The meaning of RSSI levels, as discussed previously, is that they indicate the signal strength after accounting for all possible losses due to interference or physical obstructions (Table 11.1). The higher the RSSI value (closer to zero), the stronger the signal. Since measurements are in the negative range, the closer we are to zero, the better the signal quality. For example, a value of −50 represents strong signal strength, −75 is moderate, and −100 indicates a weak, barely usable signal.

RSSI values are generally expressed in decibel milliwatt [dBm]. Interestingly, RSSI values aren't tied to any standard physical parameter, although observations and measurements show they depend on the relative distance between the transmitting and receiving devices. RSSI values can theoretically be modeled as a function of distance. This characteristic model can be applied in theory and under ideal conditions.

Table 11.1 Categorization of RSSI Values

Values [dBm]	Signal strength	Characteristics
−50	Excellent	Perfect user experience
−60	Very good	Not perfect, but provides a seamless user experience
−70	Good	Usable with minimal issues; most users operate in this range
−80	Low	Some disruptions and minor interruptions may occur
−90	Very low	Frequent disconnections; unacceptable for user-level applications
−100	No signal	Connection exists, but data exchange is not possible

11.2.2 Triangulation

Geometric Triangulation aims to divide a topological surface into triangles in such a way that they do not intersect within their interiors. This is achieved by using arcs that partition the surface into ideal triangles with vertices and edges that are consistent with each other [12]. WiFi Triangulation seeks to map RSSI values as a function of distance. This method requires a steep, linear characteristic curve for accurate implementation. Functions describe these relationships, and by inputting real-time RSSI values, they generate a speculative (x, y) coordinate pair.

11.2.3 Fingerprint Localization

A radio map database can be created using various methods, with three main techniques prevalent for building a fingerprint database [13]. The most effective approach for local mapping relies on measurements of the mapped area. The goal is to create a heat map, graphically representing RSSI values, which is advantageous as it bases the system on real measurements rather than estimates (Fig. 11.1). Unlike parametric methods, this approach uses actual values, which are collected into a multidimensional matrix. Once enough data is gathered, the database (from the multidimensional matrix) allows retrieval of specific values, forming the basis of the fingerprinting technology. During real-time RSSI measurements, the system compares the measured result to the database to find the closest match. Once the best approximation or exact match is found, it identifies the precise cell location in the multidimensional matrix.

WiFi Fingerprinting creates a radio map for a specific area. Based on data from multiple access points, it generates the probabilistic distribution of RSSI values for

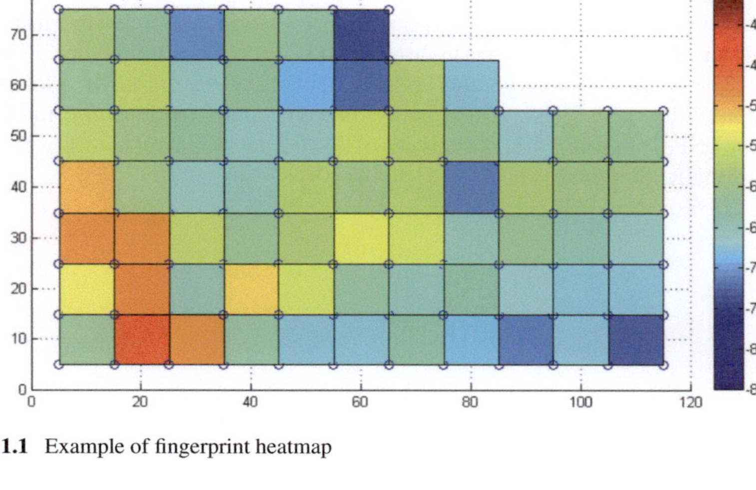

Fig. 11.1 Example of fingerprint heatmap

a given (x, y) location. Live RSSI values are then compared with the fingerprint to find the closest match, producing an estimated (x, y) coordinate pair. For illustration, Fig. 11.1. shows an image of expected results [14].

11.2.4 RSSI Characteristic

In a technical context, a characteristic refers to the graphical representation of a function, which shows the behavior of a machine, device, or metric as a function of another variable. A typical example is the current characteristic as a function of voltage. As previously mentioned, Received Signal Strength Indicator (RSSI) values are non-characteristic and not dependent on physical quantities. However, RSSI values do change with distance. While this change is not deterministic, it does follow a certain functional pattern. In my work, I examined RSSI values as a function of distance. The characteristics of RSSI depend greatly on the measurement environment and the properties of the equipment (such as router strength, ESP32 signal, antenna quality, obstacles, walls, and interference).

11.2.5 ESP32 Development Board

The **ESP32** is a development board similar to the Arduino (the term "board" is widely used in the context of microcontrollers and refers to the circuit board, or PCB, that houses the essential components of analog electronics in a structurally interconnected manner (Fig. 11.2).

With these boards, a variety of diverse tasks can be accomplished). The ESP32 is a microcontroller with a low-cost, low-power chip, featuring integrated WiFi and dual-mode Bluetooth. The ESP32 series utilizes the Tensilica Xtensa LX6 microprocessor, available in both dual-core and single-core versions. Both designs use either the dual-core Xtensa LX7 or the single-core RISC-V, and each includes a built-in antenna. The board also incorporates an RF balun and power amplifier, a low-noise receive amplifier, and filters. Additionally, power control modules are integrated within the ESP32. In this work, these types of ESP32 development boards were used for recording RSSI signals.

Fig. 11.2 ESP32 development board

11.3 Analysis and Measurement Methodology of RSSI Characteristics

The exploration and mapping of the RSSI characteristic was crucial for gaining an accurate understanding of how the Received Signal Strength Indicator behaves when used with the ESP32 microcontroller. Experience has shown that the RSSI characteristic can be quite unstable, which is evident in the measurements. The goal was to obtain the most precise dataset possible. To achieve this level of accuracy, numerous measurements had to be conducted. The graphical representation of these measurements and numerical results allowed for insights into measurement resolution. Determining the appropriate resolution is a fundamental building block for accurate positioning. This task was approached with meticulous planning. Preliminary measurements were required to ensure accurate results. These preliminary tests helped to prepare for the actual measurements, allowing potential errors to be identified and mitigated more effectively. For characterizing RSSI, I applied the assumption that RSSI values vary as a function of distance. Although this relationship is not strictly linear, it does exhibit a degree of functional dependency. As radio waves propagate in open space, they continuously lose strength, with the amount of loss dependent on distance. This attenuation is defined by the Free Space Path Loss (FSPL) model. Signal propagation is influenced by numerous factors, including temperature, object reflection, and air humidity. This propagation is described by the outdoor transmission Eq. 11.1 [15].

$$P_{rec}(d) = \frac{P_{tx} \cdot G_t \cdot G_r \cdot \lambda^2}{(4 \cdot \pi)^2 \cdot d^2 \cdot L} \tag{11.1}$$

The measurement setup included two ESP32 microcontrollers and a computer. One of the ESP32s was equipped with an external antenna to amplify the WiFi signal, a necessary enhancement since the built-in WiFi antennas of the ESP32 are insufficient for accurate measurements. Only one ESP32 required an external antenna, as strengthening the transmitter allows the receiver to detect the signal more reliably without disconnecting. This indicates that the built-in antennas of the ESP32 are less suited for transmission tasks, though they perform effectively as receivers.

11.3.1 Experimental Setup and Measuring Process

The measurement process operates on the Access Point (AP) and Station principle, where the AP functions as the transmitter and the Station as the receiver. The AP provides the WiFi connection to which the Station connects, enabling two-way communication for RSSI measurements. The ESP32 serving as the AP is equipped with the external antenna. This ESP32 runs a code written in the Arduino environment to establish the network. Three levels of resolution were tested during the measurements, with each trial conducted in a static environment to ensure repeatability. Multiple samples were taken at each measurement point due to the dynamic nature of RSSI values, which fluctuate over time and may introduce errors. Averaging these samples helped to reduce the impact of these variations, thus enhancing measurement accuracy.

The measurement was conducted using two ESP32 microcontrollers connected via WiFi. The connection setup followed the Access Point (AP) and Station model, advantageous for fixing one measurement point. The Access Point served as the origin of the measurement system, and measurements were taken at increasing distances from this origin. RSSI values were recorded in decibel-milliwatts as a function of distance, with samples collected at predetermined intervals. These intervals represented resolution values, and five samples were taken at each distance. The median of these five samples was used to determine the RSSI value for each specific distance.

Measurements were conducted along a pre-measured straight line, with the Access Point ESP32 positioned at the origin. The other ESP32 device was mobilized on a breadboard. Measurements were triggered by a push button, serving as the actuator to initiate each reading. To eliminate contact errors associated with the mechanical push button, debouncing was implemented in the code. Debouncing provided a software solution to prevent erroneous readings from partial presses or multiple readings from a single press. It ensured that the button was fully pressed before sending a signal.

The measured values were then exported to a ".txt" file, with each value indexed and separated by a comma for ease of data management. This format enabled easy importation, making the dataset manageable in CSV (comma-separated values) format, which is well-suited for structured data handling.

11.3.2 Results of the RSSI Measurements

During the first measurement, data points were recorded while moving away from the AP at intervals of 15 cm, with a maximum distance of 150 cm from the AP. At each point, five samples were collected, and the median of these values was used. This approach resulted in a total of 55 measured RSSI values (Fig. 11.3).

In the first measurement, the characteristic curve is clearly visible, with a single anomaly observed at the 120 cm mark. This outlier is likely due to signal reflection or insufficient environmental stability. In the second measurement, five samples were again taken at each measurement point. For this round, the intervals between measurement points were reduced to 10 cm, resulting in a total of 80 measured RSSI values (Fig. 11.4).

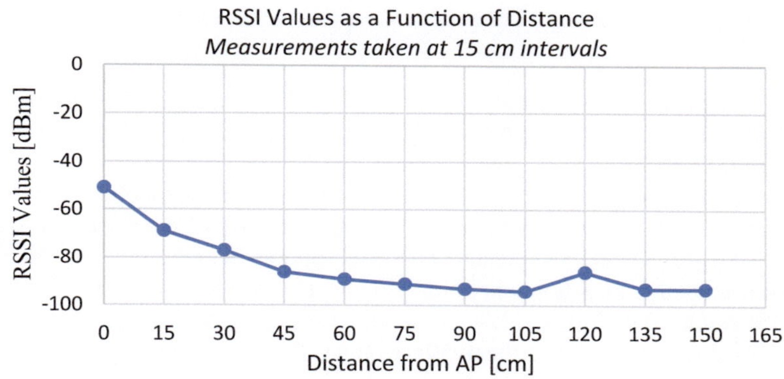

Fig. 11.3 RSSI values as a function of distance measurements taken at 15 cm intervals

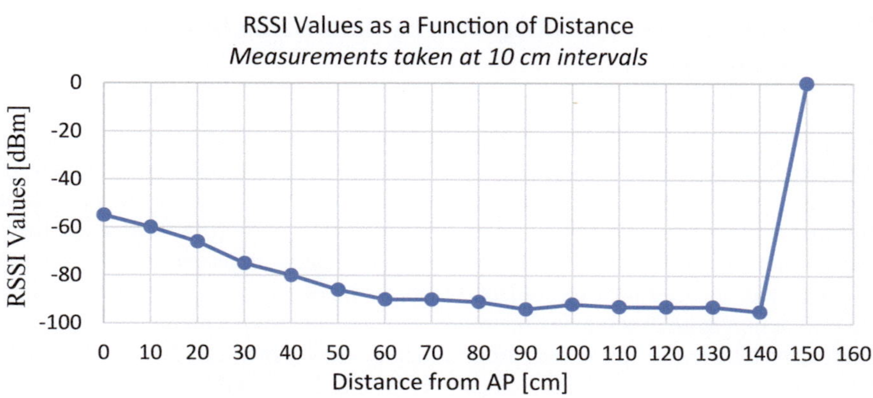

Fig. 11.4 RSSI values as a function of distance measurements taken at 10 cm intervals

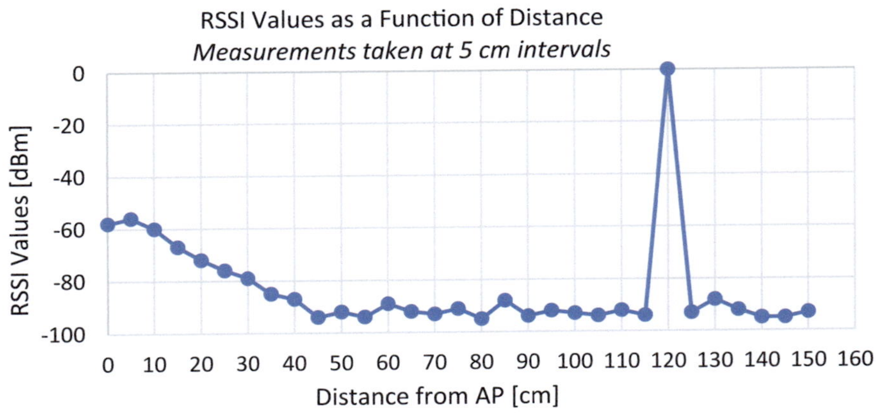

Fig. 11.5 RSSI values as a function of distance measurements taken at 5 cm intervals

In the third measurement, data points were taken at intervals of 5 cm, with five measurements performed at each point. This resulted in a total of 155 measured RSSI values (Fig. 11.5).

In the third measurement, the phenomenon of the ESP disconnecting from the AP was again observed. Notably, with measurements taken every 5 cm, the data proved to be highly noisy and heavily dependent on the measurement environment. Comparing the curves at notable measurement points reveals relatively similar values, indicating that RSSI is indeed well-defined as a function of distance. Additionally, it is evident that the curve is steepest within the 0–50 cm range. This distance range is thus optimal for positioning based on RSSI measurements.

11.4 Realization of a Navigation System Using Sensor Networks

The primary objective of this work is to achieve robot navigation using a sensor network. Navigation is performed within the workspace of a Universal Robot arm, leveraging the precision capabilities of the robot arm to map the workspace with millimeter accuracy. The core goal of this navigation system is to enable object localization within the robot's workspace without reliance on cameras or other optical sensors. Once the location of an object is identified, decisions can be made regarding the next action to be executed by the robot, such as triggering a predefined program, routine, or even adjusting speed. Additionally, safety zones can be established to detect the presence of external elements, allowing the workflow to be slowed for safe operation. The research aims to assess the reliability of position determination based on RSSI values, as accurate localization within the workspace is essential for the robot to position itself relative to an object and subsequently execute actions, such as grasping or manipulating the object. Sensor network-based localization is advantageous as it provides a highly cost-effective solution compared to camera or light-gate systems. Another notable benefit is the potential to establish safety zones. Such a system eliminates the need for cameras or light barriers for detection and relies solely on Access Points and Stations, or, more simply, a sensor network.

11.4.1 Fingerprint Heatmap of the System

The applied technique was the RSSI fingerprinting method. The workspace was divided into points, each equidistant from its neighbouring points, allowing an RSSI value to be obtained at each location and stored in a matrix. The matrix structure mirrors the segmentation of the workspace. Since four receiver stations were used to enhance localization accuracy, the data matrix is four-dimensional, with each entry containing four RSSI values. This structure enables quick and easy access to data at any point. Each location in the matrix represents four measured values, which are compared to the RSSI values of an object within the workspace. The closest match is sought within the matrix, and the point of closest correspondence is likely to represent the object's actual position in the space.

The resolution, or spatial segmentation, was set at 5-cm intervals. This configuration provided eleven measurement points per row across a total of nine rows, resulting in a measurement cycle consisting of 99 points. The measurement setup included five ESP32 microcontrollers: four as receivers and one as the transmitter. The four receiver microcontrollers were positioned at the four corners of the table using breadboards, while the access point (AP) was mounted on the end of the robotic arm. A schematic of this arrangement is shown in Fig. 11.6. This configuration was chosen to simplify movement, as only the AP needed to be repositioned rather than four separate points. By moving the AP, the relative distance to each of the four fixed points changes uniquely with each adjustment.

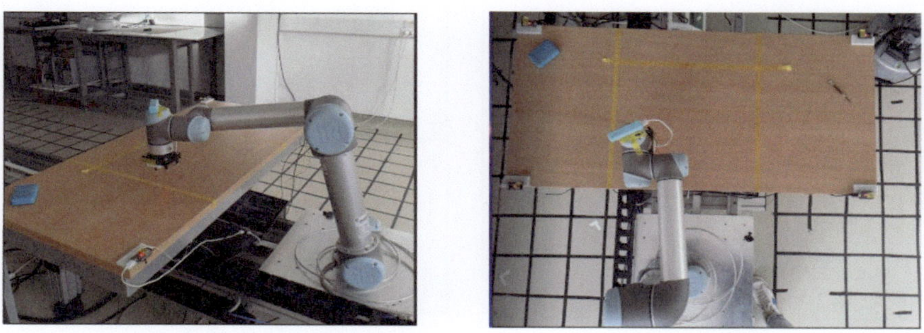

Fig. 11.6 Layout of the system with robot arm

Following the necessary preparations, the measurement process commenced. During each measurement, the robot traversed all 99 points within the workspace, pausing for three seconds at each location to record data. The microcontrollers saved the indexed data to a.txt file. At a speed of 75%, the measurement cycle took approximately 7 min to complete. The process was repeated three times, providing 15 measurements per point, from which an average value could be calculated for enhanced accuracy.

The measurement results were visualized using MATLAB simulation software (Fig. 11.7). The data was uploaded into a matrix and subsequently plotted, resulting in a total of 16 heat maps. These maps are organized into groups of four, each representing the results from Station 1 through Station 4. The final group of four heat maps displays the averaged values across the three measurement cycles.

The average results reveal a significant presence of noise. It is also observable that the ESP32 microcontrollers do not operate uniformly, or that the WiFi signal does not reach each receiver in the same way. Potential sources of this noise include the quality of the ESP antennas, environmental factors such as signal reflection, and possible interference from other devices—such as the robot arm acting as a shield or disruptions from other WiFi-connected devices like mobile phones or computers.

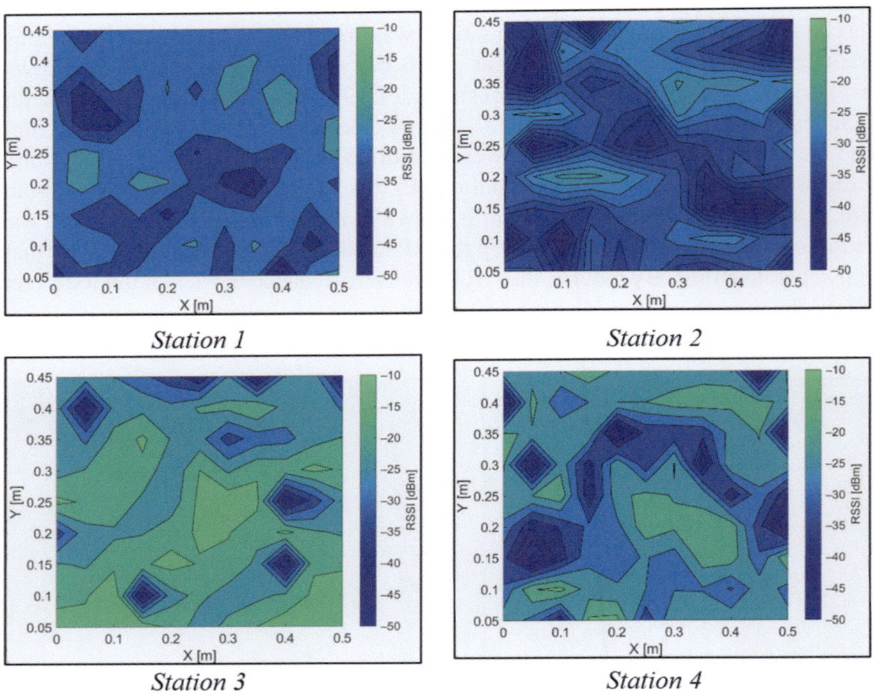

Fig. 11.7 RSSI heatmap of all 4 stations

11.4.2 Position Determination Within the Workspace

The results from the measurements discussed previously were analyzed using MATLAB simulation software. Multi-level matrices were created, specifically four-level matrices to correspond with the four measurement stations. Each matrix level contained data from one measurement station, yielding four RSSI values associated with each position.

Four points were selected within the workspace for measurement, designated as "R" to indicate their random selection. Figure 11.8. illustrates the workspace, clearly indicating the positions of these randomly placed points.

Each numbered R position corresponds to a set of four measured RSSI values, which were then compared to the RSSI values obtained during mapping. The goal was to identify the mapped set of four RSSI values that most closely resemble the measured set. The best-matching RSSI set thus indicates a specific position within the area, with this matched position being the determined location.

For each comparison, the four RSSI values from the sampled position were subtracted, level by level, from the mapped RSSI values across all positions. Specifically, the Station 1 RSSI from the sampled position was subtracted from the mapped Station 1 RSSI at each position, and this process was repeated for the other stations. This produced a difference for each position, quantifying the deviation in RSSI values. The smallest difference, closest to zero, represents the best match. The Eq. 11.2 describes this difference, where n denotes the station number, and this was applied at each of the 99 positions.

$$\text{mapped}\,\text{RSSI}_n - \text{Position}\,\text{RSSI}_n = \text{Difference}_n \tag{11.2}$$

Next, the differences calculated for each station at each position were summed, as shown in Eq. 11.3, where n represents the position number (each position has four differences in this case):

$$\sum \text{Difference}_n = \text{Diff}_{\text{total}} \tag{11.3}$$

89	90	91	92	93	94	95	96	97	98	99
78	79	R4	81	R5	83	84	85	86	87	88
67	68	69	70	71	72	73	74	75	76	77
56	57	R3	59	R6	61	62	63	64	65	66
45	46	47	48	49	50	51	52	53	54	55
34	35	R2	37	R7	39	40	41	42	43	44
23	24	25	26	27	28	29	30	31	32	33
12	13	R1	15	R8	17	R9	19	20	21	22
1	2	3	4	5	6	7	8	9	10	11

Fig. 11.8 Workspace positions with the randomly placed points

The closer the total difference is to zero, the more similar the RSSI values are between the positions. By identifying the position n with the smallest total difference, the most likely corresponding position in the workspace is determined.

11.4.3 Results of the Localization

Comparing the figures and results, the following table was created to summarize and quantify the findings, converting the graphical results into numerical form.

The results indicate that the area mapped based on the first measurement yielded more accurate outcomes, with both total and average deviations consistently lower compared to subsequent measurements. This suggests that the first measurement setup provided a more stable and reliable basis for determining the robot arm's position (Table 11.2).

The improved accuracy in the first measurement could be attributed to various factors, such as optimal environmental stability or minimized interference, which contributed to more consistent RSSI readings. This stability is essential for applications that demand high precision, particularly in tasks requiring exact positioning within the workspace. Lower deviation values not only enhance confidence in the position estimates but also reinforce the reliability of RSSI-based localization as a viable method for real-time robotic navigation.

In conclusion, the first measurement approach demonstrates the potential of RSSI fingerprinting for high-accuracy robotic positioning, highlighting the importance of careful calibration and environmental control to optimize results in practical implementations.

Table 11.2 Positions obtained based on RSSI localization

Random position	Original position	Position determined based on 1st measurement	Distance from original point [cm]	Position determined based on 2nd measurement	Distance from original point [cm]
R1	14	3	5	53	33.5
R2	36	11	42.7	35	5.0
R3	58	20	36.1	3	25
R4	80	91	5.0	21	46.1
R5	82	89	20.6	15	30.4
R6	60	40	14.1	75	20.6
R7	38	92	25.5	4	15.8
R8	16	58	22.4	73	26.9
R9	18	74	25.5	31	11.2

11.5 Conclusion

In this study, the potential of RSSI-based positioning was investigated through its application in coordinating with a robotic arm. This approach enabled the mapping of a defined area and the subsequent determination of the robotic arm's position within that area, achieving an average margin of error. Results indicate that improvements to the sensor network are required to enhance accuracy. For instance, equipping each receiver with an additional antenna or integrating RSSI with alternative positioning technologies—such as ultra-wideband (UWB) or Bluetooth Low Energy (BLE)—could yield more reliable results. The findings suggest that, in isolation, RSSI-based positioning lacks the precision required for some applications.

Nevertheless, RSSI-based localization technology presents considerable potential for further refinement. A sensor network could facilitate comprehensive robotic navigation, enabling tasks such as workpiece identification and autonomous movement. Additionally, it could support the establishment of safety zones that respond to the proximity of objects or personnel, enhancing operational security.

Security issues also warrant attention in the context of RSSI-based navigation. Sensor networks that rely on RSSI and Bluetooth technologies may be vulnerable to interference or unauthorized access, potentially compromising both accuracy and system integrity. Implementing robust security measures, such as encrypted communication protocols and continuous signal monitoring, is essential to mitigate these risks and maintain system reliability. Despite these challenges, RSSI-based robotic navigation offers significant promise as a cost-effective localization solution with broad applications, with ample opportunities for enhancement and integration with supplementary technologies.

References

1. Koyuncu H, Yang SH (2010) A survey of indoor positioning and object locating systems. IJCSNS Int J Comput Sci Netw Secur 10(5):121–128
2. Petropoulos GP, Srivastava PK (2021) GPS and GNSS technology in geosciences. Elsevier, Amsterdam
3. Li CH (2023) Ultra-wideband technology: characteristics, applications, and challenges. In: Proceedings of IEEE 2023. IEEE, New York, pp 1–12
4. Azaddel M (2023) Energy-efficient design and cost advantages of BLE. IEEE Trans Commun 71:330–335
5. Gomes A (2023) Principles and applications of RFID technology. Int J Wirel Netw 15:112–120
6. Turgut Z (2023) WiFi-based indoor positioning using RSSI: enhancing localization in WLAN systems. In: Proceedings of IEEE international conference on IoT and smart systems, pp 45–52
7. Simon J et al (2013) Navigation of mobile robots using WSN's RSSI parameter and potential field method. Acta Polytechnica Hungarica 10(4):111–116
8. Gogolák L, Fürstner I (2021) Wireless sensor network aided assembly line monitoring according to expectations of industry 4.0. Appl Sci 11(1):25
9. Ismail S, Dawoud DW, Reza H (2023) Securing wireless sensor networks using machine learning and blockchain: a review. Future Internet 15(6):200

10. Celik MF, Isik MS, Yuzugullu O, Fajraoui N, Erten E (2022) Soil moisture prediction from remote sensing images coupled with climate, soil texture and topography via deep learning. Remote Sens 14:5584
11. IEEE (2022) Indoor positioning techniques using RSSI from wireless devices. In: IEEE conference publication, 9038591
12. Tahar G (2019) Geometric triangulations and flips. C R Math Acad Sci Paris 357(7):620–623. https://doi.org/10.1016/j.crma.2019.07.001
13. Sarcevic P, Csik D, Odry A (2023) Indoor 2D positioning method for mobile robots based on the fusion of RSSI and magnetometer fingerprints. Sensors 23(4):1855
14. Navarro E (2010) WiFi fingerprinting for indoor positioning systems. In: Indoor positioning techniques, pp 22–29
15. Ramesh A, Patel M, Singh K (2020) Analysis of outdoor signal propagation using transmission equations. J Wirel Commun Syst 12(3):102–110

Chapter 12
Food Terrorism as a Real Threat

Szilvia Veress Juhaszné

Abstract The study identifies food security as a key element of critical infrastructure and highlights the potential for deliberate attacks on it, i.e. food terrorism. The author describes the concept and components of food security and some of the stages of the food supply chain that are most vulnerable to a possible deliberate attack. It describes the importance of the RASFF and TRACES alert systems for monitoring food contaminated by omission or intent and the importance of prior risk assessment for food processing and food retail outlets. The author describes these alert systems and mentions the difficulties they face in the food trade in third-world countries. In the final chapter of the paper, recommendations are made on the directions for the development of alert systems.

Keywords Food safety · Food security · Food terrorism

12.1 Introduction

The term terrorism usually brings to mind dramatic and shocking footage of bombings, with dozens of injured and dead, where the authorities make superhuman efforts to save what and who can be saved.

Terrorist acts are designed to cause panic, intimidation, economic damage and undermine trust in state governance. In our modern world, advanced technical devices and technological systems have become part of the critical infrastructure, which extremist organisations recognise and target in their actions [1]. In addition to the forms of terrorism that are familiar to all (such as armed attacks and bombings), it is necessary to draw attention to a more subtle form of terrorism, less capable of causing 'shock' [2], but all the more dangerous in that it can cause masses of people to become ill or even die. This is food terrorism, which is the deliberate, malicious damage to the food supply chain. The phenomenon has been described from the time

S. V. Juhaszné (✉)
Óbuda University, Budapest, Hungary
e-mail: juhaszne.szilvia@uni-obuda.hu

T. A. Kovács and I. Fürstner (eds.), *Critical Infrastructure Protection: Advanced Technologies for Crisis Prevention and Response*, NATO Science for Peace and Security Series C: Environmental Security, https://doi.org/10.1007/978-94-024-2308-2_12

of ancient wars when enemies used the carcasses of dead animals to destroy drinking water supplies [3].

Our globalised world offers, in addition to local attacks, a clear opportunity for attacks that affect several territories/countries simultaneously and over large distances [4]. In today's modern societies, it is the duty and responsibility of political, economic and scientific elites to monitor the risks to the security of supply. The food supply chain is the most vulnerable of the critical infrastructure elements, as it includes all aspects of animal and plant production, processing, packaging, storage systems, commercial outlets and even agricultural and food research institutes [5]. The need to protect the food supply chain as a critical infrastructure element requires a complex approach and a response based on a set of principles [1].

12.2 The Food Supply Chain as a Critical Infrastructure Element Food Terrorism

12.2.1 Terrorism and Its Model

The path to the execution of terrorist acts is long and can be described in terms of stages of radicalisation. Radicalisation is a process whereby an individual or a group increasingly adopts extremist political, social or religious ideas and aspirations [6]. According to Moghaddam's radicalisation model (see Fig. 12.1), members of radical groups go through several (precisely five) stages to carry out a specific terrorist act [7]. The last, sixth stage is the actual execution of the terrorist act. Each of these phases allows counter-terrorism to intervene and prevent the planned operation from being carried out. Still, as the phases progress, the challenges become more serious as all the tools necessary to carry out the planned operation are at the agents' disposal. The more complex the specific operation is, the more complex it becomes to prevent a terrorist act.

Given that terrorist organisations are not indiscriminate in the means they use to achieve their goals, are willing to make personal sacrifices and are constantly using new means, attacks against the food supply chain and drinking water networks must also be expected [4]. In many cases, attacks in the current form of terrorism are not directed against a specific target group but against people who are randomly targeted in order to achieve a particular objective, in accordance with the 'principle of randomness'. Hence, in the case of a terrorist attack on a food supply chain, the victims of the attack would be people who were in the wrong place at the wrong time and who, by their own choice, consumed food that they could not have known was contaminated [4]. An attack on the food supply chain could cause masses of people to fall ill and die, as international trade in food is rife in the age of globalisation. Globalisation has dramatically facilitated the activities of extremist terrorist groups at the international level [8]. Bridging the geographical distance is not a challenge for extremist radical groups today, as cooperation between different groups can be

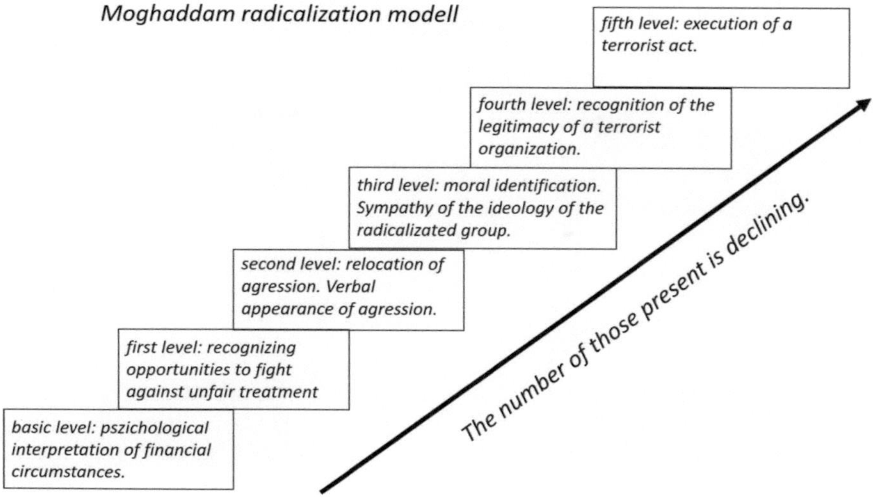

Fig. 12.1 Moghaddam radicalisation model (*source* own editing based Moghaddam, 2011) [7]

observed in the preparation and execution of individual actions. Some organisations operate as part of networks, using all possible means of modern communication [9]. Some organisations prefer to employ a person with military experience as their leader [10].

The extreme complexity of the food supply chain and the complexity of its protection are inversely proportional to the difficulty of carrying out deliberate actions against it.

12.2.2 Food Terrorism: A Deliberate Attack on the Food Supply Chain as a Critical Infrastructure Element

The food supply chain as a critical infrastructure element

As humans evolve, they are constantly building and shaping their environment, and as a result, infrastructure is constantly changing. Infrastructure has been defined from several perspectives as the set of means and institutions related to production that are not part of the direct production process but are indispensable to it [11]. In the Green Paper, the European Union defined critical infrastructure as "physical assets, services, information technology facilities, networks and real estate ….. whose disruption or destruction would have a serious impact on the health, peace, security or economic well-being of Europeans or on the effective functioning of the EU and its Member States' governments" [12].

Critical infrastructures are fundamentally characterised by dual interdependence: the interdependence of critical infrastructure networks and society's dependence

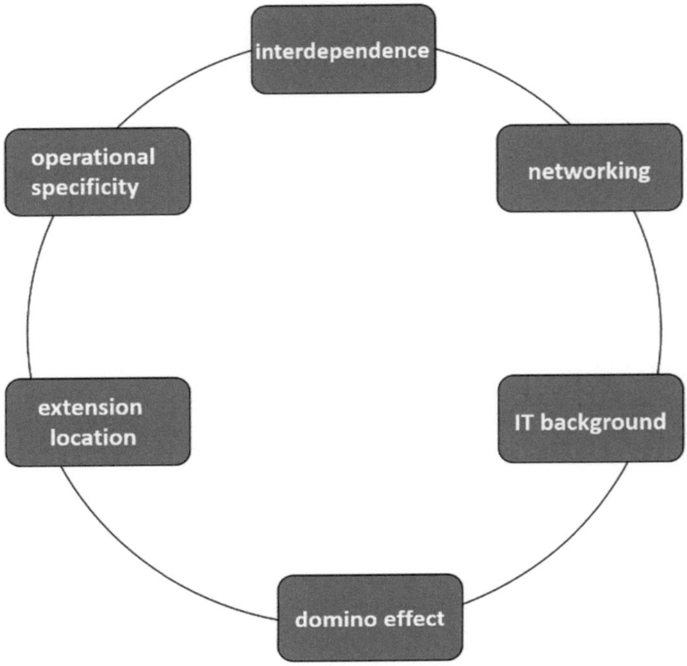

Fig. 12.2 Characteristic of critical infrastructure (*source* Bonnyai, 2014) [13]

on the infrastructure. Interdependence is only one of the characteristics of critical infrastructures. The other characteristics are shown below (see Fig. 12.2).

Of all the properties, the "domino principle" is the simplest way to describe the consequences of damage to critical infrastructure. The most important critical infrastructures include health care, energy supply, utilities, transport, telecommunications and information technology networks, drinking water and food supply, banking, public administration and law enforcement [13].

The vulnerability of the drinking water and food supply chain is extreme, as in the age of digitalisation, physical presence is no longer necessary to cause emergencies [14], and often the mere transmission or blocking of information can disrupt the infrastructure. Any element of a critical infrastructure can be the target of an information threat [15], but a disruption in just one system can have serious consequences for other sectors [16].

The stages of the food supply chain extend from 'farm to table'. A deliberate attack on this complex and sophisticated system can lead to a real disaster, as the food scandals of recent decades have highlighted. Threats to the food chain can be natural disasters, artificial emergencies, industrial disasters or deliberate abuses [17]. In this paper, we focus on deliberate acts.

Ensuring the smooth functioning of the food supply chain is essential, as threats to food security can lead to food insecurity, which in turn causes instability in society.

Discontent at different levels of society creates tensions that can lead to civil strife and radicalisation. The process of radicalisation creates a basis for the organisation of extremist groups. Acts of terrorism can then lead to political instability, with obvious long-term social consequences: refugee crisis, extreme poverty and food crisis. Food insecurity and lack of access to clean drinking water lead to further discontent, which ultimately leads to wars and armed conflicts [18]. In the following, we address the possibility of deliberate abuses that threaten food security.

Food terrorism and its purpose

Food terrorism is a tool of terrorist organisations with the same objective as any other terrorist action: '*...to* achieve *political, ideological or religious goals by creating fear* ' [19]. Terrorist attacks against the food supply chain are not necessarily carried out in all cases; often, the panic that accompanies the threat is sufficient to achieve the terrorist organisations' objectives, which involve disruption of the economy and instability of society. Food terrorism and, more broadly, agro-terrorism can cause significant devastation, according to experts' calculations [3].

A terrorist attack, regardless of the number of incidents and fatalities, can cause serious conflict between countries as trading partners. An example is perhaps one of the best-known but still controversial cases of food terrorism, which occurred in 1978 and involved the contamination of Israeli citrus fruit with mercury. A consignment of fruit arriving at the port of Rotterdam reached consumers' tables in Western European countries and caused several people to fall ill. The action was claimed by the Arab Revolutionary Army, which said it was carried out in the name of disenfranchised Palestinians and with the deliberate intention of damaging the Israeli economy. They succeeded in the latter aim, as Western European countries lost confidence in Israeli citrus fruits, which took years to restore. During this period, Israel's economy was indeed damaged by a sharp drop in citrus exports [4].

It is still unclear exactly at what point in the supply chain the oranges were deliberately contaminated with mercury: in Israel, before they were shipped (a possibility refuted by the investigations of British expert Dr Paul Wix, who concluded that the oranges had already arrived in Europe in a contaminated state), or in European ports, distribution centres or shops. In the latter case, the terrorist organisation probably had helpers [4]. The case described above is just one of many deliberate and malicious interventions in the food supply chain. The deliberate contamination of food is not a new phenomenon, a method known from the past [17], but in today's globalised world, the concept of food terrorism needs to be redefined. "For the life of me, I cannot understand why the terrorists have not, you know, attacked our food supply. Because it is so easy to do."—believes *Tommy Thompson, American Secretary of Health and Human Services* [20].

According to the above-quoted statement, carrying out a planned terrorist attack on the food supply chain is "easy". Preventing it is more difficult, as can be inferred from the definition of food terrorism: "*...the act or threat of intentional contamination of food intended for human consumption* with *chemical, biological or radionuclear agents, intended to cause injury or death to civilians and/or to disrupt social, economic or political stability*" [21].

Pillars of food safety

Food terrorism is, therefore, any deliberate intervention that prevents access to food of sufficient quality and quantity. And the FAO's 1996 World Food Summit Plan of Action defines food security as "...physical and economic access at all times to sufficient safe and nutritious food to meet the dietary needs and nutritional preferences of all people for an active and healthy life" [22].

In line with this, the pillars of food security are: access, availability, utilisation and stability [23]. The pillars are closely interlinked, and this close relationship is the basis of food security. The two definitions provide some insight into how food supply chains and drinking water supplies can be deliberately interfered with as elements of critical infrastructure. It would be a mistake to focus exclusively on food production plants, transport and storage units within the food supply chain when considering the potential for food misuse. Other targets of terrorist attacks may include research bases, plant and/or animal testing facilities, and water supply bases. The risks associated with these facilities should also be strictly considered.

In the present study, we focus on the food supply chain sectors directly linked to consumers.

Stages of the food supply chain

The food supply chain runs from the production of raw food materials (crops, livestock, and feed) to the consumer's table ("farm to table"), with the following key stages: production, processing, packaging, storage, and transport. Each of these stages represents a potential entry point for deliberate contamination, and the earlier the stage of contamination, the greater the risk [24].

The supply chain's vulnerability is further increased by the fact that its stages are separated in space and time in today's globalised world [4]. The effects of food contamination in a subtropical country can be felt weeks later by citizens in a central or western European country, and vice versa. In many cases, food supply in developing countries is determined by food imports, and consequently, masses of people can be put at risk if there is deliberate malicious interference at any point of entry in the food supply chain or even if food supplies to the destination country are sabotaged.

The most vulnerable stages of the food supply chain in terms of food contamination are the processing, packaging and storage stages [25], as these are the points where direct contact with food or its raw materials is possible.

The processing stage. During processing, foods undergo various procedures to make them fit for consumption and/or to preserve their quality and/or flavour for longer [25]. Food contamination can occur at any point in the processing chain. Contamination can be physical, biological and chemical [26]. Physical contamination is for example human hair, glass fragments; biological is the presence of various microorganisms; chemical is most often contamination with some chemical substances such as pesticide residues, antibiotics, heavy metals, other toxins [27]. At the food production and processing stage, contamination, including deliberate contamination, can occur during the cleaning of food raw materials. If we consider the water used for cleaning and washing the products to be processed, it is clear that it can be a source of introduction of many biological and/or chemical contaminants into food. For this reason (and also to ensure a safe drinking water supply), it is of paramount importance to prevent possible water-related attacks [25].

Storage and transportation. The storage and transport phases should be considered from two perspectives—transport/storage of the raw food material and transport/storage of the processed/finished product. Storage facilities for raw and processed food take different forms, depending on the type of food, where it is transported from and to, and how long it is stored. The rules for the storage/transport of fruit and vegetables differ significantly from those for the storage/transport of products of animal origin (e.g. meat and dairy products). However, some basic rules are generally applicable, such as maximum hygiene and the exclusion of unauthorised persons from coming into contact with food [27]. For food businesses, strict control of staff is essential. The installation of alarm and detection systems, surveillance monitoring [25], risk assessment and crisis management can be effective tools for this purpose.

12.3 Means to Fight Food Terrorism

12.3.1 Risk Management

Risk assessment as a preventive procedure

Risk assessment and analysis should be carried out at all stages of the food supply chain. Various models are available for risk assessment. According to the so-called 'Irish model,' the risk classification of a given food establishment is based on the following criteria: type of unit, quality of unit design, hazard of the product produced, technology, number of consumers exposed to the hazard and their health sensitivity, hygiene experience from previous inspections, and internal quality system of the manufacturer [27].

Risk classification therefore provides an excellent opportunity to develop and implement prevention methods and warning and alert systems. Their importance is paramount both for companies involved in food production and trade and for countries exporting and importing food.

12.3.2 Crisis Management

Crisis response or crisis management is in fact a plan for dealing with a crisis that has already occurred, which in most cases involves not only recovery tasks but also the prevention of similar events in the future. Crisis management plans are used by food production and distribution companies to respond as quickly and effectively as possible to a problem that has already occurred and to minimise the negative consequences.

The analysis of the event that has taken place and the conclusions to be drawn from the crisis that has occurred and its management are essential, following the damage repair and rehabilitation that are the main steps in crisis management. The lessons learned should be incorporated into the company's crisis management action plan [28].

12.3.3 Alarm Systems as Preventive Devices

Effective crisis management is inconceivable without the cooperation of public authorities in both damage prevention and damage response [29]. Given the increasingly global nature of the food supply chain today, the operation of international alert systems is crucial.

Among the two systems directly involved in ensuring food and feed safety at the international level are the Rapid Alert System for Food and Feed (RASFF) and the Trade Control and Expert System (TRACES), both of which aim to protect the public from food-related hazards and play a critical role in detecting and responding to food incidents.

The EU's ultimate goal is to create a single European crisis management system [17], which it intends to achieve through the use of alert systems.

The RASFF is a notification system operated by the European Union (EU) that informs EU Member States and the European Commission in real time of food and feed risks detected in the EU. The system allows for rapid and coordinated action between countries, provides timely alerts and exchanges information on measures taken to address food safety hazards. RASFF works in cooperation with other countries and regions around the world to ensure global food security.

TRACES plays a vital role in ensuring food safety within the European Union. It is a web-based platform developed by the European Commission to facilitate the exchange of information on the import, export and transit of animals, animal products and certain types of food. The primary objective of TRACES is to enhance traceability and transparency throughout the food supply chain, thereby protecting public health and ensuring effective control of food safety risks.

Direct and immediate communication with third countries is ensured by a new development of the RASFF alert system, the RASFF Window online database, but this system provides limited access to third countries [30]. Limited access is not a perfect solution for products entering or intended for export to third country markets. The African Union has therefore set itself the goal and plans to create a continuously updated information database specifically for African countries, similar to the RASFF system. The system is intended not only to ensure food safety, but also to ensure that food exports from African countries reach European markets and ultimately contribute to the economic stability of African countries [31]. Currently, the global food safety regimes developed by the European Union are so strict that food from third countries is often rejected or destroyed at the border, with a sharp increase in the period 2008–2018. According to the RASFF online database, the number of rejections of oilseeds and fruits at the border from Sudan to Greece increased again in 2019 compared to the previous year, with Greece rejecting these products 368 times in 2018 and 650 times in 2019 [32].

Although the strictness of RASFF and TRACES gives a sense of security against possible attacks on the food supply chain, the economic impossibility of third countries should not be a 'side effect' of the system.

Based on interviews with businesses involved in food trade at European level, the author also concludes that, although the rigour of the system is indeed effective and allows for precise monitoring, it makes trade in certain products considerably more difficult. This is particularly true for products with a short expiry date (e.g. meat and dairy products), where the verification of the origin of the products and their ingredients is a complex and time-consuming task, and the practical administration (e.g. obtaining an official veterinary certificate) can take several days.

Alarm systems have proven to work perfectly in many cases, preventing products from reaching consumers in the event of contamination above a certain limit, but it must also be clear that they cannot prevent contamination from occurring. Accordingly, it would not be able to prevent deliberate contamination of the food chain, i.e. an act of food terrorism.

In order to ensure the efficient functioning of economic processes and food trade, the harmonisation of food standards between African countries and the European Union, and even the global market, is essential for the future. In addition, it is worth developing the systems that have already been developed and are highly effective in terms of monitoring, so that they can prevent contamination from occurring and, if it does occur, identify it as early as possible in the food chain.

12.4 Preparing Against Food Terrorism

Terrorist organisations are not indiscriminate in their means, as mentioned several times in the study, so any action against the food supply chain is a real risk. Potential attacks must be prepared for and, where possible, prevented.

The systems described above (based on experience from other incidents) are in most cases capable of detecting contamination that has already occurred in the food supply chain and preventing the final consumer from coming into contact with contaminated food. The key to preventing food terrorism is to ensure broad international coordination of these systems and to focus on the protection of critical infrastructure, including through further development of the systems. In addition, raising awareness in society about critical infrastructure and its protection options is of paramount importance [14]. Critical infrastructure protection programmes should be extended to all areas of the food supply chain, involving all stakeholders - including final consumers as a significant part of society.

The aim is to build and maintain prevention, protection and response capabilities at international level [4]. Terrorist organisations are constantly using new tools to organise and conduct operations. Consequently, it is a realistic proposal that, in the case of high-risk target groups in terms of terrorist threat, the relevant leadership should organise regular simulation exercises involving the authorities or even all the parties concerned, such as the two simulation exercises carried out by the United States of America in the early 2000s, Black Winter and Atlantic Storm [33].

Based on simulation exercises in developed countries, the tasks for preparing for terrorist attacks on the food supply chain are as follows:

In times of terrorist threat: census of potential targets; identification of persons particularly vulnerable to possible use of biological agents (e.g. pregnant women, immunocompromised persons); continuous liaison with local community leaders, with particular attention to liaison with specific socio-cultural minorities.

After the first signs of a terrorist attack: informing the public in a credible way about the emergency; taking advantage of the fact that people naturally adopt a risk-averse behaviour and try to reduce their movements; encouraging the use of special helplines for those who notice a suspicious change in their own health or that of people around them.

Where there is already evidence of a terrorist attack: if vaccination is justified, those at greatest risk of infection (e.g. health workers) should be vaccinated first; every effort should be made to minimise turnover; and regular information should be provided to those affected about the risks of avoiding vaccination [17].

To perform these tasks to the required standard, it is essential to have highly qualified, dedicated professionals, trained by security engineering, security science, military and other public service educational institutions.

12.5 Conclusion, Recommendations

The food supply chain, as an element of critical infrastructure, deserves particular attention in the fight against terrorism. By its very nature, it is one of the most vulnerable critical infrastructures and is, therefore, subject to an increased threat, which is not negligible for terrorist organisations, given the constant use of new means by terrorist groups to achieve their goals.

Food and drinking water are basic human necessities, and if these are threatened, the consequences can be unpredictable - mass disease, deaths, and, with or without this, distrust of the companies producing and/or distributing the food and/or the importing/exporting country and its government, which can lead to an economic crisis. This is the aim of terrorist organisations. They seek to achieve their own political, ideological and religious goals by creating a crisis situation.

Specific food scandals in recent decades have highlighted the particular vulnerability of the food supply chain as a critical infrastructure element. The "farm to table" principle has led to the development of international alert systems capable of detecting accidental or deliberate food contamination at any stage of the food chain and preventing contaminated food from reaching the final consumer. These alert systems, working in conjunction with each other, can indeed effectively manage an incident and prevent it from causing harm to the wider community. However, preventing food contamination itself is a much more complex task. In this area, the cooperation and responsibility of the competent authorities bodies and companies cannot be questioned. Attacking the food supply chain and drinking water supply is not yet a common tool among terrorist organisations. However, as the study has repeatedly pointed out, since the risks must be taken into account, all the institutions, bodies and authorities involved in food security must initiate a dialogue to improve existing systems or develop a new system capable of taking preventive action against food terrorism.

References

1. Radvanovsky R, Mcdougall A (2023) Critical infrastructure: homeland security and emergency preparedness, 5th edn. CRC Press. https://doi.org/10.4324/9781003346630
2. Monke J (2007) Agroterrorism: threats and preparedness (CRS Report No. RL32521). Library of Congress Washington DC, Congressional Research Service
3. Eslami E, Siamian H, Rezaei Orimi J et al (2024) Pattern of bioterrorism in ancient times: lessons to be learned from the microbial and toxicological aspects. Wien Med Wochenschr 174:288–298. https://doi.org/10.1007/s10354-023-01029-1
4. Jurica K, Vrdoljak J, i Brčić Karačonji I (2019) Food defence systems as an answer to food terrorism. Arhiv za higijenu rada i toksikologiju 70(4):232–255. https://doi.org/10.2478/aiht-2019-70-3357; Meulenbelt S (2001) Assessing chemical, biological, radiological and nuclear threats to the food supply chain. Glob Secur Health Sci Policy 3(1):14–27. https://doi.org/10.1080/23779497.2018.1509675; Foxell JW (2001) Current trends in agroterrorism (antilivestock, anticrop, and antisoil bioagricultural terrorism) and their potential impact on food security. Stud Conflict Terror 24(2):107–129. https://doi.org/10.1080/10576100151101623
5. Azizsafaei M, Sarwar D, Fassam L, Khandan R, Hosseinian-Far A (2021) A critical overview of food supply chain risk management. In: Jahankhani H, Jamal A, Lawson S (eds) Cybersecurity, privacy and freedom protection in the connected world. Advanced Sciences and Technologies for Security Applications. Springer, Cham. https://doi.org/10.1007/978-3-030-68534-8_26
6. Moghaddam FM (2011) The staircase to terrorism—a psychological exploration Georgetown university. J Strateg Secur 4(4):37–62
7. Szeremley Cs, Juhász PG (2021) The DNS educational system as a device helping to prevent the spread of radicalisation in Malawi. J Cent Eastern Eur Afr Stud 1(3):119–134
8. Nnani NE (2023) The dynamics of the globalization of terrorism in West Africa: probing the case of Nigeria's Boko Haram. Soc Sci Res 9(1). https://journals.aphriapub.com/index.php/SSR/article/view/2048
9. Besenyő J, Sinkó G (2021) The social media use of African terrorist organizations: a comparative study of Al-Qaeda in the Islamic Maghreb, Al-Shabaab and Boko Haram. Insights Reg Dev 3(3):66–78. https://doi.org/10.9770/ird.2021.3.3(4)
10. Sinkó G, Besenyő J (2021) Comparison of the secret service of al-Shabaab, the Amniyat, and the National Intelligence and Security Agency (Somalia). Int J Intell Counter Intell 36(1):220–240. https://doi.org/10.1080/08850607.2021.1987143
11. Velasco-Munoz JF, Mendoza JMF, Aznar-Sanchez JA, Gallego-Schmid A (2021) Circular economy implementation in the agricultural sector: definition, strategies and indicators. Resour Conserv Recycl. https://doi.org/10.1016/j.resconrec.2021.105618
12. European Commission. https://home-affairs.ec.europa.eu/pages/page/critical-infrastructure_en. Accessed 20 Oct 2023
13. Boruzs H (2024) Physical protection characteristics of critical infrastructures Hungary. Armed Secur Guard Saf Secur Sci Rev 6(1):41–52. https://oda.uni-obuda.hu/handle/20.500.14044/25503

14. Jasiūnas J, Lund PD, Mikkola J (2021) Energy system resilience—a review. Renew Sustain Energy Rev 150:111476. https://doi.org/10.1016/j.rser.2021.111476
15. Djenna A, Harous S, Saidouni DE (2021) Internet of things meet internet of threats: new concern cyber security issues of critical cyber infrastructure. Appl Sci 11(10):4580. https://doi.org/10.3390/app11104580
16. Rajnai Z, Fregan B (2016) Kritikus infrastruktúrák védelme (jogi szabályozás). Műszaki Tudományos Közlemények 5:349–352
17. Szendrő ÉM (2021) Az élelmiszerlánc, mint kritikus infrastruktúra és védelme (Doctoral Dissertation). Nemzeti Közszolgálati Egyetem, Budapest
18. Juhász PG (2017) Az élelmezés-biztonsági helyzetre ható tényezők Malawiban (Doctoral Dissertation). Szent István Egyetem, Gödöllő
19. Schmid AP (2023) Defining terrorism, ICCT Report, p 21. https://doi.org/10.19165/2023.3.01
20. Food Navigator. https://www.foodnavigator.com/Article/2004/12/20/New-food-security-solution-to-beat-terrorist-threat/. Accessed 13 Nov 2024
21. Rasco BA, Bledsoe GE (2004) Bioterrorism and food safety. CRC Press, Boca Raton
22. Food and Agricultural Organization. https://www.fao.org/fileadmin/templates/faoitaly/documents/pdf/pdf_Food_Security_Cocept_Note.pdf. Accessed 1 Dec 2024
23. Calloway EE, Carpenter LR, Gargano T et al (2023) New measures to assess the "Other" three pillars of food security–availability, utilization, and stability. Int J Behav Nutr Phys Act 20:51. https://doi.org/10.1186/s12966-023-01451-z
24. Alvarez MJ, Alvarez A, De Maggio MC, Oses A, Trombetta M, Setola R (2010) Protecting the food supply chain from terrorist attack. In: Critical infrastructure protection IV: fourth annual IFIP WG 11.10 international conference on critical infrastructure protection (ICCIP 2010), Washington, DC, USA. Revised Selected Papers 4. Springer, Heidelberg, pp 157–167
25. World Health Organization. https://iris.who.int/bitstream/handle/10665/42619/9241545844.pdf?sequence=1. Accessed 1 Dec 2024
26. Center for Disease Control and Production. https://www.cdc.gov/foodborne-outbreaks/foodproductionchain/?CDC_AAref_Val=https://www.cdc.gov/foodsafety/production-chain.html. Accessed 1 Dec 2024
27. Szeitzné Szabó M (2008) Élelmiszer-biztonsági helyzetelemzés és kockázatértékelés. Agroinform Kiadó, Budapest, pp 1–238
28. Kasza Gy, Surányi J, Lakner Z, Bódi B, Deák F, Faludi G, Horváth A, Mészáros L, Szántó A, Danczák I (2012) Rendkívüli helyzetek és kezelésük az élelmiszer-kereskedelemben–irányelvek és tapasztalatok. Élelmiszervizsgálati közlemények 58:3–4
29. Szeitzné Szabó M (2011) Élelmiszerbiztonság: tények, tendenciák, teendők: a Magyar Tudományos Akadémia és a Magyar Élelmiszer-biztonsági Hivatal tanulmánya az Új nemzeti élelmiszerbiztonsági program megvalósításához. Agroinform, Budapest
30. Tarimo HH (2014) Rapid alert system for food and feed (RASFF). In: SADC SPS regional food safety meeting, Gaborone
31. Agritrade. https://agritrade.cta.int/Agriculture/Topics/SPS-Food-safety/African-Union-moves-to-establish-a-region-wide-food-safety-network.html. Accessed 1 Dec 2024
32. Traoré OZ, Latouche K, Tamini LD (2020) European's RASFF border rejections, African countries's reputation and exports of edible vegetables and fruits. Cahier de recherche/Working Paper 5
33. Smith BT, Inglesby TV, Brimmer E, Borio L, Franco C, Gronvall GK, Kramer B, Maldin B, Nuzzo JB, Schuler A, Stern S, Henderson DA, Larsen RJ, Hamilton DS, O'Toole T (2005) Navigating the storm: report and recommendations from the Atlantic Storm exercise. Biosecur Bioterror 3(3):256–267. https://doi.org/10.1089/bsp.2005.3.256. PMID: 16181048

Chapter 13
Device for Testing the Thrust of Small Rocket Engines

Dražan Kozak, Antun Brnić, and Wei Li

Abstract Breakthroughs in satellite technology enabled the gathering of enormous quantities of data regarding our planet, from weather predictions to sea-level and underground water levels measuring. Today, the only method of putting satellites into orbit is via rockets. The thrust curve of rocket engines is one of the key parameters of successful spaceflight. Providing a reliable way of measuring the thrust of rocket engines in discrete time intervals is therefore of paramount importance. This work covers the design and manufacturing process of a simple device built for that purpose using a load cell for measuring the force and an ESP32 microcontroller for gathering and transmitting the measured thrust data via Bluetooth. A Python program with a user interface collects the transmitted data and organizes it, in the end producing a thrust curve diagram.

Keywords Rocket engines · Thrust · Satellites · Measuring

13.1 Introduction

For successful spaceflight knowing the thrust engines generate is of paramount importance. For that reason, thrust measuring system have existed for as long as rockets have. With the prominence of model rocketry today, many authors have explored the idea of building a test bench for measuring the thrust of small rocket engines used in model rocketry. As stated in [1], after designing and constructing a solid-fuel

D. Kozak (✉) · A. Brnić
Mechanical Engineering Faculty in Slavonski Brod, University of Slavonski Brod, Slavonski Brod, Croatia
e-mail: dkozak@unisb.hr

A. Brnić
e-mail: abrnic@unisb.hr

W. Li
School of Mathematics and Statistics, Xidian University, Xi'an, P. R. China
e-mail: liweilw@mail.xidian.edu.cn

T. A. Kovács and I. Fürstner (eds.), *Critical Infrastructure Protection: Advanced Technologies for Crisis Prevention and Response*, NATO Science for Peace and Security Series C: Environmental Security, https://doi.org/10.1007/978-94-024-2308-2_13

rocket motor its integrity and performance must be evaluated on the ground multiple times to ensure its functionality and safety and to verify that the engine performance is in line with design parameters. Such tests are done on static fire test benches ae. devices for measuring the thrust force engines produce. The authors designed a thrust measuring device, measuring the produced thrust via a loadcell connected to a computer with a USB cable. This is not ideal for one would want to be as far away from the test bench as possible, no matter how small the rocket engines being tested are. The authors also stressed an important fact of having shrapnel protection around the tested engine in the case of uncontrolled explosion. Babayomi et al. in [2] developed a rocket engine test bench with a much larger load cell able to withstand up to 5000N of force, therefore able to measure much larger rocket engines. They also used a professional data acquiring system which is quite expensive and as such unavailable to most people interested in model rocketry. In [3] authors tested various models of small rocket engines used in model rocketry and compared the obtained data with the data provided by the engines manufacturers. Four different engines were measured three times each which showed little variation in thrust in each subsequent measurement of the same engine. For obtaining the thrust data they used an apparatus with a load cell in which engines were placed horizontally. The authors emphasized the importance of rocket flight for national security, and the scientific development of the nation.

The device for measuring the thrust of rocket engines developed in this work tries to combine the best of both worlds from all the authors. It uses an inexpensive load cell sensor, rated up to 200 N, for measuring the thrust force, which can easily be upgraded for a more capable one, with measuring capabilities up to 2000 N. Barriers are placed around the tested motor ensuring protection form uncontrolled explosions and small legs are added to the platform which enable devices calibration in the horizontal plane. For data collection a wireless Bluetooth system was designed with an ESP32 and Python which provides additional safety for the test bench operator. For controlling the whole process, a Python program with a UI ("User interface") was developed which helps visualize and compare the obtained data.

13.2 Earth Observation Satellites

The rapid advancement of technology, while yielding numerous benefits, has also brought significant challenges for humanity, particularly in the form of environmental degradation and resource depletion. Among the most pressing consequences are the depletion of the planet's finite resources and the speeding up of climate change, both of which are critical threats to life as we know it. One resource of paramount importance is water, an essential component for sustaining life on Earth. Given its criticality, monitoring and managing water resources have become a primary focus of scientific and technological innovation.

Earth observation satellites represent one of the most sophisticated tools developed for this purpose. Operating from the heights of their orbit, these satellites provide a unique ability to monitor Earth's surface on a global scale. Their high-altitude orbits allow them to capture vast regions of the planet within a single day, making rapid data acquisition across many terrains and climates possible. This capability is particularly important for tracking hydrological patterns, including the distribution, quality, and availability of freshwater resources. These satellites are equipped with advanced sensors capable of measuring a range of variables, such as surface water levels, soil moisture content, snow cover, and precipitation rates. By using technologies like synthetic aperture radar (SAR) and multispectral imaging, they can gather data with exceptional precision. Furthermore, satellites observations contribute to climate modeling by monitoring changes in ice caps, glaciers, and sea levels, providing critical insights into the interconnectedness of water cycles and climate systems.

One of the satellites missions that observes the Earth´s water resources is GRACE: Gravity recovery and Climate experiment. Developed as a joint venture between NASA and the German Aerospace center, it measured Earths gravitational anomalies and its changes over time. Gathered data was used to determine the amount of underground fresh water supplies, pressure on the bottom of oceans, loss of ice sheets and more [4, 5].

Copernicus is the Earth observation component of the European Union's Space program. It provides accurate, timely and easily accessible information to improve the management of the environment, understand and mitigate the effects of climate change and ensure civil security. These missions carry a range of technologies, such as radar and multi-spectral imaging instruments for land, ocean and atmospheric monitoring [6].

NASA Landsat is a multi-satellite mission that continuously monitors water resources, offering comprehensive data on critical parameters such as water quality, temperature, and physical or chemical changes within water bodies. This ensures real-time insights into aquatic ecosystems, supporting effective management, conservation, and sustainable use of water resources. Researchers used Landsat images of Jakobshavn Glacier on Greenland to determine that in 37 years it has lost 88 billion metric tons of ice [7].

Currently the only way to put satellites into Earth's orbit is to launch them on top of rockets. For such a complex engineering system to work properly one must have a lot of data for all individual components. Data is used to model and understand their behavior in various situations. One of many important parameters in rocketry is the thrust curve of rocket engines which describes how the magnitude of the force generated by the engines varies with time.

13.3 Rocket Engines

Rocket engines are devices that generate a high-temperature gas stream through fuel combustion. When this high-temperature gas is expelled at high speed through a nozzle, according to Newton's third law, a reaction force is generated. This reaction force is the thrust that propels the rocket in the opposite direction of the expelled gas mass. Because of this working principle, Rocket engines are known as reaction engines, and the fuel that burns is the reaction mass. There are two main types of rocket engines: those that use liquid propellant ("Liquid-propellant rocket motors") and those where the propellant is solid ("Solid rocket boosters," or SRB). Rocket propellant consists of fuel and an oxidizing agent.

Engines using liquid propellant have separate tanks for the fuel and oxidizer, which are delivered to the combustion chamber through a complex system of pumps and pipes. This makes these engines extremely complex with many different components. The Rocketdyne F1 engines, which used liquid propellant and took humans to the Moon in 1969.

On the other hand, SRB engines are much simpler. The combustion chamber also serves as the propellant tank. The propellant is ignited at the top of the chamber and burns downward, producing a stream of high-pressure hot gas that is further accelerated by the nozzle geometry at the chamber's exit.

SRB engines are mainly used as additional thrust sources to assist rockets in reaching Earth's orbit, after which they are discarded. They are practical for rocket modeling because of their simplicity. Rocket model engines are typically solid-fuel motors and are commercially available in Europe in a wide range of thrusts, classified from A to O, depending on the total impulse they produce. The total impulse is the integral of thrust over time or, more simply, the product of burn time and average thrust force.

13.4 Device Construction

The construction began with conceptualizing the function the device needs to perform. The device's function is to ignite a small rocket engine and collect data on the thrust it produces at specific time intervals. A sensor is required to collect the data, and a microcontroller is needed to read the data from the sensor and send it to a personal computer. A program on the computer controls the microcontroller's actions. Figure 13.1 shows a prototype model of the device.

The sensor chosen for measuring the thrust is a "Load cell" sensor, shown in Fig. 13.2. Load cells are designed as a cantilever beam equipped with three strain gauges. An object is placed at the end of the cantilever that is not fixed, and as the beam bends, the length of the wires within the strain gauges changes, which consequently alters their electrical resistance. This change in resistance is converted into the force that the object exerts on the end of the cantilever. At the end of the cantilever, in this case, there is a platform permanently attached to the sensor. This platform allows for the easy placement of an interchangeable mold that holds the engine securely in place during testing. The ease of changing the mold is crucial because the engines themselves are manufactured with different casing diameters.

The sensor, its supporting structure, and the mold in which the engine is secured are mounted on a horizontal platform. The engine ignition area is surrounded by plates, shown in Fig. 13.3, designed to ensure the safety of the surroundings in the

Fig. 13.1 CAD model of the prototype device

Fig. 13.2 Load-cell and the mounting platform

Fig. 13.3 Safety barriers added

event of uncontrolled engine explosions. The platform itself is supported by legs that allow for leveling of the platform, as seen in Fig. 13.4.

The microcontroller chosen is the ESP32 due to its low cost and extensive capabilities. The ESP32 comes with built-in Wi-Fi and Bluetooth modules, which simplify the construction of the device, as no additional Bluetooth module is required for transmitting data from the microcontroller to the personal computer where the data is processed.

The microcontroller is controlled via Bluetooth using a Python program, seen in Fig. 13.5, on a personal computer. In the program, it is necessary to enter the name of the engine being tested and select the COM port through which the wireless Bluetooth communication between the computer and the microcontroller is established. Once the required data is entered, the testing can be initiated. After pressing the "Start Test" button, a 5 s countdown begins. At the end of the countdown, the program sends a command to the microcontroller to initiate the engine ignition and start data collection. Additionally, the application creates an Excel file where the received data on force and time are stored.

Fig. 13.4 Added legs

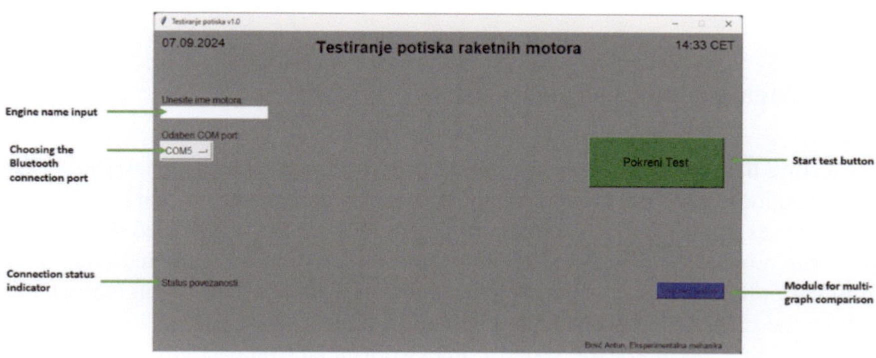

Fig. 13.5 Python program user interface

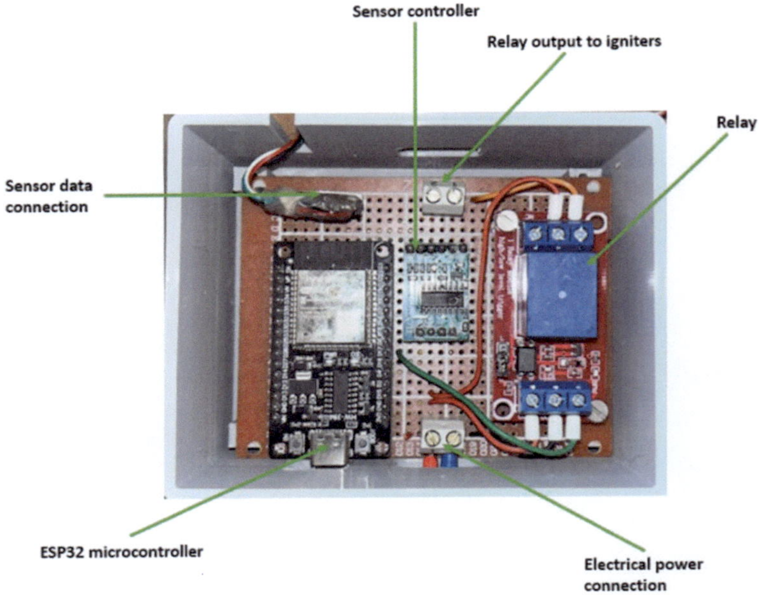

Fig. 13.6 Electronics assembly

Figure 13.6 shows the electronics assembly. The power supply for the electronics is an adapter that provides 5 V and 3 A of current and can be plugged into any household socket. Microcontroller is powered separately to negate the voltage drop effect while igniting the engines.

13.5 Measurements and Results

For testing the functionality of the device two solid propellant engines were used. These engines are widely used in model rocketry and therefore there are engine thrust curves available from the manufacturers. Manufacturer provided thrust curve diagrams provide us with an additional way of validating the gathered data, apart from multiple tests of the same engines. Figure 13.7 shows the two engines used in testing, the Estes D12-7 (brown) and Klima C2 (red).

The Klima C2-0 motor was tested 4 times to ensure the repeatability of the experiment. Collected data and the graphs it produced were compared to the manufacturers thrust curve. The experiments concluded that the device works as intended, producing the same results each time. Figure 13.8 shows the experiments being done.

Although the thrust was measured to be a bit less than expected values, based on manufacturers data, the difference between the expected value and the recorded value is constant at approximately 0.8 N. That is to say the graph produced with testing

Fig. 13.7 Estes D12-7 (brown) and Klima C2 (red)

the engines follows the general outline of the manufacturers graph. The manufacturers graph is shown on Fig. 13.9, while Fig. 13.10 shows the graph produced via experimental testing of the engines.

Recorded graph also indicates that the tested engine, being a hobby model rocketry engine, is not built at a particularly high standard. After the initial high thrust of the first second the thrust varies quite a bit from 0.8 to 1.5 N for the next four seconds. That would indicate that the solid rocket fuel does not burn equally or is not pure enough because of the contaminants. The next figure, Fig. 13.11, shows the comparison of the data collected from multiple measurements of the same engine.

Fig. 13.8 Testing the engines

Comparing the graphs made from multiple experiments with the one provided by the manufacturer a few things can be deduced. Firstly, all the tested engines give roughly the same performance and roughly the performance indicated by the manufacturer. Secondly, that the device produces reliable data which can be used for further study. After the reliability and the repeatability of the device was concluded, it was used to measure another kind of engine to compare it to the Klima C2-0. Using the graph comparing module of the Python application, Fig. 13.12 was produced, showing the difference in performance of the D12-7 and Klima C2 engines.

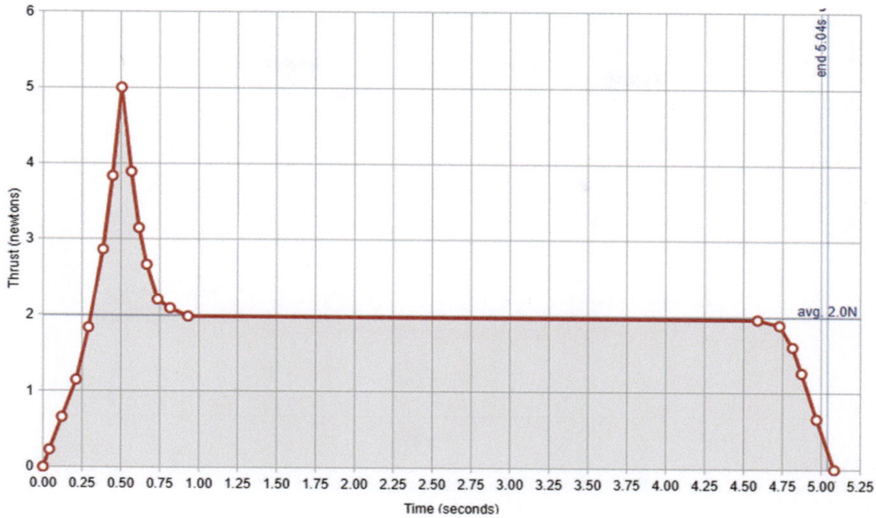

Fig. 13.9 Klima C2-0 thrust curve (from manufacturer)

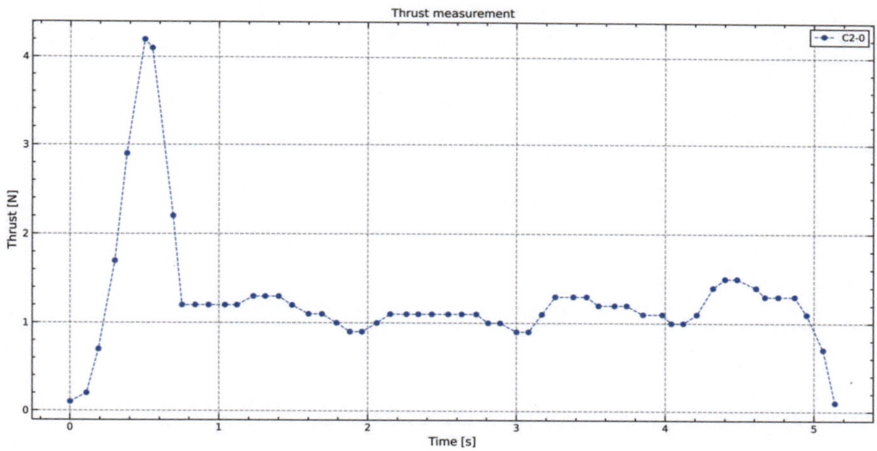

Fig. 13.10 Klima C2-0 thrust curve obtained by testing

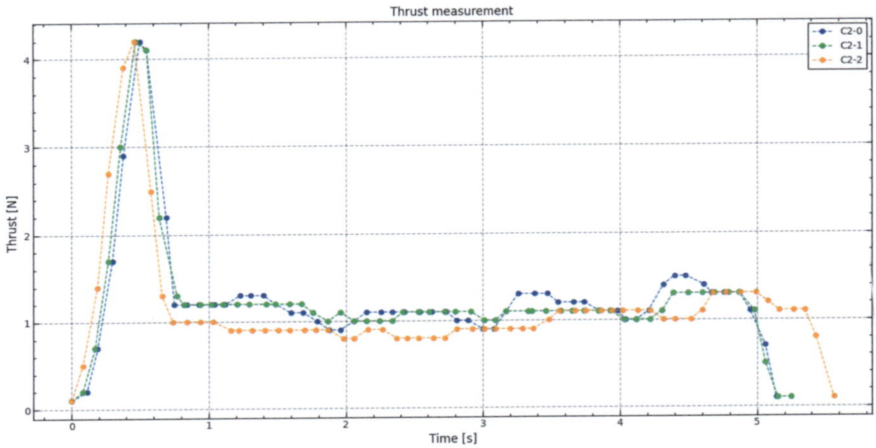

Fig. 13.11 Comparison of three tests of the same engine Klima C2-0

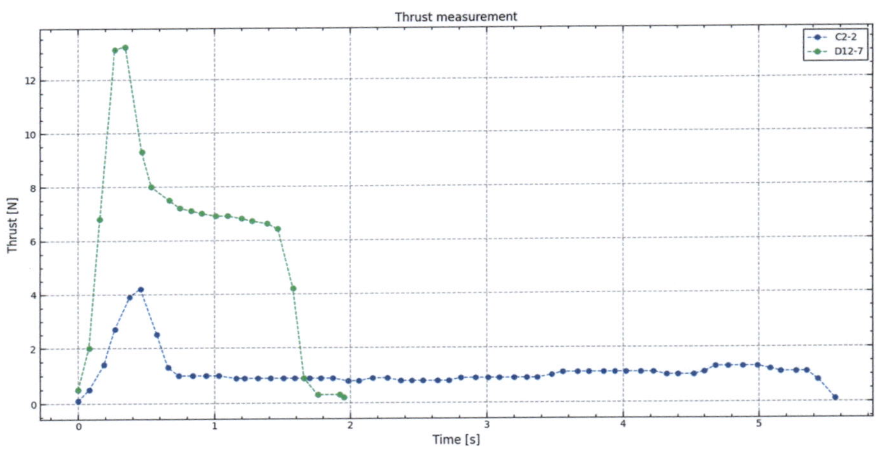

Fig. 13.12 Thrust comparison of D12-7 and Klima C2 engines

13.6 Conclusion

In this work a device for testing the thrust of small rocket engines was constructed using simple and cheap electronic components and a Python application. That the device provides good data was established comparing the engine thrust data gathered with the data available from engine manufacturers. Repeatability of the measurements was also proved with repeated measurements of the same engines.

References

1. Soares D, Teloken F, Severo T, Poletto M (2018) Development of a vertical static test bench for amateur rocket engines. Int J Adv Sci Tech Res 5. https://doi.org/10.26808/rs.st.i8v5.05
2. Babayomi O, Osheku C, Adetoro M, Okon A, Opasina A (2013) Low cost ground test equipment for atmospheric rocket flights using embedded systems, pp 26–29. https://doi.org/10.1109/NIGERCON.2013.6715635
3. Harridon M, Sa'aid HB, Noh HM, Wardiah PNS, Abas MN, Karim MA, Hilmi MZ, Amran MA (2021) Thrust measurement of different rocket engines. Int J Sci Res Publ (IJSRP) 11(1). ISSN: 2250-3153. https://doi.org/10.29322/IJSRP.11.01.2021.p10973
4. Frappart F, Ramillien G (2018) Monitoring groundwater storage changes using the gravity recovery and climate experiment (GRACE) satellite mission: a review. Remote Sens 10:829. https://doi.org/10.3390/rs10060829
5. https://www.jpl.nasa.gov/spaceimages/images/largesize/PIA04235_hires.jpg. Accessed 15 Dec 2024
6. https://www.esa.int/Applications/Observing_the_Earth/Copernicus/The_Sentinel_missions. Accessed 15 Dec 2024
7. https://landsat.gsfc.nasa.gov/. Accessed 15 Dec 2024
8. https://en.wikipedia.org/wiki/Rocketdyne_F-1#/media/File:F-1_rocket_engine_at_KSC.jpg. Accessed 15 Dec 2024
9. https://sh.wikipedia.org/wiki/Raketa_na_kruto_gorivo#/media/Datoteka:SolidRocketMotor.svg. Accessed 15 Dec 2024
10. https://eurospacetechnology.eu/index.php?id_product=1151&id_product_attribute=0&rewrite=klima-c2-0&controller=product. Accessed 15 Dec 2024

Chapter 14
Construction and Operation of the Hungarian Technical Border Barrier

Milán Majer

Abstract In the following study, I aim to describe the legal background which enabled the construction of the Hungarian technical border barrier, the construction process, and how the completed structure and systems are operated. The structures and systems I present are currently operational and imperative in the protection of Hungary against illegal immigration, therefore no specificities can be shared beyond the presented. The goal of my study is to decide whether the border lock can be improved, and its efficiency increased. I aim to present my answer based on personal experiences, statistics, and the analysis of available literature. It can be concluded that in order to operate effectively, you have to remain highly responsive to constantly changing challenges. There is debate among researchers about whether the Hungarian border lock is a protection system suitable for critical infrastructure.

Keywords Hungarian technical border closure · Legal background · Construction · Operation · Illegal migration

14.1 Introduction

Types of barriers that cause no damage, e.g., fences are an important group of technical border locks. In this case, the fence has been erected by utilising the defence resources of the country, resulting in security level increase and reduction of endangerment [1].

As a member of NATO and the European Union, Hungary created its national program for the protection of critical infrastructures in 2008.

M. Majer (✉)
Bánki Donát Faculty of Mechanical and Safety Engineering, Óbuda University, Budapest, Hungary
e-mail: majermilan@gmail.com

T. A. Kovács and I. Fürstner (eds.), *Critical Infrastructure Protection: Advanced Technologies for Crisis Prevention and Response*, NATO Science for Peace and Security Series C: Environmental Security, https://doi.org/10.1007/978-94-024-2308-2_14

According to the concept adopted by NATO, critical infrastructure comprises of: "…those facilities, services and information systems that are so vital to nations that their inoperability or destruction would have a debilitating effect on national security, the national economy, public health, public safety and government for its effective operation."

14.2 The Legal Background Behind the Construction

The number of immigrants who attempted to enter Hungary rapidly rose in January 2015. Around that time, it became crucial to create a border lock on the southern border of Hungary in order to stop immigrants who wanted to enter the country illegally. In response to illegal immigration rising to extraordinary levels, on June 17, 2015, the Hungarian government decided to shut down the green border between Serbia and Hungary. The purpose of this decision has been to direct immigrants through the legal border crossing points instead of crossing the green border illegally.

The 1401/2015. (VI. 17.) government decision [2] declared a 175 km long, 4 m-high temporary fence for border control purposes was to be installed on the Hungarian side of the border.

The Serbian border has been closed completely on September 15, 2015. Illegal immigration immediately increased through the Croatian border, therefore on October 16, 2015, the construction of the Temporary Security Border Barrier (referred to as IBH) has been completed in the Croatian border, as well. With its construction, the closure of the green border has been realised by presenting continuous physical obstacle to crossing on the length of the entire southern border section.

The Parliament adopted Act LXXXIX of 2007 on the State Border addition to the law [3], and later "…compensation related to public interest rights of use…" was regulated in a government decree. Moreover, the aforementioned government decree has been amended twice in August 2015: the name "fence" has been amended; the new name of the structure became "temporary security border closure".

The creation of the electric fence protection system is regulated by 1165/2016. (IV. 5.) [4] which has been prescribed by a government decision. This decreed it necessary to build a manoeuvre road on the Hungarian-Serbian and Hungarian-Croatian border sections in order to ensure the necessary supply logistics of the system. It has been determined that the budget will cover all the necessary financial requirements for this.

The legitimacy of the technical border fence has been also ensured by the fact that Act C of 2012 on the Criminal Code (Btk.) [5] defined the prohibited crossing of the border fence in Article 352/A. §, 352/B. §, 352/C. §. In addition, human trafficking in § 353. § 5 of the State Border Law stipulates that the Hungarian authorities "…may detain a foreigner who is illegally present in Hungary within the 8 km zone of the external border line or from the border mark, and may escort them through the nearest gate of the establishment according to paragraph (1)…" from July 5, 2016, and "…in the entire territory of the country, any foreigner illegally present may be detained and escorted through IBH" from March 28, 2017.

This concludes the establishment of the legal background for the construction of the border fence, in full compliance with domestic and international standards (action against illegal migration [6], prevention of terrorism) [7].

The first version of the IBH system has been built in in the summer of 2015. The Military of Hungary [8] as well as BV Holding [9] took part in the construction of the fence. The construction has been completed in August 2015. The name of this first system was Quick Installation Wire Barrier (GYODA). This unit, which can be considered a rudimentary technical border barrier, has been further reinforced in the fall of 2015 by a wire fence several meters high.

Illegal immigration continued to be increase. The government responded with the 1165/2016. (IV. 5.) [4] government decree, in July 2016. Installation of a high-current network ensuring energy supply began with the aim to deliver electricity to the so-called "intelligent" components of the technical border barrier over a length of more than 100 km. At the same time, the construction of the communication network along the entire border section commenced. The construction and installation have been completed in September 2016. The installed electrical network became responsible for the operation of the later-planned Intelligent Signalling System (IJR).

In August 2016 the government decided to build the second version of the IBH, which is the so-called Temporary Security Border 2 (IBH2). Metalcom Zrt. [10] conducted a feasibility assessment and found that the border closure would only work according to requirements if IBH would be supplemented with an intelligent signalling system. In this way, the protection will be provided by the physical barrier and the electronic system combined [11]. The human resources necessary for the operation of this system are no subject of this study, however they may form the content of further publications. The first elements of the border closure were erected on a 10 km-long path between Bácsszentgyörgy and Bácsborsód. The construction lasted about a month and half; it has been handed into operation in December 2016.

The completed IBH section lived up to expectations, meaning that in its existence it has been able to reduce the illegal migration threat on the given border section. Based on the result and conclusions drawn from its construction, the decision has been made in January 2017 that the IBH and IJR system should be installed on the entire border section between Serbia and Hungary. In the first phase of construction, the section between Csikéria and Röszke has been installed, taking a total of 60 days to complete. In the second phase, construction has been carried out on the remaining sections, which took a total of 90 days to complete. It can be concluded that the system has been rapidly deployed.

The overall conclusion is that Hungary has been able to respond speedily and effectively to the threat caused by the severe increase in migration. All construction work concluded by March 2017, and the entirety of the system has been successfully handed over to operation.

14.3 Physical Design

The fence is supported by steel columns, as this method was the most effective technical solution due to its height, stability, necessary protection requirements, and the subsequent attachment of other elements to each other [12]. It can also be defined as a continuously upgradeable physical protection. The steel columns are placed 5 m apart, the stiffening of the columns is ensured with inclined steel supports per every 10 columns. The wire braid has been fixed on the top plane of the Hungarian side, on tension wire with several strands, several millimetres in diameter per strand. On the top of the columns, Y-shaped necks extend in the direction of Hungary and the neighbouring country; these are over 100 cm long in both directions.

The neck elements were provided in both directions with a blade-sharp wire cylinder guide wire with a diameter of between 30 and 80 cm, placed in several rows. I am unable to include further data of the physical designs in this study, given that the technical border lock is still in operation and therefore sharing of its specifications present a security risk. A rebar mesh has been attached to the wire mesh in the higher parts to prevent cuts. A razor-sharp wire cylinder has been also attached to the bottom of the stems. Other fastenings of the braces have also been ensured. The vehicle and the gates allowing the passage of people were built in accordance with the requests of the police and Gemenc Ltd [13]. The border closure has been since reinforced in some areas which have seen repeated attacks.

Reinforcement should be understood as reinforcing steel mesh over the entire height of the fence, which are fixed with high-quality clips. The manoeuvring route along the entire length of the fence has been designed to take into account the characteristics of the area and to avoid natural obstacles. It has been designed as close as possible to the border lock for maximum efficiency.

The IBH system provides physical protection at the country's border, however, it is important to note that the construction did not take place entirely along the lines of the border marks, but on average at a distance of between 0 and 100 m within the actual national border, still in Hungarian territory.

This is a guarantee of legality, since in this way Hungary does not violate the territorial integrity of neighbouring countries during works outside the fence, or its measures against migrants. The terrain conditions had to be taken into account, as well.

It can be concluded that the physical lock [14] on the one hand inhibits and, on the other hand, hinders the entry of illegal migrants, therefore it fulfils the expectations attached to it in terms of efficiency.

The IJR provides the IBH with the ability to detect and select the attacking penetration points, and it also has the ability to deter entry since electricity has been introduced into the fence. Due to the energization, the manpower involved in the defence had to be specifically trained in the operation of the system. The IJR system is securely attached to or near the fence, and it receives its electrical supply from the IBH electrical network. IBH is directly connected to IJR's monitoring system.

As system elements exist outside the fence, we should mention the medium-voltage overhead network, as well. The low-voltage underground cable networks completely cover the affected border sections. Their task is to supply the active elements of the entire IBH with electricity.

System components outside the fence can also be defined as infrastructure distribution devices that ensure the system's energy supply; and infrastructure [15] devices, i.e., cameras operated by the National Infocommunications Service Provider Ltd. [16] the joint operation of elements, the joint operation of uninterruptible power supplies, routers, transmission technical components, distributors, controllers, and elements required for gate opening.

System elements within the fence are the gates (small gates and large gates), which ensure the passage of passenger and freight traffic. An internal element is also the emergency button, which enables the IJR system to be switched off and restarted on the spot in the event of an emergency. Electric security fences are identified by clearly visible, striking warning signs. The warning signs are easy to read, both from the internal, protected area and from the external, public area. At least one warning sign has been placed on both sides of the border closure element provided by the introduction of electricity.

System components have also been installed that supply the IJR system with electricity, detect sabotage, and then transmit an alarm to the monitoring system.

During the design and cabling of the system elements and structures, transformer districts were created. Along the line of the IBH, a relatively large number of transformer districts are responsible for the energy supply, and several small low-voltage connections are provided by MVM Next Energiakereskedelmi Ltd. (The former Southern Transdanubia Electricity Provider) [17]. The transformer districts consist of several pieces per side, one district supplies between 5 and 10 energy-giving devices with electricity.

An infrastructure system element has also been designed [18], to act as distributor at the border lock. The strung, high-current cable is attached to the low-voltage side of this. Three types of prefabricated elements were installed on the IBH line. They provide the energy for the operation of system components, i.e., cameras, uninterruptible power supplies and high-current distributors.

The energy source device, also known as the Druid device [19], contains several multi-zone energy sources, one of them is responsible for supplying energy to the inner fence and the other one is responsible for the outer fence. The internal structure and design of the device is considered sensitive information; therefore, it is excluded from this publication.

The devices are equipped with sensors, lightning, and overvoltage protection devices. Overvoltage protection of the IJR system was solved with II–III grade devices. The contractors installed lightning protection spark gaps on the high-voltage forward and return fibers against possible excessive voltage coming from the direction of the wires of the IJR system.

Additional functions of the energy transmitters are to supply the high-voltage fence with energy. It converts electrical power into short, high voltage pulses every second. Its output voltage can be adjusted, it can also be used in low energy mode.

The system is also capable of high-energy mode, but the voltage can be regulated here as well.

The length of the voltage pulses can be adjusted, so despite the high voltage, an approximately low amount of energy still reaches the fence. In terms of human life [20], the standard and the literature indicate an amount of energy of 5 J [21], the system operates within this value limit. The set amount of energy runs in wires that conduct electricity to the required extent [22].

As noted above, standard electricity warning signs were placed on the surface of the fence, which make it clear that the system is under voltage, and the constructor also posted signs with warnings written in several languages. Self-developed LED voltage indicators are also in operation. Their placement is possible regardless of polarity, and they work with both low and high voltage. The voltage indicators were placed in several places on both sides to make their use easier.

The energizing system also include a software user interface, on which the values of the output, return and capacity measurements are displayed. They are also able to display the average values, and it is possible to switch between them. Each display tab shows a different group of values. Among these, the system provides information about the voltage levels that could be measured in the fence going back several minutes, broken down by section.

The system energy supply has been designed and deployed in such a way that it cannot be detected by illegal migrants. In addition, the necessary steps toward protecting human life have been taken with signs warning of the operation of all elements capable of causing bodily harm posted along the scale of the deployment.

An additional signalling device has been also integrated into the fence. According to its intended operation, the device manages 50–100 sensors. Sensors are allocated to segments. The maximum length of a segment is 100 m, and the sensor also has the ability to turn, so it can follow the terrain. The sensors are differentiated based on their segment, in the event of a malfunction, the exact location of the problem can be determined, to ensure that it can be easily and quickly located, and rapid maintenance response or a reset can be performed.

In the event of an alarm due to the movement around the fence, a 7-language announcement in English, French, Turkish, Arabic, Pashto, Farsi and Urdu shall be played from the loudspeakers in the area affected by the intrusion. The loudspeakers broadcast the following: "I warn you that you are on the Hungarian state border, at the border lock owned by the Hungarian state, which is equipped with an electric protection system. Approaching and touching the border lock can cause injury and pain. If you damage the border lock, cross the border lock in a prohibited manner, or attempt to do so, you commit a crime. I urge you not to do so! You may submit your asylum application at a Hungarian embassy outside the Schengen area."

The computer control system alerts different events with a specified type of alarm. In the event of a communication error, the system is capable of retaining contact with the control device. If a problem occurs in the operation of one or more sensors, the system shall search for the possible source of the error. In general, the short circuit, breakage, contact error, misaddressing, power supply failure, or bus cable short circuit of the sensor is found responsible for the error, this can be repaired quickly and efficiently.

A controller outage error may be alerted as the system itself becoming unable to establish connection. In such cases, a power supply, network, or controller error may occur, which can be easily repaired and detected.

In the event of a maintenance signal error, if the sensors have been moved by thermal expansion [23] or other external influences [24], the operator has to wait for the sensor to return to its initial position, adjust and then the signal shall disappear without further action required. In most cases the operators are able to apply a staged method to troubleshooting error messages' root causes, which allows for rapid localisation and issue resolution.

The internal structure and design of the signalling device is deemed to be sensitive information; therefore, it cannot not be detailed further in this publication.

The system is also able also to detect any physical change that may occur during an illegal penetration attempt including that the location of the change shall be precisely presented to human defence resources. The border lock is capable of its own physical protection, as well as the precise localization of possible damage.

A camera system operates along the entire length of the Hungarian technical border barrier. In the case of cameras in a constant state, continuous colour image analysis works in specific monitoring zones or lines. If someone or something passes through their areas or makes continuous movement, this shall generate a video analytics alarm. The evaluable daytime range of the laser cameras is more than 1000 m, and the night-time range is more than 100 m. Thermal cameras have an evaluable viewing distance of 1000 m.

The protection of the Hungarian technical border barrier is supplemented using drones [25]. Using a drone, the territory of the state border is mapped by the patrol service, and the flying drone performs the detection of possible illegal routes, resting places, and possible tunnels. It searches for persons who have committed illegal border crossings and is also able to perform other tasks.

The main function of the opening sensor is to detect the opening of the gates through their inputs, and to detect the opening of the above-mentioned energy-supplying devices. The entire control unit receives energy through the energizing device, the voltage required for its operation here is 12 V. This control board can control 6 inputs and 5 relay connections.

To detect openings, the system uses Reed relays [26], magnets [27] and other components operating on a similar principle, based on international experience and the operation of border locks [28] in other countries.

The most essential elements of the Ethernet communication network were placed in the infrastructure devices. A router was installed in each device. The large number of routers enable the localization method of effective troubleshooting.

Each device, such as energy transmitters, controllers, input cards, thermometers, etc. is connected to the network of the Hungarian National Police Headquarters (ORFK) via the routers. ORFK's network is operated by NISZ Ltd; they provide the connection with the servers located in the control rooms.

The task of the communication network is to ensure stable and safe communication between all the devices that make up the IJR and IBH system. Alarms and signals reach the monitoring system through the correct operation of the communication network, and the network is designed to allow for the remote switching off and on of the devices, as well as possible configuration. This network consists of active devices [29], i.e., routers, switches, firewalls and security modules, SFP modules, and the data transmission cables that connect them, i.e., optical, UTP, or copper cables. The optical, UTP, or copper data transmission cables of the network are connected to the system elements performing various tasks through constructions made in safe places.

The cables were installed in LPE protective tubes with a diameter of 2–4 cm. A data transmission backbone was built parallel to the border, within the territory of Hungary, and active data transmission devices were installed at the low voltage feeding points, which are under the supervision and operation of the NISZ. Furthermore, the signalling and control elements of the IJR and IBH systems are also connected to this network.

The communication network is Internet Protocol address [30] (IP address) based. An IP address is a unique network identifier used to identify computers that communicate with each other using a network protocol. IP addresses according to IPv4 are 32-bit integers, which can traditionally be defined as four one-byte decimal numbers between 0 and 255, separated by dots for easier readability. Each device of the IJR and IBH systems was assigned its own private IP address.

In addition to the design noted above, there is also a remote monitoring software that continuously collects data on the operation of the border lock. External and internal fence elements are displayed on the map with icons. There is also a separate server interface for selecting the various stations and displaying the data of connections and disconnections. In the user interface, the application shows each object with a different icon and marks its exact location on the map. The colours and pictograms of the icons have specific meanings.

A location icon can be a county, site, etc., and the interface shows the available information about the current status of the devices under the icon.

Electric fence sections are made up of various measuring and controllable devices. These appear in the interface as fence sections along with the functions of all devices installed on the section.

Devices that generate electrical impulses and detect cutting, short-circuiting, and electrical loading of the fence can also be installed on the fence sections. Stretch sensors indicate the displacement of electric fence cables. Gate opening sensors show the opening of the gates located on the fence section.

The fence sections also indicate the current status of the devices. Additional information can also be obtained, such as login and logout details, generation and management of PDF reports, management of devices, management of signals, log files, error reports. From the operational point of view, the main task of the network is the monitoring and operation of the communication system, the software, and the infrastructures of the technical border barrier.

In the IJR system, its role is to monitor the availability of devices, examine the error signals of individual devices, monitor measured values (for example, temperature), and issue alarms according to re-configured alert settings for the different types of events that may occur. Multiple services are deployed which shall alert when such a pre-set event or an emergency occur or when a limit value is exceeded.

The system extracts the data from the IJR server, not from the previously mentioned devices installed close to the border lock. The monitoring elements of the signalling system is suitable for continuous monitoring of the operation, as well as the immediate recognition and localization of possible malfunctions or attacks, thereby ensuring the effective operation of the technical border barrier [31].

14.4 Conclusion

In this article, I described how the Hungarian technical border barrier complies with domestic and international legal requirements, and what makes its construction and operation legitimate. From an engineering point of view, the construction of the fence is suitable for preventing and mitigating attacks caused by illegal migration [32]. The building can be considered a "smart facility" due to the advanced communication between the different elements of the signalling system and the user interfaces.

The continuous development of the border lock is deemed necessary, as new methods and tools are constantly being developed by attackers.

Such a development possibility is, for example, the software synchronization of the face recognition system and the available camera system, because if the attacker's [33] face is recorded, it can even be automatically compared with photos in the available databases.

It would be expedient to declare the border fence a critical infrastructure, since illegal migrants, including terrorists [34], would become able to enter the country and through it, the European Union, unchecked in the event of the fence's destruction. Therefore, "in a densely populated area of 300 people/km^2 or more, a serious disturbance of public peace, a harmful psychological and public health effect affecting the population" would be created. These effects are defined in the Hungarian regulatory system as social impact criteria, and as public and state administrative impact.

References

1. Besenyő J. Fences and border protection: the question of establishing technical barriers in Europe. Acad Appl Res Mil Public Manag Sci 16(1):77–87. https://doi.org/10.32565/aarms.2017.1.7
2. 1401/2015. (VI. 17.) Government decision on certain measures necessary to manage the extraordinary immigration pressure
3. LXXXIX of 2007. Act on the state border
4. 1165/2016. (IV. 5.) Government decision on the provision of budget resources necessary for the construction of the maneuver road to be built on the Hungarian-Serbian border section and to ensure a reliable power supply

5. C of 2012. Act on the criminal code
6. Resperger I (2013) International and domestic crisis management. Teaching aids. Ludovika University of Public Service, pp 3–11
7. Besenyő J, Prantner Z, Speidl B, Vogel D (2016) Terrorism 2.0—the Islamic state. Kossuth Publishing House, pp 22–50
8. Hungarian Army Homepage. https://honvedelem.hu. Accessed 10 Dec 2024
9. General information about workers in Hungarian prisons. https://www.bvholdingkft.hu. Accessed 10 Dec 2024
10. General information about Hungarian border barrier builders. https://metalcomzrt.eu. Accessed 10 Dec 2024
11. Horváth S, Kósa C (2010) Technical communication. Textbook. Óbuda University, pp 12–32
12. Berek L, Berek T, Berek L (2016) Personal and property security. Textbook. Óbuda University, pp 20–88
13. Hungarian Forest Economy Homepage. https://gemenczrt.hu. Accessed 10 Dec 20214
14. Padányi J (2015) Technical lock at the border. Tech Mil J 25(3):21–33
15. Bognar B, Bonnyai T, Vámosi Z (2019) Protection of critical infrastructures I. Dialog Campus Publishing House, pp 48–124
16. Hungarian National Information Communication Service Homepage. https://nisz.hu. Accessed 10 Dec 2024
17. Hungarian Energy Provider Homepage. https://www.mvmnext.hu. Accessed 10 Dec 2024
18. Bagossy B (2021) Thesis, Óbuda University, pp 1–53
19. General information about the druid energy transmitter. https://druid.apache.org/docs/latest/design/architecture. Accessed 10 Dec 2024
20. Zalessky A (1968) The electric arc. Hungarian Technical Book Publisher, pp 5–12
21. General information about joule experiments. https://www.netfizika.hu/tudas/node/6372. Accessed 10 Dec 2024
22. Balla S, Bán K, Rider A, Szabó A (2011) Material knowledge. Textbook. Budapest Technology and Economics University, pp 52–72
23. Szakmány S, Hömöstrei M, Wiener C, Molnar A, Pesthy S, Jenei P (2020) Physical experiments. Note. Lóránd Eötvös University, pp 15–22
24. Restás Á (2021) Combustion and fire fighting theory. Textbook. Ludovika University of Public Service, pp 23–35
25. Óbuda University (2023) Drone pilot training. https://uni-obuda.hu/szakok/pilota-nelkuli-legijarmu-uzemelteto-szakmernok-szakember. Accessed 10 Dec 2024
26. General information about converters operating on a special principle. https://tudasbazis.sulinet.hu/hu/szakkepzes/gepeszet/gepeszeti-szakismeretek-3/kulonleges-elven-mukodo-atalakitok/reed-csovel-torteno-erzekeles. Accessed 10 Dec 2024
27. Patocskai T (2011) Testing the performance and energy efficiency of networks supplemented with mobile relays. TDK thesis, Budapest University of Technology and Economics, pp 2–10
28. MTA Migration Working Group (2015) Factors, trends and prospects of migration to Europe, which has accelerated since 2015. Analysis, pp 8–31
29. Metalcom Ltd. Educational material, pp 1–32
30. Molnár P (2014) Multicast stream reception with ethernet controller. TDK thesis, Óbuda University, pp 12–20
31. Baukó L (2023) Summary report on the Hungarian technical border closure, pp 1–10
32. Besenyő J, Kármán M (2022) Changing Nigerian migration trends and its Hungarian context. J Cent East Eur Afr Stud 2(1):148–166. https://doi.org/10.59569/jceeas.2022.2.1.90
33. Wertz R, Stryckman B, Besenyő J, Jasani G (2024) A comparative study of terrorist attacks targeting healthcare in North America vs Europe, vol 18, no 4. Cambridge University Press. https://doi.org/10.1017/dmp.2024.235
34. Besenyő J, Issaev L, Korotayev A (2024) New perspectives on North Africa and the Sahel Region. In: Terrorism and political contention, pp 1–17. https://doi.org/10.1007/978-3-031-53429-4. Accessed 17 Dec 2024

Chapter 15
Side-Channel Analysis for Hardware Trojan Detection: A Deep Learning Approach

Piroska Stanić Molcer and Ahmed Nour Abdesselam

Abstract Hardware Trojans pose a significant threat to the security and integrity of integrated circuits, particularly in the context of complex System-on-Chip designs that incorporate third-party intellectual property. This paper presents a novel approach to hardware Trojan detection using deep learning techniques applied to side-channel analysis. We employ a one-dimensional convolutional neural network to classify electromagnetic side-channel signals from AES encryption cores, aiming to detect the presence of hardware Trojans. Our method leverages the Trust-Hub dataset, focusing on AES-T1300 and AES-T1400 benchmarks with varying trigger conditions and payloads. The proposed CNN architecture demonstrates high accuracy in distinguishing between Trojan-free and Trojan-infected circuits. This research contributes to the field of hardware security by offering a non-invasive, machine learning-based approach to Trojan detection that can be applied in real-world scenarios where golden chip references are unavailable. The success of our method underscores the potential of deep learning in enhancing the security of integrated circuits against sophisticated hardware-level threats.

Keywords Deep learning · Supervised machine learning · Hardware trojan detection

P. S. Molcer (✉)
College of Applied Sciences Subotica Tech, Subotica, Serbia
e-mail: piroska@vts.su.ac.rs

A. N. Abdesselam
Free University of Bozen-Bolzano, Bozen-Bolzano, Italy
e-mail: AhmedNour.Abdesselam@student.unibz.it

T. A. Kovács and I. Fürstner (eds.), *Critical Infrastructure Protection: Advanced Technologies for Crisis Prevention and Response*, NATO Science for Peace and Security Series C: Environmental Security, https://doi.org/10.1007/978-94-024-2308-2_15

15.1 Introduction

15.1.1 Embedded System Vulnerability

While software threats mainly impact the embedded systems used in IoT, hardware threats are an issue in all types of embedded products: consumer, IoT, mobile, automotive, medical, and industrial. System on Chip (SoC) is the main building block of embedded devices. It is complex in its architecture, may contain an ARM processor, memory, a memory protection unit, a Direct Memory Access (DMA) engine, a nested interrupt vector controller, a debug JTAG, a wake-up controller, a Coarse Grain Reconfigurable Architecture (CBRA) accelerator, and depending on the SoC sometimes a core link which is interfacing an FPGA application processor. The SoC design flow is complicated and comprises numerous stages that incorporate both application-specific hardware design and software design processes. Hundreds of engineers are designing one SoC and multiple parts are acquired from integrators.

To reduce the design costs, as well as the time to market, SoC integrators acquire pre-verified hard or soft Intellectual Property (IP) blocks from a large pool of untrustworthy vendors. These IPs largely serve as black boxes, since the third-party IP vendors do not reveal the design information to maintain competitive advantage in the market. Third-party IP vendors often make use of untrustworthy and sometimes unverified automation tools for the design and verification of their IP blocks. Use of unreliable software tools can lead to design discrepancy and open the door for malicious backdoors. Due to the rising verification costs, SoC integrators rarely reverify the procured IP blocks.

ARM TrustZone is a security technology that provides hardware support throughout a system, allowing secure processing environments to be virtualized on top of the normal system. Introduced with the ARMv6 architecture [1], TrustZone is widely implemented in devices such as smartphones, tablets, wearables, and IoT-embedded devices. Despite its advantages, TrustZone is susceptible to inadequate hardware and software implementations [2]. TrustZone includes security extensions to the ARM System-On-Chip (SoC) architecture, covering processors, memory, and peripherals. It has been extended to the ARMv8-M architecture for the Cortex-M processor family, maintaining similar features to those for application processors. Context switching between the secure and normal worlds is performed in hardware for faster transitions. TrustZone's security extends beyond the processor, encompassing SoC peripherals like the DRAM controller, DMA, secure boot ROM, Generic Interrupt Controller, and is implemented using special hardware components such as the TrustZone Address Space Controller, the TrustZone Protection Controller, and the Dynamic Memory Controller. All the components communicate via the AXI bus, with the SoC interacting with peripherals through the AXI to APB bridge. Third-party vendors often implement SoC peripherals, and some may not fully comply with TrustZone specifications to cut costs [2]. TrustZone creates two partitions: a "normal world" for typical tasks and a "secure world" for sensitive operations, such as encryption or handling sensitive data. The normal operating system (OS), also known as

Rich OS, runs in the normal world, while a smaller Secure OS runs in the secure world. A hypervisor manages shared hardware resources, including processor registers, memory, caches, and the Memory Management Unit (MMU), allowing both environments to operate in a time-sliced manner. Switching between the secure and normal worlds involves Monitor mode, triggered by the Secure Monitor Call (SMC) instruction or hardware interrupts. Monitor mode saves the state of the current world and restores the state of the switched-to world during transitions. Introduced Non-Secure (NS) bit in the configuration register indicates whether the current operation is in the normal (non-secure) or secure world, enabling the memory controller to distinguish between memory accesses.

In [3] it is described how an accelerator with a hardware Trojan inside can break the TrustZone. The Secure Configuration Register (SCR) allows the hypervisor to configure CPU responses to fast interrupts, normal interrupts, or external data aborts when the non-secure world is active. Interrupts are channeled through Monitor mode, which determines whether they should be handled by the normal or secure world based on their configuration. Each world has its own interrupt vector routine to service the corresponding interrupts. The non-secure world can explicitly enter Monitor mode through a software-generated exception using the SMC instruction. The Accelerator Coherency Port (ACP) connects a hardware accelerator to the ARM memory subsystem. To ensure security, the hardware accelerator's address space should be restricted to prevent compromising the processor's software. This paper describes how a Hardware Trojan (HT) within a third-party IP interfacing with memory through the ACP can access the entire processor memory using physical addresses. This vulnerability allows for DMA attacks originating from an HT embedded in a hardware accelerator.

To prevent illegal peripheral access, ARM TrustZone Controllers provide access control mechanisms by monitoring bus transactions for violations. However, Trojans described in [4] do not modify or create valid bus transactions, allowing them to evade detection by TrustZone. These Trojans exploit idle bus cycles, where no properties or assertions are checked, to communicate without violating protocol checks. The scenario assumes at least one system bus component has a Trojan that uses channel information, and another Trojan that leaks data or snoops bus data. These Trojans do not interfere with valid bus transactions but repurpose existing bus signals to create a "Trojan" bus protocol, facilitating covert communication across the SoC. Trojan data uses the same lines as normal bus traffic, with additional signaling, labeled as Control, to indicate when Trojan data is being transmitted. The Leakage Conditions Logic, which is protocol-dependent, determines when it is safe to replace original bus signals with Trojan values. For the AMBA AXI4/AXI4-Lite protocol, this logic exploits the VALID/READY handshake signals, which indicate when data is valid or the receiver is ready. When the VALID signal is low, master-driven signals, like WSTRB, have no functional meaning, allowing Trojan data insertion. Preventing bus Trojans requires refining the specification to address these vulnerabilities.

The issue of hardware Trojans has garnered global attention from governments and industry due to the significant risks associated with deploying Trojan-infested ICs in sensitive applications. Traditional IC testing methods are inadequate for detecting hardware Trojans, as they focus on modeled defects and fail to uncover unmodeled malicious modifications, which are often hidden within the allowed process variations without altering the IC's functionality. Conventional post-manufacturing tests struggle to detect hardware Trojans due to their complexity, diversity, and rare triggering conditions. Trojans can be designed to activate only under very rare conditions, which are unlikely to occur during standard tests but may arise during prolonged in-field operation. Combinational Trojans activate when specific conditions occur simultaneously at certain internal nodes, while sequential Trojans act as time bombs, triggered by a sequence of rare events after extended use. Destructive testing techniques, which involve physically examining the IC layer by layer and reverse engineering the design, are well-documented and highly accurate. These methods involve chemical treatment and various scanning techniques, such as scanning electron microscopy and scanning optical microscopy. Despite their accuracy, and automation techniques [5, 6] these methods are expensive, require skilled personnel, and necessitate a dedicated laboratory [7].

15.1.2 Related Work

Protection against hardware Trojans has been widely explored by researchers through multiple different approaches. The first approach is specialized functional testing. This method focuses on triggering unknown Trojans and observing their effects at the output ports. It involves generating test vectors to activate a Trojan circuit and observe its malicious effects on the payload at the primary outputs. Both functional and structural test vectors can be applied in this approach. A novel test pattern generation technique termed multiple excitation of rare occurrence is introduced for logic testing in [8]. Structural netlist based analysis is used in [9]. Special structural checking tool was developed for tracking data flow at gate level and detect hardware Trojan in [10]. A new method based on directed graphs for extracting structural gate-level features is proposed in [11].

The second approach is side-channel analysis. This method relies on observing the effects of Trojans in physical parameters such as supply current or path delay. The analysis includes parameters like power, temperature, radiation pattern, time delay, noise. Researchers compare the physical parameter profiles of Trojan-free reference designs with those of the Design-Under-Test. Significant deviations in the latter parameters suggest possible modifications. Due to process variations, careful threshold limits must be set to avoid false positives or negatives. All side-channel analyses are based on observing the effect of an inserted Trojan on physical parameters. For example, extra gates will consume leakage power, additional switching activities will consume more dynamic power, and added gates and capacitance will increase path delay. Additionally, electromagnetic radiation due to switching activity is another parameter observed. Statistical analysis of side-channel measurements has become a prominent method for detecting hardware Trojans [12]. This technique relies on the idea that hardware Trojans will subtly distort the side-channel parametric profile of an integrated circuit (IC), even if the IC's functionality remains unchanged. These distortions, although small and hidden within design margins, are systematic and detectable through statistical analysis. One advantage of side-channel analysis is its effectiveness in detecting Trojans that do not cause observable malfunctions in circuits. However, a significant drawback is that large process variations in modern nanometer technologies and measurement noise can mask the effects of Trojan circuits. Modeling of process drift for HT detection without a golden model is described in [13]. A deep learning model is developed for HT detection and identification in [14]. Research on HT detection in cryptographic circuit is described in [15]. Principal component analysis combined with multi-variant execution environment as well as support vector machine is used on side channel data to differentiate between Trojan-free and Trojan-infected chips. A golden model free neural network assisted approach for HT detection is described in [16]. In [17] feature scoring was accomplished after feature extraction from gate level netlist. The best features were feeded for training to an XGBoost classification model.

15.2 Methods

15.2.1 Dataset

The hardware Trojan dataset on Trust-Hub.org is a valuable resource for researchers and professionals in the field of hardware security. The dataset includes a wide range of hardware Trojans with different characteristics, behaviors, and insertion methods. This variety allows researchers to study different types of threats and develop countermeasures. The dataset contains benchmark circuits that have been intentionally infected with hardware Trojans. These benchmarks are widely used in the research community to evaluate the effectiveness of detection and mitigation techniques. Trust-Hub encourages collaboration among researchers by providing a platform to share findings, techniques, and tools. The research described in this paper uses the dataset provided with [18] hosted on IEEE DataPort (TM) [19] for hardware Trojan (HT) detection which consists of power and electromagnetic (EM) side-channel signals from a series of benchmark tests. These benchmarks, derived from Trust-Hub, focus on the AES (Advanced Encryption Standard) encryption core circuit. The data collection conditions include measurements of both power consumption and EM radiation, with twelve different benchmarks used, each featuring specific trigger conditions and payloads. Data is collected under three conditions: with the Trojan disabled (present but not enabled), enabled (present and enabled but not triggered), and triggered (present, enabled, and triggered). Input vectors for the AES core are generated in two ways: using a fixed value repeatedly or using the output of one encryption cycle as the input for the next, creating a sequence of varied input vectors. Measurements are taken at various temperatures (25, 35, 45, 55, 65, 75, and

85 °C) to account for temperature effects on side-channel emissions. In this research datasets named AES-1300-1, AES-T1300-2, AES-T1400-1 and AES-T1400-2 were used. In all datasets Trojan payload was leaking the key by increasing dynamic power consumption. The difference between the used datasets is that with AES-T1300 the trigger condition was getting a predefined number as input and in case of the AES-T1400 dataset the trigger condition was getting a predefined sequence of numbers as input. Each dataset consists of 10000 time series. In each time series there are 2500 samples.

When examining the datasets differing in the last number 1 or 2, the primary difference lies in the method used to generate the input plaintext for the AES encryption process. This distinction can significantly impact the analysis of side-channel signals, especially in the context of security evaluations and the study of potential vulnerabilities. In the first dataset (labeled 1), the input plaintext is dynamically generated based on the "next input = current output" method. The initial input data is set to zero, and the input for each subsequent cycle is the output ciphertext from the previous cycle. This creates a sequence of plaintexts that evolve over time.

The difference of power signals is higher when the input is dynamically generated (Fig. 15.1) than in the case when the input signal is constant (Fig. 15.2). In Fig. 15.3 power signals of HT triggered and HT enabled are shown for a dynamically generated input revealing a very slight difference in amplitude.

Histograms of power capture using AES-T1400 dataset with dynamically generated input for cases Trojan enabled, disabled and triggered state is shown in Fig. 15.4.

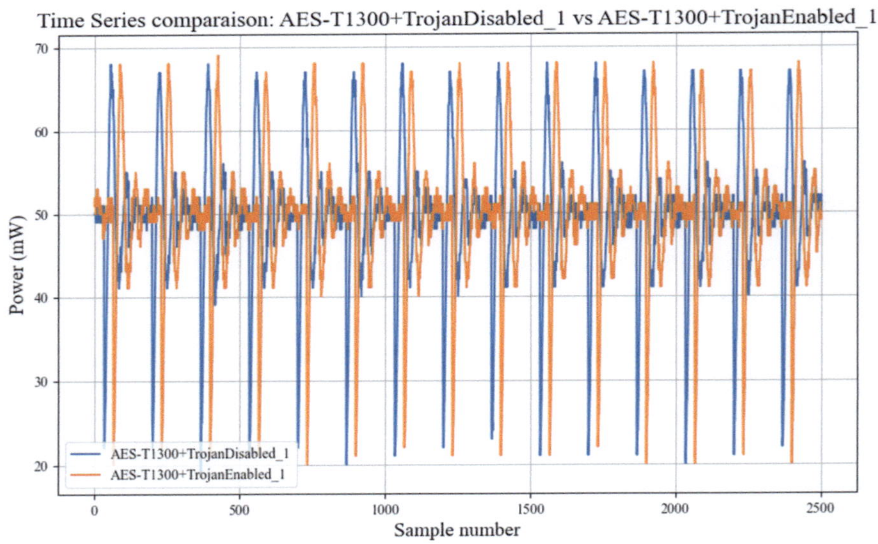

Fig. 15.1 Time series data of the encryption chip power while HT is disabled and enabled, with a dynamically generated input and specific number as a trigger condition

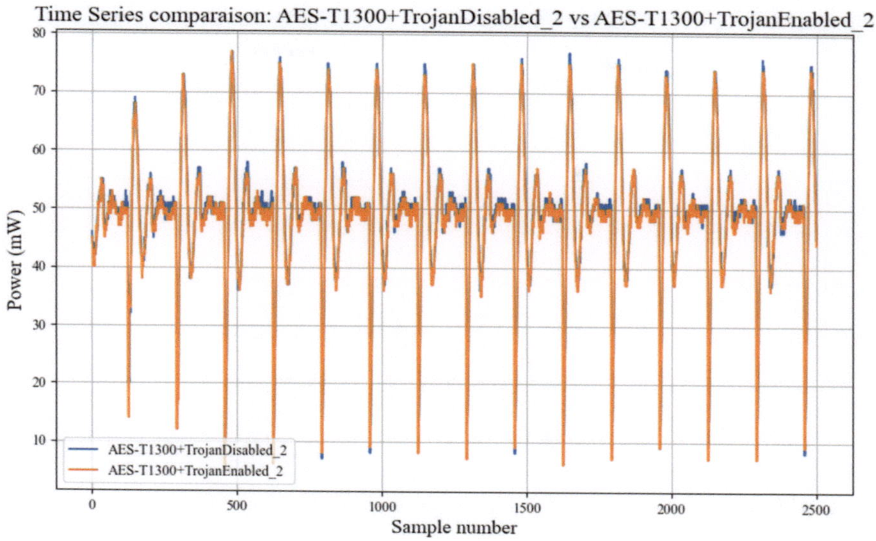

Fig. 15.2 Time series data of the encryption chip power while HT is disabled and enabled, with a constant input and specific number as a trigger condition

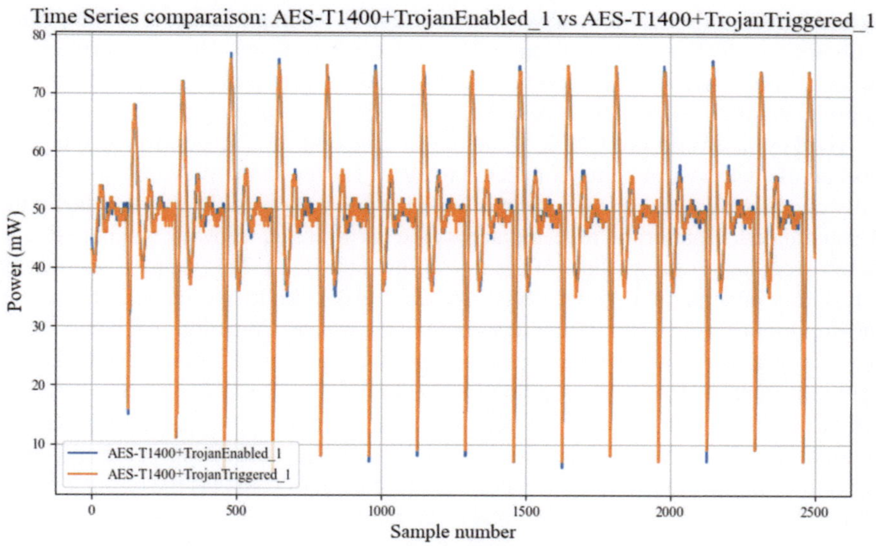

Fig. 15.3 Time series data of the encryption chip power while HT is enabled and triggered, with a dynamically generated input and specific sequence as a trigger condition

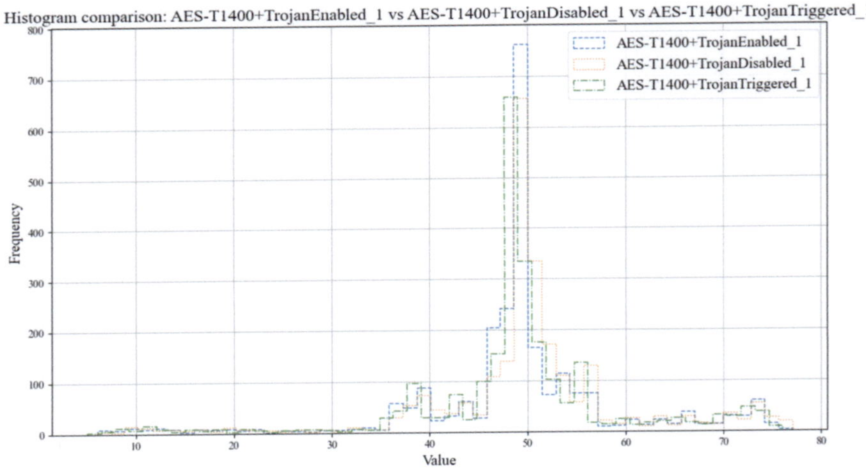

Fig. 15.4 Histograms of the power values for trojan enabled, disabled and triggered states

15.2.2 *Convolutional Neural Network*

While basic neural networks have proven effective for many tasks, they have limitations when dealing with high-dimensional data like images. This is where convolutional neural networks (CNNs) come into play. CNNs are a specialized kind of neural network designed to process data with a grid-like topology, such as images.

The key innovation of CNNs is the use of convolutional layers. In these layers, instead of each neuron being connected to every neuron in the previous layer (as in fully connected layers), each neuron is connected to only a local region of the input. This connection pattern is repeated across the entire input, allowing the network to detect local patterns regardless of their position in the input. This property, known as translation invariance, is particularly useful for tasks like image recognition where the important features can appear anywhere in the image. Convolutional layers work by sliding a small window—kernel—across the input, performing a convolution operation at each step. This operation involves element-wise multiplication between the values in the kernel and the corresponding values in the input, followed by summation. The result is a feature map that highlights the presence of specific patterns or features in the input. Another important component of CNNs is the pooling layer, which typically follows a convolutional layer. Pooling reduces the spatial dimensions of the feature maps, making the network more computationally efficient and helping to make the detected features more robust to small translations in the input. Common pooling operations include max pooling, which takes the maximum value in each local neighborhood, and average pooling, which takes the average. Usually several convolutional and pooling layers are followed by one or more fully connected layers. This architecture allows the network to progressively learn more abstract and complex features. The training process for CNNs is similar to that of standard neural networks, using backpropagation and gradient descent. However, the shared weights in the convolutional layers mean that the same feature detector is learned and applied across the entire input, significantly reducing the number of parameters compared to a fully connected network of similar depth.

CNNs have revolutionized the field of computer vision, achieving state-of-the-art performance on tasks such as image classification, object detection, and facial recognition. Their success has also led to applications in other domains, including natural language processing and speech recognition. CNNs are also used to classify time series. Authors in [20] propose using one-dimensional convolutional neural network (1D CNN) for classifying EEG time series data to diagnose epilepsy. The study introduces a sequential 1D CNN model and state-of-the-art architectures like the 1D inception and ResNet modules. These models automate and enhance the accuracy of epilepsy diagnosis, with the sequential model achieving a high ROC-AUC score of 0.98, indicating its strong classification performance and potential clinical utility. Time series classification by converting time series data into 2D texture images and utilizing deep CNN for classification is presented in [21]. Initially, the study employs the synthetic minority oversampling technique for data enhancement, specifically on arrhythmia data. The time series data is then transformed into two-dimensional texture images using recurrence plots. These images are subsequently classified using a deep CNN classifier. The proposed method effectively processes time series data by treating the classification task as an image recognition problem, demonstrating its potential through experiments on arrhythmia data. Authors in [22] introduce a hybrid model combining 1D CNN and bidirectional long short-term memory (Bi-LSTM) networks for classifying cardiac arrhythmia beats. The proposed method preprocesses electrocardiogram (ECG) signals using stationary wavelet transforms and a two-stage median filter with a Savitzky–Golay filter, followed by segmentation and z-score normalization. A six-layer 1D CNN extracts features from the preprocessed ECG signals, and a Bi-LSTM network classifies the arrhythmias. In [23] a novel approach to HT detection using Siamese neural networks (SNN) and side-channel analysis is presented. The authors propose a non-destructive, golden chip-free method that analyzes electromagnetic radiation data to detect the presence of hardware Trojans in integrated circuits. The study implements and compares three SNN architectures—Siamese CNN, Siamese GRU, and Siamese LSTM—using the Trojan Power and EM Side-Channel dataset.

In this research a supervised deep convolutional network is used to be trained to detect if the chip is infected with hardware Trojan.

The same architecture of a convolutional neural network is trained to detect if there is a hardware Trojan in the chip in a use case when the Trojan is triggered (Fig. 15.5). The architecture is built from 3 convolutional layers with pooling layers. Feature extraction is accomplished by the 3 convolutional layers, one with 64 neurons and two with 128 neurons. The classification layer is a feedforward neural network with 128 neurons in the hidden layer. Dropout technique is used to avoid overfitting at the feedforward layer.

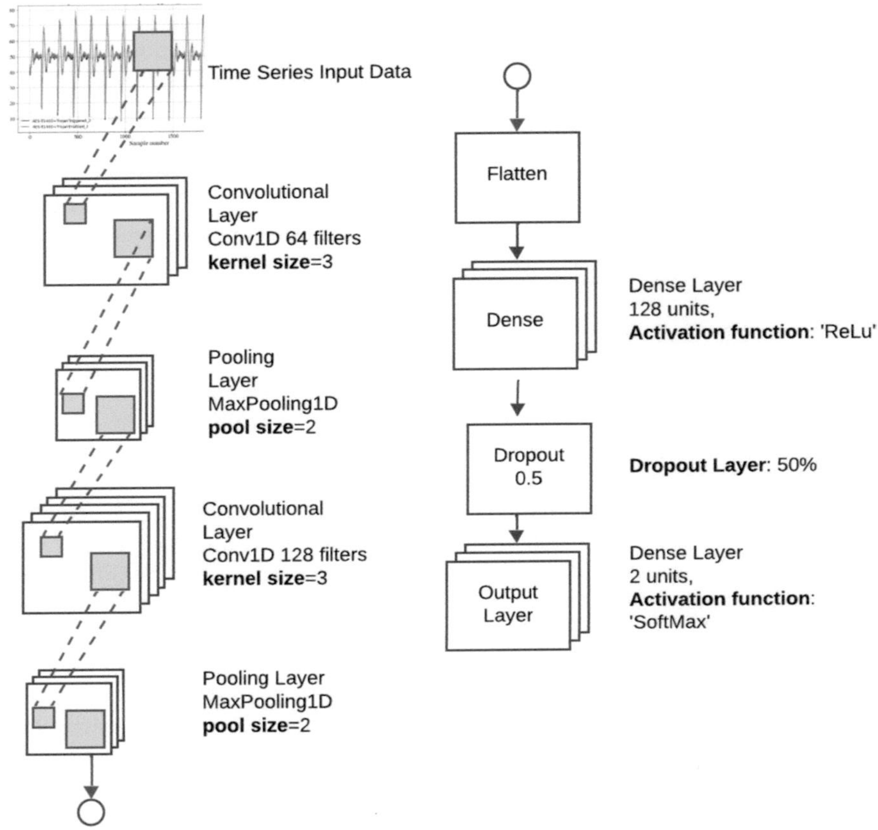

Fig. 15.5 Diagram of a neural network architecture for processing time series input data. The network consists of two convolutional layers with Conv1D, the first with 64 filters and a kernel size of 3, followed by a max pooling layer with a pool size of 2. The second convolutional layer has 128 filters and a kernel size of 3, followed by another max pooling layer with a pool size of 2. The output is flattened and passed through a dense layer with 128 units using the ReLU activation function. A dropout layer with a 50% rate is applied before the final dense output layer with 2 units and a SoftMax activation function

15.3 Results

Performance results were obtained on the test data which has never been shown to the model before. The dataset was split into 60% of training data, 20% of validation data for use while training, and the remaining 20% of data for testing to calculate the performance parameters. Validation data is used to prevent overfitting during the training phase.

The proposed 1D CNN architecture is assessed using five statistical measures. Accuracy represents the fraction of correctly classified instances in the dataset, indicating the model's overall performance in distinguishing between Trojan-free and Trojan-present instances. False Positive Rate (FPR) denotes the proportion of instances that are incorrectly classified as Trojan-present among all Trojan-free instances, highlighting the model's tendency to produce false alarms. True Positive Rate (TPR) measures the ratio of correctly identified Trojan-present instances to the total number of actual Trojan-present instances, reflecting the model's ability to detect the presence of Trojans. True Negative Rate (TNR) represents the proportion of correctly classified Trojan-free instances among all Trojan-free instances and reflects the model's ability to correctly identify Trojan-free cases. False Negative Rate (FNR) denotes the proportion of Trojan-present instances that are incorrectly classified as Trojan-free among all actual Trojan-present instances, highlighting the model's tendency to miss detecting Trojan-present instances. Precision is the ratio of correctly predicted Trojan-present instances to the total instances predicted as Trojan-present, indicating the accuracy of the model's positive predictions. Recall, also known as sensitivity, measures the proportion of actual Trojan-present instances that are correctly identified by the model, indicating its effectiveness in capturing true positives. F1 Score is a balanced metric that combines precision and recall, representing the harmonic mean of both, and accounts for both false positives and false negatives, providing a comprehensive view of the model's performance. ROC Curve, or Receiver Operating Characteristic curve, plots the TPR against the FPR or TNR against FNR at various threshold settings, providing a visual representation of the model's performance across different classification thresholds and helping to evaluate the trade-offs between sensitivity and specificity. Table 15.1 contains performance values of all the models trained by specific dataset pairs. Performance parameter ROC area under curve (ROCAUC) gives a single scalar value that summarizes the model's performance.

Table 15.1 Macro average performance values

Dataset pair	Accuracy	Precision	Recall	F1 score	ROC AUC
T1300TrojanEnabled_2 T1300TrojanDisabled_2	0.89	0.89	0.89	0.89	0.96
T1300TrojanTriggered_2 T1300TrojanDisabled_2	1	1	1	1	1
T1400TrojanTriggered_2 T1400TrojanDisabled_2	0.92	0.92	0.92	0.92	0.97
T1400TrojanTriggered_1 T1400TrojanDisabled1	0.78	0.79	0.78	0.78	0.85
T1400TrojanEnabled_1 T1400TrojanDisabled1	0.64	0.64	0.67	0.65	0.7
T1300TrojanEnabled_1 T1300TrojanDisabled_1	0.99	0.99	0.99	0.99	0.99
T1300TrojanTriggered_1 T1300TrojanDisabled_1	1	1	1	1	1

AES-T1300 benchmark employs a straightforward trigger mechanism. The hardware Trojan embedded within the AES encryption circuit is designed to activate when it receives a predefined number as input. This means that the Trojan continuously monitors the input data, and upon detecting the specific number, it initiates its malicious activity. In case of AES-T1300, the payload involves leaking the cryptographic key by increasing the dynamic power consumption of the circuit. This increase in power consumption can be observed and analyzed through power side-channel signals, providing an indirect means of extracting the key.

The simplicity of the single number trigger mechanism in AES-T1300 makes it relatively easy to implement. However, this simplicity also makes it more susceptible to detection. Security mechanisms that monitor for unusual input patterns or power consumption anomalies can potentially identify the presence of such a Trojan. In contrast, the AES-T1400 benchmark employs a more complex and stealthy trigger mechanism. Instead of a single predefined number, the hardware Trojan in AES-T1400 is triggered by a predefined sequence of numbers. This sequence must be received in a specific order for the Trojan to activate. Similar to AES-T1300, the payload in AES-T1400 involves leaking the cryptographic key by increasing the dynamic power consumption. The dataset pair of T1400TrojanEnabled2 and T1400TrojanDisabled2 was too big challenge for the model, since with T1400 HT is harder to detect, and moreover, Trojan was not triggered in this dataset, only enabled. ROC curve of TNR versus FNR for AES-T1400 for model trained with pair of data measured when Trojan is disabled and Trojan is triggered (dashed) and when one of the dataset pairs was Trojan enabled is shown in Fig. 15.6.

The use of a sequence of numbers as the trigger condition significantly enhances the stealthiness of the Trojan. Detecting a specific sequence is more challenging than detecting a single number, as the sequence can be intricate and less likely to occur accidentally. This complexity reduces the likelihood of unintended activation and makes the Trojan harder to detect using conventional security monitoring techniques.

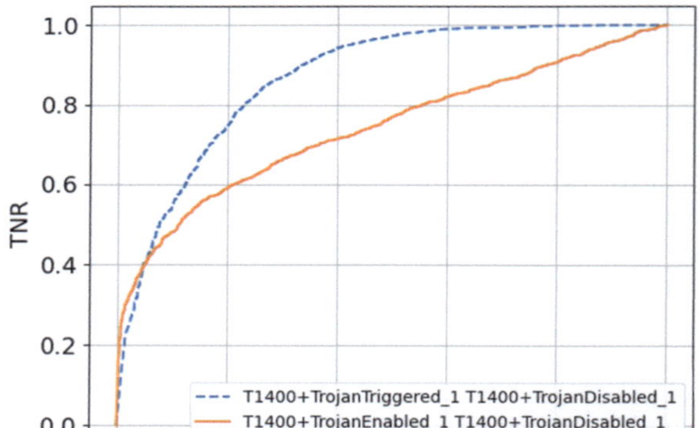

Fig. 15.6 ROC curves of models trained with trojan triggered and trojan disabled data pairs and trojan enabled and trojan disabled data pairs using the T1400 dataset

While both AES-T1300 and AES-T1400 share the same payload—leaking the cryptographic key by increasing dynamic power consumption—their trigger mechanisms set them apart. The AES-T1300's reliance on a single predefined number makes it simpler but more detectable. Security systems designed to monitor for specific input values or anomalous power consumption patterns may be able to identify and neutralize such a Trojan. On the other hand, the AES-T1400's trigger mechanism involving a predefined sequence of numbers enhances its stealthiness. The complexity of the sequence makes it less likely to be accidentally triggered and harder to detect. This makes the HT behavior captured in AES-T1400 dataset a more sophisticated and insidious threat, posing a greater challenge to security mechanisms aimed at detecting and mitigating hardware Trojans.

In Fig. 15.7. ROC curves of TNR versus FNR for AES-T1400-1 and AES-T1400-2 are shown, while Fig. 15.8 shows plots of TPR versus FPR for datasets for AES-T1400-1 and AES-T1400-2.

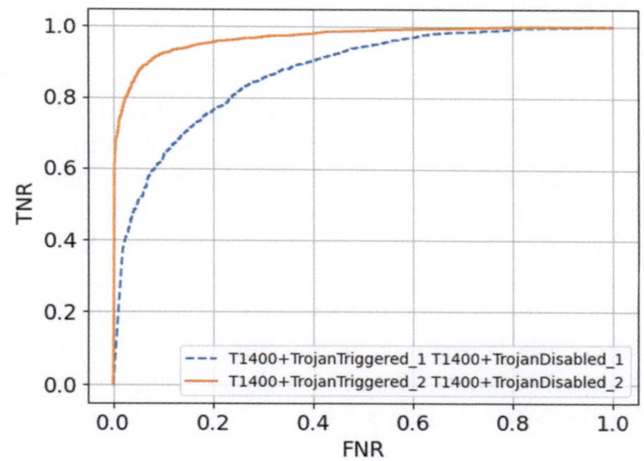

Fig. 15.7 ROC curves of models trained with trojan triggered and trojan disabled data pairs for use case 1 and use case 2 using the T1400 dataset

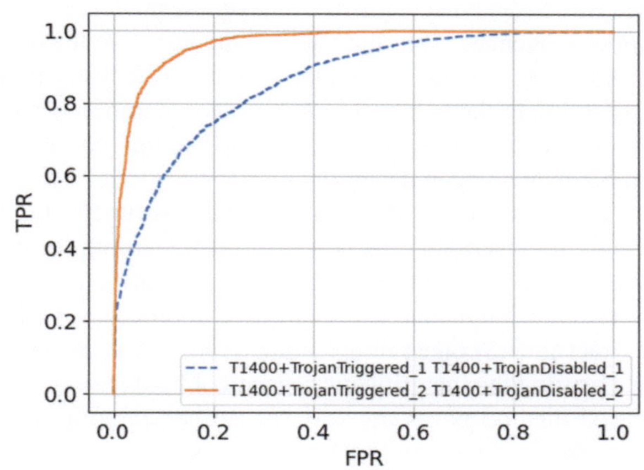

Fig. 15.8 ROC curves of FPR and TPR for models trained with trojan triggered and trojan disabled data pairs for use case 1 and use case 2 using the T1400 dataset

Using dynamically generated plaintext introduces variability into the power side-channel signals. This method reflects more realistic encryption scenarios where the input data changes over time, providing insights into how the AES encryption handles varying data. This variability can help in studying the resilience of the encryption process to differential power analysis attacks, as each cycle's power consumption will differ based on the evolving input. In the second dataset labeled 2, the input plaintext to the AES encryption code is a fixed value. Using a fixed plaintext means that the same data is repeatedly encrypted in each cycle, resulting in consistent power side-channel signals. This consistency is useful for establishing a baseline for side-channel analysis without the influence of variable inputs. It allows for easier comparison and detection of anomalies solely based on hardware behavior rather than input variability.

15.4 Conclusion

This study demonstrates the efficacy of deep learning techniques, specifically one-dimensional convolutional neural networks, in detecting hardware Trojans through side-channel analysis. Our approach, which utilizes power and electromagnetic side-channel signals from AES encryption cores, proves highly effective in distinguishing between Trojan-free and Trojan-infected circuits.

The success of our method hinges on the CNN's ability to automatically extract relevant features from the raw side-channel data. This feature extraction process is particularly powerful in the context of hardware Trojan detection, where the subtle differences between clean and infected circuits may not be immediately apparent in the raw signals. The convolutional layers in our model effectively learn to identify these distinguishing characteristics, enabling accurate classification without the need for manual feature engineering. Our results across different benchmarks (AES-T1300 and AES-T1400) highlight the model's robustness in detecting Trojans with varying trigger mechanisms, from simple predefined input numbers to more complex input sequences. This adaptability suggests that the features learned by the CNN are sufficiently general to capture a range of Trojan behaviors, yet specific enough to differentiate them from normal circuit operations. The high performance achieved by our deep learning model opens up exciting possibilities for further advancements in hardware Trojan detection.

In the broader context of hardware security, this research contributes significantly by leveraging the power of deep learning for hardware Trojan detection. As the complexity of integrated circuits continues to increase and supply chain threats persist, such machine learning-based approaches, with their ability to automatically extract and leverage subtle features from side-channel data can play an important role in ensuring the integrity and security of hardware systems. The success of our CNN-based method paves the way for further exploration and refinement of AI-driven techniques in the ongoing battle against hardware-level threats.

References

1. https://developer.arm.com/documentation/102418/0101
2. Stajnrod R, Ben Yehuda R, Zaidenberg NJ (2022) Attacking TrustZone on devices lacking memory protection. J Comput Virol Hacking Tech 1–11
3. Gross M, Jacob N, Zankl A, Sigl G (2022) Breaking TrustZone memory isolation and secure boot through malicious hardware on a modern FPGA-SoC. J Cryptogr Eng 12(2):181–196
4. Fern N, San I, Koç CK, Cheng KT (2016, March) Hardware trojans in incompletely specified on-chip bus systems. In: 2016 Design, automation and test in Europe conference & exhibition (DATE). IEEE, pp 527–530
5. Nasr AA, Abdulmageed MZ (2016) An efficient reverse engineering hardware Trojan detector using histogram of oriented gradients. J Electron Test 33(1):93–105. https://doi.org/10.1007/s10836-016-5631-z
6. SVM to reverse engineering-based hardware Trojan detection. In: Fifteenth international symposium on quality electronic design. IEEE, pp 47–54
7. Bao C, Forte D, Srivastava A (2014, March) On application of one-class SVM to reverse engineering-based hardware Trojan detection. In: Fifteenth international symposium on quality electronic design. IEEE, pp 47–54
8. Govindan V, Chakraborty RS (2018) Logic testing for hardware Trojan detection. In: The hardware Trojan war: attacks, myths, and defenses, pp 149–182
9. Liu Q, Zhao P, Chen F (2019) A hardware Trojan detection method based on structural features of Trojan and host circuits. IEEE Access 7:44632–44644
10. Le T, Di J, Tehranipoor M, Forte D, Wang L (2016, May) Tracking data flow at gate-level through structural checking. In: Proceedings of the 26th edition on Great Lakes symposium on VLSI, pp 185–189
11. Hashemi M, Momeni A, Pashrashid A, Mohammadi S (2022) Graph centrality algorithms for hardware Trojan detection at gate-level netlists. Int J Eng 35(7):1375–1387
12. Ismari D, Plusquellic J, Lamech C, Bhunia S, Saqib F (2016, November) On detecting delay anomalies introduced by hardware trojans. In: 2016 IEEE/ACM international conference on computer-aided design (ICCAD). IEEE, pp 1–7
13. Gubbi KI, Saber Latibari B, Srikanth A, Sheaves T, Beheshti-Shirazi SA, PD SM, Rafatirad S, Sasan A, Homayoun H, Salehi S (2023) Hardware Trojan detection using machine learning: a tutorial. ACM Trans Embed Comput Syst 22(3):1–26
14. Wang H, Panoff MK, Wang S, Forte D (2023, June) HT-EMIS: a deep learning tool for hardware Trojan detection and identification through runtime EM side-channels. In: Proceedings of the Great Lakes symposium on VLSI 2023, pp 51–56
15. Liu Y, Jin Y, Nosratinia A, Makris Y (2016) Silicon demonstration of hardware Trojan design and detection in wireless cryptographic ICs. IEEE Trans Very Large Scale Integr (VLSI) Syst 25(4):1506–1519
16. Vakil A, Mirzaeian A, Homayoun H, Karimi N, Sasan A (2021) AVATAR: NN-assisted variation aware timing analysis and reporting for hardware Trojan detection. IEEE Access 9:92881–92900
17. Dong C, Chen J, Guo W, Zou J (2019) A machine-learning-based hardware-Trojan detection approach for chips in the internet of things. Int J Distrib Sens Netw 15(12):1550147719888098
18. Faezi S, Yasaei R, Barua A, Al Faruque M (2021) Brain-inspired golden chip free hardware Trojan detection. IEEE Trans Inf Forensics Secur (IEEE TIFS'21)
19. Hardware Trojan Power and EM Side-Channel Dataset ieee-dataport.org/3599. A dataset of power and electromagnetic side-channel signals collected for hardware Trojan benchmarks
20. Manocha G, Rustagi H, Singh SP, Jain R, Nagrath P (2021) Epilepsy seizure classification using one-dimensional convolutional neural networks. Data Manag, Anal Innov
21. Cao W, Zhang C, Xiong Z, Wang T, Chen J, Zhang B (2021) Image classification of time series based on deep convolutional neural network. In: 2021 40th Chinese control conference (CCC), pp 8488–8491

22. Rahul J, Sharma LD (2022) Automatic cardiac arrhythmia classification based on hybrid 1-D CNN and Bi-LSTM model. Biocybern Biomed Eng
23. Nasr A, Mohamed K, Elshenawy A et al (2024) A Siamese deep learning framework for efficient hardware Trojan detection using power side-channel data. Sci Rep 14:13013

Chapter 16
Expert Application for the Explosive Cladding Process Specification

Zoltán Nyikes and Tünde Anna Kovács

Abstract The explosive cladding process is a unique welding process. It can find some research results in the literature to support engineers in determining the welding parameters. In this research, the authors analysed the method of Kriventsov et al. and built an expert application for the explosive cladding process specification. The cladding process parameters, such as the setup of the cladding (parallel setup), the stand-off distance of the setup, and the mass of the explosive material, are a function of the explosive material density and explosion speed. The expert application is a good tool for specifying the explosive cladding process parameters based on the physics of this process and the database of the materials by a suitable algorithm. The algorithm uses basic calculation models. Based on the parameters of metals and explosive materials (density, size), the stand-off distance is determined in the case of a parallel setup. The expert application for the explosive cladding process is useful for mild steel, stainless steel, copper, aluminium, titanium and other metals.

Keywords Explosion · Cladding · Expert application · Welding · Explosive material

16.1 Introduction

Welding is a traditional joining technology that has been used by industrial applications for a long time. There are two groups of welding technologies. One of the fusion welding groups, and the other one is the solid state welding [1]. The most common fusion welding processes are gas welding and gas metal arc welding processes. In the case of these processes, the base metal and the filler metal need to be melted to

Z. Nyikes (✉)
Milton Friedman University, Budapest, Hungary
e-mail: nyikes.zoltan@uni-milton.hu

T. A. Kovács
Bánki Donát Faculty of Mechanical and Safety Engineering, Óbuda University, Budapest, Hungary

T. A. Kovács and I. Fürstner (eds.), *Critical Infrastructure Protection: Advanced Technologies for Crisis Prevention and Response*, NATO Science for Peace and Security Series C: Environmental Security, https://doi.org/10.1007/978-94-024-2308-2_16

establish a metallic joint between the components. In the case of fusion, the welding process can be done without or with filler metal, which also melts and composes the joint in the welding process [2]. The other group of welding processes are solid-state welding. Solid-state welding is made without filler metal. The most common solid-state welding processes are friction welding, friction stir welding, forge welding, diffusion bonding, pressure welding, cold welding, roll welding, ultrasonic welding and explosion welding [1, 3]. In the case of these processes, the joint results from pressure alone or a combination of heat and pressure [1]. The plastic deformation makes the a solid-state welding joint. Plasticity is an essential property of metals that creates a metallic joint without melting it. As a function of temperature and the deformation rate, the microstructure of the metallic material may change. The cold working (plastic deformation under the recrystallisation temperature) causes hardening, increasing the dislocation density [4]. The high deformation rate (shock wave) in the case of some face-centred cubic crystal structure metals can cause the deformation of the crystal structure (e.g. austenitic structure change to martensite) [5]. The change in the crystal structure shows increasing in hardness [6, 7]. The Hadfield steel railway frog surface hardening is made by explosive hardening [8].

The solid-state welding over the recrystallisation temperature can't cause hardening because of the dynamic recrystallisation process (e.g. forge welding). The mechanical properties of solid-state welded joints depend on the process temperature, deformation rate, and degree of deformation [9].

The solid-state welded joints are made without melting, which is the reason for these several advantages, as they are free of solidification defects, gas porosity and other inclusions established from the melting and solidification [10]. Cold welding, forge welding, and friction stir welding require high-performance machines, which limit the size and productivity of the process. In the case of the explosion welding process, the sizes are not limited because it doesn't use any machines. The collision pressure earned from the explosion established gas pressure, which depends on the quality and quantity of the explosive material used [11]. Even the explosion welding process is a solid-state process, and in this case, it is possible to use for butt welding, but it is not a standard process [12]. Explosion welding is suitable to join metal sheets with each other. The process is also called explosion cladding technology [13]. The advantages of the explosion welding process are the cheap cost of the process and applied materials, the variety of the weldable metal pairs, and the freedom of size. Some difficulties of explosion welding are the limited availability and wide variety of properties of explosive materials, the mechanical properties of various weldable sheet metals and the required safety condition of the process. In the case of the explosive welding technology specification, the safety requirements and conditions need to be taken into account. The handling of explosive materials needs to be carried out by a certified explosive process expert [14].

Because of the welding quality assurance, a welding engineer must define the welding process parameters for the used process and materials. Explosive welding is a rarely used process. The explosive materials and the effect of the welding setup need to be known to determine suitable welding parameters.

In this work, the authors wanted to introduce the rules and equations of explosive welding parameter determination based on the literature and experimental knowledge. This work aims to make a valuable algorithm for the specification of welding process parameters based on the material properties of the wanted joint components and the available explosive materials.

16.2 The Setup of the Explosive Welding

For explosive welding, you can use oblique and parallel setups. A parallel setup is suggested for big-size sheets. The oblique setup is suitable for small components and unique metal pairs. The determination of the oblique setup parameters is the most complex and depends on several factors. For the determination of parallel setup parameters, theoretical and practical knowledge is defined clearly [15–17]. The parallel setup is shown in Fig. 16.1.

The parameters of the parallel setup are the next (Fig. 16.1)

- h (mm) thickness of the explosive material,
- s (mm) thickness of the flyer plate,
- l (mm) the stand-off distance.

The size of the stand-off distance (l) is a key parameter of the setup because it is the flyer plate impact length. The thickness of the explosive material (h) is proportional to the quantity of the explosive material. In practice, the explosive material quantity is determined by the thickness of the explosive material layer. The flyer plate thickness (s) depends on the thickness of the required cladding layer thickness. During the explosive welding process, the flyer plate thickness decreased as a function of the explosion resulting in plastic deformation.

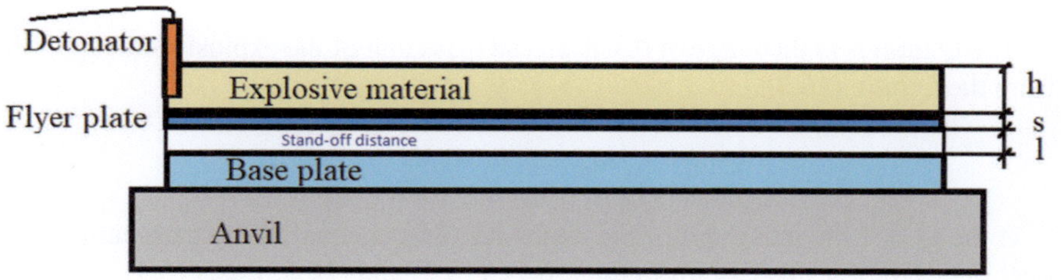

Fig. 16.1 Explosive welding parallel setup (h is the thickness of the explosive material, s is the thickness of the flyer plate, and l is the stand-off distance)

16.3 Determination of the Setup Parameters

The parallel setup is shown in Fig. 16.1. The base plate and the flyer plate thickness are defined by the required result of the welding process. It needs to be noted that the thinning of the flyer plate depends on the explosion-established pressure and the flyer plate's plasticity. In the case of a very thin flyer plate, must be used a buffer sheet between the flyer plate and the explosive material. The buffer sheet absorbs some of the pressure-caused impact energy and causes smaller deformation of the flyer plate. An important aspect also of this process is the selection of the explosive material. Common explosive materials detonation velocity (6000–7000 m/s). For explosive welding, the suitable detonation velocity is lower than the sound velocity in the used flyer metal. The velocity of the sound in the metals is between (2000–7000 m/s) [18]. The suitable explosive velocity for explosive welding is in the range of 2000–3000 m/s [15, 19–21]. On the base of the empirical tests, it needs to be noted that the used explosive material quantity (layer thickness) needs to be enough to hold the explosion along the entire length of the plate.

For the calculation of the setup parameters, the following equations:

Flyer plate mass units M_f (g/cm^2) (16.1)

$$M_f = \rho_f \cdot s \tag{16.1}$$

where

ρ_f density of the flyer plate (g/cm^3)

s flyer plate thickness (cm)

For the explosive material unit M_e determination we use the Q (-) ratio (16.2)

$$Q = \frac{M_e}{M_f} \tag{16.2}$$

The Q ratio is in the range of 0.5–1.5. The mass unit of the explosive material Me from the Q ratio (16.3)

$$M_e = Q \cdot M_f \tag{16.3}$$

If the Q = 1 the mass unit of the explosive (M_f) is equal to the mass unit of the flyer plate (M_e). If this case from the (16.1), (16.2) and (16.3) Eqs. (16.4)

$$Q \cdot \rho_f \cdot s = \rho_e \cdot h \tag{16.4}$$

In this case, the thickness of the explosive material (h) can be calculated with the Eq. (16.5)

$$h = \frac{Q \cdot \rho f \cdot s}{\rho_e} \tag{16.5}$$

where

ρ_e density of the explosive material (g/cm^3)

For the stand-off distance (l) determination it needs to use the range of $0.5 \cdot s < l < 1.6 \cdot s$, where s (cm) is the thickness of the flyer plate [22]. In practice, the stand-off distance depends on the explosion speed of the explosive material. In the case of a low-speed explosive, it can use a bigger stand-off distance and it is true for the opposite event when the explosive speed is higher the stand-off distance is smaller.

16.4 The Algorithm of the Welding Process Specification Data

Based on the introduced parameter determination, the application algorithm can be used (Fig. 16.2.).

16.4.1 Input Data

The input data are the material and geometrical parameters of the flyer plate and the parameters of the explosive material. Important to define for the user the units of the data, because the calculation can be false in the case of unsuitable data.

The input data are the next:

- ρ_e explosive material density (g/cm^3)
- s flyer plate thickness (cm)
- ρ_f flyer plate material density (g/cm^3)
- Q ratio in the range of 0.5–1.5

16.4.2 Calculation Process

For the parameter calculation, the algorithm uses the introduced questions: (1), (2), (3), (4), (5).

16.4.3 Output Parameters

The output parameters are the setup parameters for the explosive welding process (Fig. 16.1). The output parameters unit is (cm) which is useful for the process realisation.

The output data are the next:

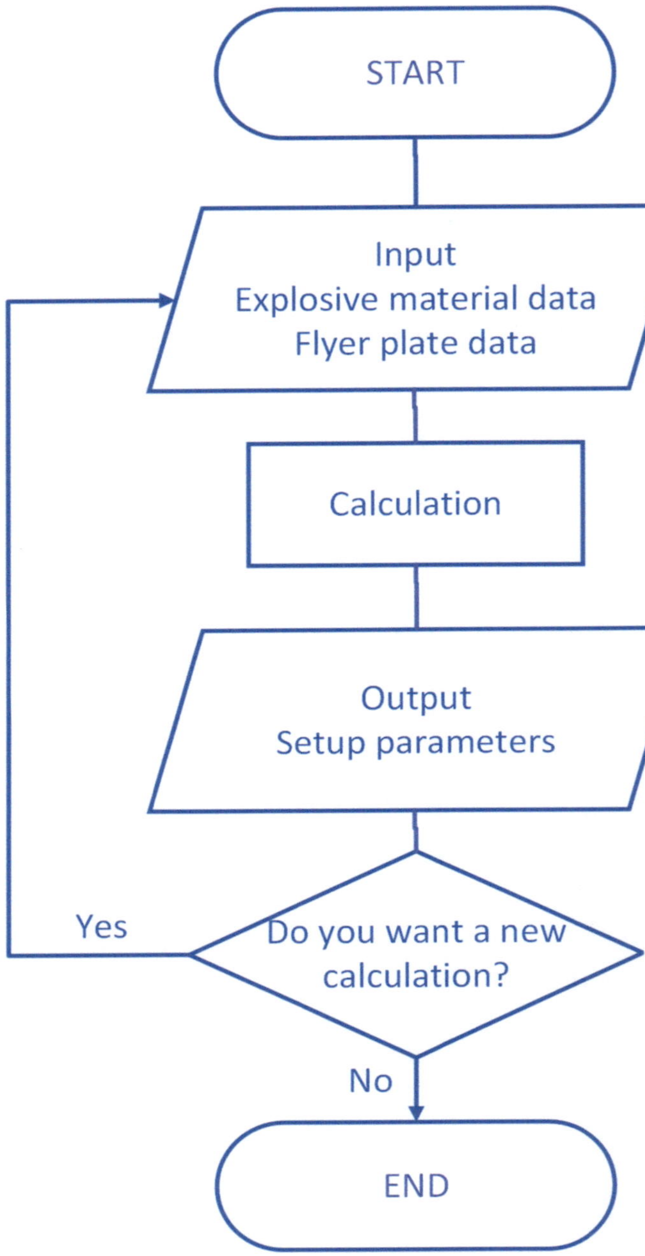

Fig. 16.2 Algorithm of the explosive welding process parameter determination

- h thickness of the explosive material (cm)
- l stand-off distance (cm)

16.4.4 New Calculation Option

An optional step in the algorithm is that the user can start a new calculation, to determine new setup parameters in the case of any modification of the parameters or to calculate new welding process parameters.

16.4.5 End of the Process

End of the algorithm, when the user wants to finish the calculation.

The algorithm is useful for the parallel setup parameters calculation, the setup shown in Fig. 16.1. The algorithm (Fig. 16.2) based application helps to calculate the setup parameters. It needs to be noted that they are useful for the test welding process because it is well known that the material parameters are only constant with a tolerance. In the case of an explosive welding process specification, it needs to follow the required steps to assure the welded joint quality [23].

16.5 Test Welding Process Specification (WPS)

Welding technology is designed using calculations and practical knowledge. When designing a technology for series production, we must follow the steps of WPS design [24]. The technology specification steps are the next: Preliminary Welding Procedure Specification (pWPS), Welding Procedure Qualification Record (WPQR) and the results of the Welding Procedure Specification (WPS). The introduced application is useful for the pWPS preparation. After the pWPS determination, it can do the test welding. The test welding is followed by the quality control. The quality control includes material testing of the test welded sample. In the case of the explosive welding process, the useful quality tests are the destructive tests (DT) and the non-destructive tests (NDT) [25].

Destructive tests [26]:

- Metallographic test
- Hardness test
- Tensile test
- Ram-tensile test
- Share test
- Bending test
- Charpy test

Non-destructive tests [27]:

- Visual test
- Eddy current test
- Penetration test
- Ultrasonic test

For the qualification of the welded joint, it needs to specify the suitable test as a function of the clad plate's required properties. In the case of the test process specification needs to take into account the application of the clad plates and the required mechanical properties.

The results of the material tests shall be evaluated and if the joint meets the requirements of the WPQR, the technological parameters are considered to be appropriate. The WPS is ready to apply for the explosive process.

The WPS for the explosive welding process needs to include all process parameters [28]. The WPS datasheet for the explosive welding process (according to the EN ISO 4063 standard the welding process designation number is 441) [29] is found in Fig. 16.3. On the base of the datasheet the welding process can be done. Important that the setup parameters and the explosive material quantity and quality fitting. It can summarize the instructions in the Notes line to give information to the welders. The sheet's surface preparation before the welding process is a key step of the process, in the note it needs to define it. The post-weld straightening technology required and the post-weld heat treatment required shall also be defined. The welding process setup parameters are determined by the algorithm but the attached data is also needed for the WPS. The datasheet includes the other needed information.

16.6 Conclusion

In this work, the authors wanted to make a useful algorithm for the determination of the parameters of the explosive welding process. Explosive welding is an inexpensive and easy-to-use process for plate cladding. The advantage is that identical and dissimilar metals can be combined. The number of possible material combinations is very large. However, the process of joining requires the use of appropriate parameters that can be determined by calculation and practical knowledge. A very wide range of explosives is available, which means different parameters.

We prepared an algorithm for the parallel explosive welding setup parameters determination. We used the mathematical models from the literature and also used the empirical knowledge for the parameter determination.

For the successful process, we prepared a suitable WPS datasheet to collect all data which are required to implement a successful high-quality explosive welding process. It is important to address the advantages and dangers of explosive welding, and these are also discussed in this paper. We plan to develop the algorithm and add data on commonly used metals and explosives available today.

Welding Procedure Specification (WPS)		***Company name***	***Date***
Base Material		**Work material**	
Base plate material Flyer plate material		Explosive material specification Holder specification	
Joint Details		**Joint Position**	
Welding process: 441 (EN ISO 4063) Joint type: cladding Manual/Mechanised: manual Base plate thickness (b) Flyer plate thickness (s) Length of the sheets		Parallel	
Joint Sketch		**Welding Setup**	
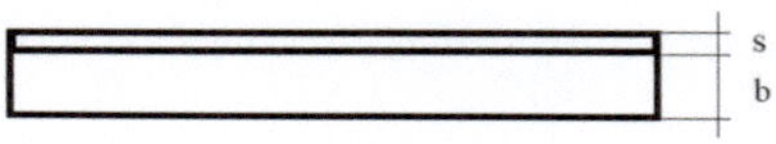		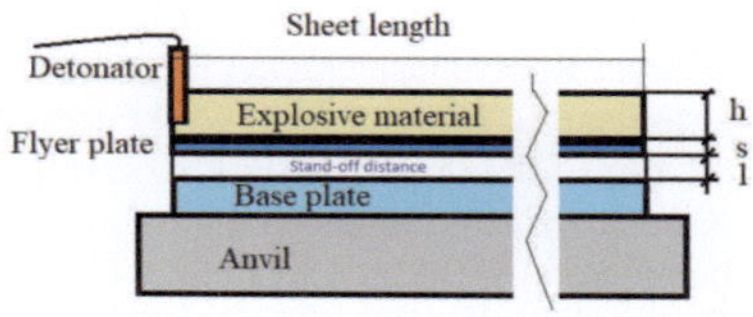	
Welding Details			
Explosive material thickness (h) cm	Stand-off distance (l) cm	Detonator ID	Explosive material specification
Post heat treatment Post straighten Quality control VT 100 %		**Notes**: 1. Holder made before welding 2. Surface preparation before welding 3. Detonator use by expert 4. etc.	

Fig. 16.3 Explosive welding WPS datasheet

References

1. Olson DL, Siewert TA, Liu S, Edwards GR (1993) ASM handbook: Welding, Brazing, and Soldering, Vol 6, 1st edn. ASM International, USA
2. Weman K (2003) Welding Processes Handbook, 1st edn. Woodhead Publishing Ltd., England
3. Khedr M, Hamada A, Järvenpää A, Elkatatny S, Abd-Elaziem W (2023) Review on the Solid-State welding of steels: Diffusion bonding and friction stir welding processes. Metals 13:54. https://doi.org/10.3390/met13010054
4. Kovács I, Zsoldos L (1973) In International Series in Natural Philosophy, Dislocation and Plastic Deformation, Vol 60, 1st edn. Pergamon, Budapest. https://doi.org/10.1016/B978-0-08-017062-6.50001-3
5. Li T, Zheng J, Chen Z (2016) Description of full-range strain hardening behavior of steels. Springerplus 5:1316. https://doi.org/10.1186/s40064-016-2998-3
6. Kovacs-Coskun T (2016) Explosive surface hardening of austenitic stainless steel. IOP Conf Ser Mater Sci Eng 123. https://doi.org/10.1088/1757-899X/123/1/012010
7. Meyers MA, Murr LE (1980) Shock waves and High-Strain-Rate phenomena in metals. In: International Conference on Metallurgical Effects of High-Strain-Rate Deformation and Fabrication. Albuquerque, NM, pp 91–111

8. Havlíček P, Bušová K (2012) Experience with explosive hardening of railway frogs from hadfield steel, Metal 2012. Czech Republic, Brno
9. Varbai B, Pickle T, Májlinger K (2018) Development and comparison of quantitative phase analysis for duplex stainless steel weld. Period Polytech Mech Eng 62(3):247–253. https://doi.org/10.3311/PPme.12234
10. Greenberg BA, Ivanov MA, Kuz'min SV, Lysak VI (2020) Explosive Welding, Processes and Structures. 1st edn, CRC Taylor ans Francis, USA
11. Lukács L, Szalay A, Zádor I (2012) Explosive forming and the aerospace. Repüléstudományi Közlemények, vol 14/2, pp 431–446
12. Szunyogh, L.: Welding and Related Technologies. 1st edn. Gépipari Tudományos Egyesület, Hungary (2007).
13. HSFL Carvalho G, Galvão I, Mendes R, M Leal R, Loureiro A (2020) Aluminum-to-Steel cladding by explosive welding. Metals 10(1062). https://doi.org/10.3390/met10081062
14. Daruka N (2024) Explosion protection for explosives manufacturer plant work processes. In: Daruka N, Ember I, Kovács ZT (eds) Drilling-Blasting international symposium special issue, Magyar Robbantástechnikai Egyesület, Budapest, pp 254–271
15. Crossland B, Bahrani AS (1968) Fundamentals, of explosive welding. Contemp Phys 9(1):71–87. https://doi.org/10.1080/00107516808204394
16. Kriventsov AN, Baluev AV, Kuz'min VI (1999) A method of calculating explosion welding parameters. Weld Int 13(11):897–899. https://doi.org/10.1080/09507119909452069
17. Sherpa BB, Rani R (2021) Explosive welding process to clad materials with dissimilar metallurgical properties. Mater Flow Anal 1–14. https://doi.org/10.5772/intechopen.94448
18. David R (2007) Lide edited: Handbook of Chemistry and Physics, 87th Edition Taylor and Francis Group LLC
19. Ivančo M, Erdélyiová R, Figuli L (2019) Simulation of detonation and blast waves propagation. Transp Res Procedia 40:1356–1363. https://doi.org/10.1016/j.trpro.2019.07.188
20. Blazynski TZ (1983) Explosive Welding, Forming and Compaction, 1st edn. Applied Science Publisher, England, Essex, pp 23–268
21. Durgutlu A, Okuyucu H, Gulenc B (2008) Investigation of the effect of the stand-off distance on interface characteristics of explosively welded copper and stainless steel. Mater Des 29(7):1480–1484. https://doi.org/10.1016/j.matdes.2007.07.012
22. Lukács L, Rácz P (2013) High energy metal working, 1st edn. Nemzeti fejlesztési ügynökség, Budapest, Hungary, pp 115–202
23. Nyikes Z, Kovács TA (2022) Experimental study of the explosive welding process applicability for train weight decreasing. Transp Res Procedia 63:2523–2528. https://doi.org/10.1016/j.trpro.2022.06.290
24. EN ISO 15609–1 standard
25. https://www.explosionbonding.com/Testing.html. Accessed 01 Jan 2025
26. Bitay E (2021) Basic Welding (original in Hungarian: Hegesztési Alapismeretek), EME, Kolozsvár, Romania
27. Halmshaw R (1996) Introduction to the Non-Destructive Testing of Welded Joints, 2nd edn. Abington Publishing, England
28. Patterson A (1993) Fundamental of explosion welding, ASM Handbook, Welding, Brazing and Soldering, ASM International, Vol 6, pp 160–164. https://doi.org/10.31399/asm.hb.v06.a0001351
29. EN ISO 4063 Standard

Chapter 17
Military and Civilian Uses of Cyber-Physical Systems in Crisis Situations

Zoltán Nyikes and László Tóth

Abstract Cyber-physical systems (CPS) are integrated systems that combine the IT and physical worlds to perform real-time data collection, analysis and automated interventions. This paper examines the role and application of CPS in various crises, such as natural disasters, industrial accidents and military conflicts. Particular emphasis is placed on key components of the systems, such as sensors, infocommunication networks, artificial intelligence and robotic systems, which enable rapid and accurate decision-making and automated interventions. Military applications such as drones, autonomous systems and cyber defence technologies offer significant advantages on the battlefield, reducing human casualties. It highlights the benefits of CPS applications, such as rapid response time, accuracy, and efficiency, as well as the challenges of cyber-attacks, technology dependency, and high costs. Future development directions, such as the development of artificial intelligence, enhancing cyber security and improving network redundancy, will contribute to increasing the reliability and efficiency of systems in crises.

Keywords Cyber-physical systems (CPS) · Crises · Automatic intervention · Military applications · Cyber security

---

Z. Nyikes (✉)
Milton Friedman University, Budapest, Hungary
e-mail: nyikes.zoltan@uni-milton.hu

L. Tóth
Banki Donat Faculty of Mechanical and Safety Engineering, Obuda University, Budapest, Hungary

T. A. Kovács and I. Fürstner (eds.), *Critical Infrastructure Protection: Advanced Technologies for Crisis Prevention and Response*, NATO Science for Peace and Security Series C: Environmental Security, https://doi.org/10.1007/978-94-024-2308-2_17

17.1 Introduction

Cyber-physical systems (CPS) are technological solutions that rely on tight integration of IT systems and the physical world, enabling real-time data collection, processing, and automatic response. Crises, such as natural disasters, industrial accidents, or military conflicts, are particularly challenging, as every moment can be crucial in managing the situation. Cyber-physical systems play an essential role in this environment, as they can provide fast and accurate information and automate interventions, reducing the burden on human resources and improving response times.

We aim to demonstrate the relevance of applying CPSs in crises, especially in the military and civilian sectors. By examining their adaptability to different crises and the solutions they offer, we will show how cyber-physical systems can support rapid response, efficiency of field interventions, and decision-making processes. The paper also considers future directions for development that will further enable the effective use of cyber-physical systems in crisis management situations as technology evolves.

17.2 The Role of Cyber-Physical Systems in Crises

Cyber-physical systems (CPS) are fundamentally based on closely integrating the information (cyber) and physical worlds. These systems not only collect and process data but are also able to act and intervene in the physical world through active feedback. In crises, these systems are paramount, as the rapid and accurate flow of information and automated decision-making profoundly impact the outcome of events.

17.2.1 Key Components of Cyber-Physical Systems in Crises

CPS are integrated systems that digitally sense and control the physical world and automatically respond to changes. In crises, cyber-physical systems play a key role by enabling the coordination of a rapid and effective response. The main components of cyber-physical systems and their role in crises are described below

Sensors and data collection devices. Sensors and data collection devices provide the basic information cyber-physical systems need about environmental changes and critical events. These devices continuously monitor the environment and transmit the collected data to processing systems.

Monitoring environmental changes. Sensors such as temperature, humidity, and motion sensors can monitor changes in the environment in real-time. This can help, for example, identify predictions before an earthquake or detect a fire hazard early [1].

Imaging systems. Such as cameras mounted on drones, surveillance systems, and satellite monitoring, can quickly scan large areas, especially in hard-to-reach or dangerous locations. They can help document disaster sites in real-time.

Specialised sensors. Sensors such as earthquake detectors, gas leak detectors and toxic detection sensors play a critical role in rapidly detecting natural or technological disasters and identifying hazards.

Infocommunications networks. Infocommunication networks ensure the rapid transmission of data and information sharing between crisis management levels. Effective communication enables decision-makers and intervention teams to coordinate their responses [2].

Real-time data transmission. The rapid transmission of data from sensors ensures that decision-makers and operational centres always have up-to-date and accurate information. Maintaining continuous info-communication is a prerequisite for a fast response.

Emergency local area networks. Mobile ad-hoc networks (MANET) and satellite communication systems (SATCOM) are critical when traditional infocommunications channels are down due to a disaster. Such networks allow the rapid exchange of necessary information and the coordination of local operations.

Data processing and decision-making. Processing and making decisions based on large amounts of data is significant in crises. Rapid data analysis and the use of predictive models allow appropriate responses.

Big Data and Artificial Intelligence (AI). Big data analysis and AI-based models enable real-time data processing, anomaly detection and rapid decision-making. For example, AI can help predict events such as future floods or model the risk of landslides after earthquakes.

Predictive modelling. Rapid forecasting allows intervention teams to prepare for further hazards and develop evacuation plans if necessary.

Actuators and robotic systems. Actuators and robotic systems are the components of cyber-physical systems that automatically take the necessary actions in crises, minimising the need for human intervention [3].

Automated intervention devices. For example, rescue drones can search for survivors after an earthquake, while fire-fighting drones can help with water or extinguishing agents.

Automatic control of physical infrastructure. Robotic systems and actuators can automatically control critical infrastructure, such as gas distribution systems, power grids, or water supply systems, to minimise the impact of water use and water resources.

17.2.2 Crisis-Specific Applications

Cyber-physical systems (CPS) play a key role in crisis management by integrating the physical world with digital systems, collecting data in real-time, and analysing and automating responses [4]. CPS-based solutions for different crises are wide-ranging and can help prevent damage, improve response times, and coordinate interventions efficiently. Below, we show how cyber-physical systems can be applied in different crises [5].

Natural disasters. Natural disasters (e.g., earthquakes, floods, and forest fires) require rapid response and extensive coordination. Cyber-physical systems can help in disaster forecasting, rescue operations, and infrastructure protection.

Forecasting and alerting. Collecting and analysing meteorological and seismological data: Sensors and data collection systems can measure environmental changes such as temperature, humidity, water levels, and seismic activity in real-time. Tsunami detection systems can detect undersea earthquakes and automatically send warnings to coastal areas.

Tsunami warning systems: by detecting seabed movements and changes in water levels, the systems can alert the public to the danger in time.

Rescue operations. Drones and robots for mapping rescue zones: Drones can quickly assess disaster-stricken areas and inaccessible or dangerous locations. Rescue drones can search for survivors under the rubble, quickly locate traffic routes, and gather basic information about the disaster site.

Infrastructure monitoring. Real-time monitoring of the stability of bridges, buildings, and dams: Sensors used in cyber-physical systems, such as structural monitoring systems, can continuously monitor the condition of buildings, bridges, dams, and other critical infrastructure. Analysis of static and dynamic data helps determine where immediate intervention is needed [6].

Technological disasters. Technological disasters, such as industrial accidents or nuclear leaks, can pose a substantial risk to the environment and human life. Cyber-physical systems play a key role in rapid and safe intervention.

Emergency shutdown. Automated systems for the safe shutdown of industrial facilities: CPSs can automatically shut down industrial plants such as oil refineries or chemical plants during an emergency to prevent explosions, leaks or other dangerous events. This prevents serious consequences before human intervention.

Debris removal. Robots and drones can remove hazardous materials such as oil, chemicals, or radioactive materials. Automated systems can provide access to dangerous areas without endangering people.

Social crises. Social crises such as mass unrest, political instability, or terrorist attacks require special attention to maintain public safety. In such situations, cyber-physical systems can help monitor the situation on the ground in real-time and coordinate the response effectively [7].

Mass incident control. Facial recognition and crowd tracking systems: These technologies can help law enforcement quickly identify potential crowd threats and detect dangerous individuals or groups. The real-time flow of data allows for rapid response in maintaining public order.

Field reconnaissance. Use of drones and robots in the field: Drones can provide police and other authorities with real-time images and data to scout the scene and make quick decisions about the need for response [8].

Cyber security crises. Cybersecurity crises, such as cyberattacks against critical infrastructures, require rapid protection of digital systems. Cyberphysical systems protect physical infrastructures and data traffic [9].

Monitoring and protecting network traffic. Detection of cyber attacks against critical infrastructures: Cyber-physical systems provide real-time monitoring to protect infrastructures, for example, to detect attacks against healthcare institutions, financial systems, government systems, IT networks, water supply systems, energy supply, or transport systems [10]. The system can trigger immediate response actions to prevent system downtime or data theft.

Ensure redundancy. Emergency network protocols: Redundant systems maintain necessary network connections during a cyber attack. Emergency network protocols help maintain information security and avoid data loss in critical systems.

17.2.3 Benefits in Crisis

Using cyber-physical systems (CPS) in crises offers several advantages that directly contribute to a rapid, accurate and effective response [11]. The most important benefits are described in detail below

Speed. In crises, the response must be fast and efficient, as every minute can be crucial to minimising damage or saving lives. Cyber-physical systems can process data in real-time, allowing a rapid response to threats.

Real-time data processing: sensors continuously measure and transmit environmental data, such as temperature, humidity or even air pollution levels, and analyse the data almost instantaneously. For example, in an earthquake or flooding, the system can immediately signal and issue first alerts.

Immediate intervention. Response times are significantly reduced, which is particularly important in natural disasters.

Accuracy. Avoiding mistakes in crises is vital. Cyber-physical systems are powered by highly accurate sensors and advanced algorithms that can accurately identify threats to ensure the correct response.

Sensor and sensor reliability. Sensors such as earthquake sensors, gas leak detectors, or cameras can detect even the smallest environmental changes. This allows them to give a highly reliable signal, such as predicting an earthquake or avoiding a flood.

Algorithms and Artificial Intelligence (AI). AI-based analytical systems can process data quickly, predict potential threats and determine the most appropriate intervention. The models help predictive analysis, which anticipates the outcome of disasters and plans the necessary response accordingly.

Efficiency. Managing human resources in crises can be extremely challenging, as access to dangerous areas and rapid intervention are often not feasible with human resources. However, cyber-physical systems allow for automated intervention, reducing the need for human resources and thus increasing efficiency.

Reduced human resource requirements. Robotic systems, such as rescue robots or fire-fighting drones, can operate in hazardous areas without directly risking human life. For example, drones can be deployed to quickly scout a fire, a chemically contaminated area, or a radiated area, while robots can perform risky remediation tasks instead of humans [12].

Automated interventions. This increases the speed of intervention while reducing labour costs and increasing response time efficiency.

17.2.4 Constraints and Challenges

Although cyber-physical systems (CPS) have many advantages in crisis management, their application also faces limitations and challenges. These are described in detail below

Vulnerability to cyber-attacks. Cyber-physical systems are particularly vulnerable to cyber-attacks because they use network-connected devices and data. Faulty data processing or system manipulation can have serious consequences, especially in crises where time and accuracy are of the essence. In some cases, systems may be under attack in cyberspace and the physical world, which is also a source of problems [13].

Manipulation and downtime. If the security of cyber-physical systems is not properly managed, external attackers can take control of the system. This can even lead to a complete system shutdown. For example, if an attacker intentionally generates incorrect data that distorts emergency decision-making in a crisis where every minute counts, manipulation or shutdown of the system can have tragic consequences.

Cyber security vulnerabilities. CPSs require network connectivity, which increases the attack surface. Weak cyber defences, such as inadequate encryption, access privileges, or vulnerabilities, can allow cyber attackers to exploit the system.

Technological dependence. With the proliferation of cyber-physical systems, society and industry are becoming increasingly dependent on technology, which can be a particular problem if these systems fail or become inoperable.

System failure. If cyber-physical systems such as drones, robots or sensors malfunction or fail, the possibility of human intervention may be limited. Over-dependence on machines and systems means that traditional human intervention or manual tools are not always available. This can be particularly dangerous if rapid human intervention is required in disaster situations.

Systems integration problems. Coordinating and integrating different cyber-physical systems and related technologies can also be problematic. If several systems are among different vendors' products, platforms, or protocols, they may not work together seamlessly, and the whole system's efficiency may be reduced.

Costs. Developing, installing, maintaining and continuously updating cyber-physical systems requires significant financial resources, which, in many cases, not all institutions or organisations can afford.

Development costs. The development and implementation of CPSs are costly processes that require the integration of the latest technologies, sensors, data processing systems, and communication protocols. Such systems often require specialised hardware, advanced artificial intelligence, and adaptive algorithms, all contributing to the costs.

Maintenance and upgrades. Cyber-physical systems require constant maintenance to ensure reliability and long-term operation. This includes software upgrades, integration of new sensors, troubleshooting of system failures, and continuous review of cyber security protocols. All of these are ongoing costs in the long term.

Deployment costs. Deploying cyber-physical systems in crises can be extremely costly, depending on the location and its compatibility with the infrastructure. Establishing and maintaining infrastructure costs can be high in industrial plants, nuclear facilities or remote areas.

17.3 Military Cyber-Physical Systems in Crisis

Military cyber-physical systems (CPS) are key to modern warfare and crisis management. These tools not only increase the success of operations but can also reduce the loss of life and damage to property. Crises, such as armed conflicts, cyber-attacks or humanitarian crises, require rapid and accurate decision-making, for which CPSs provide key data and intervention capabilities.

17.3.1 Main Types of Military Cyber-Physical Systems

Military cyber-physical systems (CPS) play a key role in modern warfare, where precision, rapid response time and complex technological solutions are critical. Such systems provide significant advantages on the battlefield through intelligent and autonomous technological solutions [14]. Below, we describe the main types of military cyber-physical systems

Autonomous systems and drones. Drones and autonomous systems are key players in modern military operations, allowing precision attacks, real-time intelligence and mapping of dangerous areas without directly risking human lives [15].

Reconnaissance drones. Reconnaissance drones take aerial photographs and provide real-time intelligence information to troops. These drones can continuously monitor the battlefield, enemy forces, and infrastructures, aiding decision-making and strategic command. They also play an important role in tracking enemy movements and identifying hidden threats such as underground bunkers or mines.

Armed drones. Armed drones, such as precision strike drones, can carry out targeted attacks with high precision, minimising civilian damage and losses of own combat personnel. These drones can be used specifically to destroy high-value targets such as strategic infrastructures, command centres and tanks.

Robots on water and land. Aquatic and terrestrial robots play a key role in military operations, as they can be used for various tasks such as mine clearance, logistical support and mapping hazardous areas. Aquatic robots are particularly important in operations in water areas, while terrestrial robots are used for battlefield reconnaissance and identifying threats in the field.

Cyber defence systems. One of the most important aspects of military cyber defence is protecting information, systems, and infrastructure from hostile cyber attacks, which can be critical in a war environment. Without adequate cyber defences, military systems can be disrupted or manipulated [16].

IT network traffic monitoring. Continuous IT network traffic monitoring and detecting hostile cyber attacks are essential during military operations. Monitoring systems can identify suspicious activity, such as DDoS attacks or attempts to steal data, and can take immediate action to stop attacks.

Data integrity protection. Protecting data in military systems, particularly by ensuring encryption and data integrity, is critical to safeguarding military communications, command and control, and intelligence. Manipulation or destruction of data can have serious implications for the success of operations and the safety of soldiers.

Offensive cyber tools. Using offensive cyber assets enables the jamming and neutralising of enemy systems, such as radar systems, info-communications channels, or command and control systems. Cyber attacks that create confusion and chaos can give the attacking party a strategic advantage and weaken the enemy's ability to fight.

Critical infrastructure monitoring systems. In modern warfare, the essential operation of infrastructure such as power, water and fuel systems is crucial for logistics and combat operations [17]. Cyber-physical systems help to ensure the stability and protection of these related systems.

Sensor networks. Various sensor networks monitor essential infrastructures such as fuel, water, and energy supply in battlefields or war zones. These sensors provide real-time data on potential problems such as burst pipes, leaks, or system overloading, allowing for rapid response and necessary maintenance.

Automated control systems. Automated control systems ensure supply chain continuity between the frontline and the hinterland. The system continuously monitors logistics routes, the status of supplies and the needs of the various military units, ensuring efficient operations and optimal resource allocation.

Real-time decision support systems. Rapid real-time data analysis and strategic planning are essential for military decision-making success. Artificial intelligence (AI) and simulations are key in this process.

Artificial Intelligence (AI). Artificial intelligence enables the rapid analysis of data and the prediction of potential threats and combat scenarios. AI can recommend the best response strategies to command centres, assess the situation of attacking forces, and optimise defensive measures.

Simulations and forecasts. Battlefield simulations and forecasts model potential scenarios, allowing commanders to simulate different situations and develop the best response measures. Simulations help predict combat outcomes, support planning, optimise strategies, and minimise risks [18].

17.3.2 Advantages in Military Operations

In military operations, cyber-physical systems offer several advantages that can significantly increase the efficiency and security of operations:

Efficiency. Automated systems speed up decision-making and reduce the chance of human error, as data is processed in real-time using algorithms, allowing commanders to make accurate decisions faster.

Minimizing risk. Instead of sending humans into dangerous environments, such as battlefields, robots and autonomous systems can replace them. This reduces casualties for military personnel while allowing dangerous tasks such as mine clearance, reconnaissance or targeted strikes to be carried out more safely.

Data integration. Combining data from different sources (drones, sensor networks, satellite imagery) allows fast and accurate decision-making. Integrated data allows for real-time situation assessment and helps develop optimal operational strategies.

17.3.3 Challenges and Risks

Using cyber-physical systems in military operations offers many benefits but is not without challenges and risks. In this case, the role of insider perpetrators becomes particularly important, as they can damage the organisation and infrastructure through insider attacks [19]. In the following, I detail the key challenges and potential risks that may affect the reliability and effectiveness of these systems [20].

The threat of cyber-attacks. Cyber-physical systems are highly vulnerable to cyber-attacks based on digital communication, data collection and processing. The following risks may arise:

Paralysis of systems. Cyber attackers may be able to paralyse systems, making them inaccessible to decision-makers, thus hindering the management and execution of operations.

Transmitting false information. Attackers can manipulate data (e.g., information collected by drones or sensors) to mislead control personnel. This can lead to erroneous decisions that can have serious consequences, such as endangering friendly forces or removing operational objectives.

Technical failures. Due to the complexity and high degree of automation of cyber-physical systems, technical failures can have serious consequences, especially in critical situations. Examples of potential problems:

Sensor failure. Accurate data collection is the basis for situational awareness. Suppose a sensor fails, such as a ground or airborne reconnaissance device. In that case, it can lead to data loss or inaccurate information that can compromise the success of current operations and thwart the entire operational plan.

Systems failure. System failure can shut down critical infrastructure (e.g., communications networks, power management, autonomous vehicles), slowing or making responding impossible.

Ethical issues. The use of cyber-physical systems raises several ethical dilemmas, in particular for autonomous systems and weapons [21]

Responsibility of autonomous weapon systems. Autonomous weapons can make autonomous decisions, such as neutralising an enemy target. However, the question arises about who is responsible for a wrong decision or civilian casualty. Clarifying the legal and moral responsibilities of using autonomous systems is essential for safe and ethically sustainable use.

Protection of human rights and international rights. Cyber-physical systems, particularly autonomous systems, raise legal issues regarding protecting human rights. How can we ensure that automated systems do not violate human rights or international humanitarian law? Autonomous systems require appropriate legal and ethical standards to ensure they can only be used in lawful and acceptable circumstances [22].

Dependence on technology. The use of cyber-physical systems can lead to significant dependence on technology. This can be particularly problematic in emergencies, where failure or hacking of technological equipment could lead to a possible complete disruption of operations

Reduced decision-making. Excessive automation, if not properly monitored, can reduce the role of human decision-making. If a system malfunctions and the decision-maker relies only on technological responses, the need for human intervention may be reduced, with potentially serious consequences in critical situations. Human intuition and experience are essential, especially in fast-changing battlefield environments.

Automation system failures. In the case of over-automation, if the system makes a mistake, decision-makers are often unable to react appropriately, as they are no longer used to direct decision-making by machines. This can pose an even greater risk when quick decisions are required.

17.3.4 Future Directions for Development

Future developments are key to increasing military cyber-physical systems' efficiency, security, and sustainability (CPS). Three important development directions are presented below, and they are expected to significantly impact the systems' operation and applicability in the future.

Network redundancy. In military operations, reliable communication and data exchange is essential. Network downtime can have serious consequences, especially during a conflict or crisis [23]. Developing network redundancy is, therefore, a priority:

Systems that can operate in the event of partial network failures. Future developments will aim to create systems that can maintain operation even if parts of the network are damaged or fail. Using mobile ad-hoc networks, satellite-based communication systems and dynamically adaptive network structures would allow the military to operate effectively in emergencies where infrastructure may be damaged [24].

Self-contained local area networks. Military units may be able to establish regional, self-contained operational networks that ensure data flow and command and control in the event of a central network failure.

Further development of artificial intelligence. Artificial Intelligence (AI) will play a key role in the future development of cyber-physical systems, particularly for military applications. AI capabilities are continuously evolving, and significant advances are expected in the following areas:

Strategic planning. Using AI to analyse complex data patterns and battlefield dynamics will enable strategic decision-making. AI will rapidly analyse massive amounts of data (e.g., satellite imagery and sensor data) and, identify key patterns, predict potential threats, and optimise combat operations [25].

Autonomous decision-making. This can lead to faster and more accurate operations, especially when human decision-making time is insufficient.

Cybersecurity improvements. Cyber security is one of the most critical factors in protecting military cyber-physical systems. Military systems are desirable targets for cyber-attacks, as breaches of these systems can significantly impact the success of operations [26]. Future developments aim to develop more advanced and effective defence mechanisms against cyber attacks

Improved attack detection and response mechanisms. New, more advanced algorithms and machine learning-based systems will be needed to detect systematic cyber-attacks. With the help of AI, systems will be able to detect cyber-attacks in real-time and respond quickly to them, minimising the damage.

Redundant security systems. Future military CPS will be able to automatically switch to safe, emergency modes if the primary layer of defence is breached.

Encryption and data protection. Advanced encryption mechanisms will ensure that military information remains confidential even during cyber-attacks. Due to the growing threat of conflict in cyberspace, continuous improvements in encryption and quantum encryption will be significant in the future.

17.4 Summary

The paper discusses the importance of Cyber-Physical Systems (CPS) in crisis management and highlights their capabilities to support crisis management through real-time data collection, analysis, and automated intervention.

Using CPS systems is particularly important in natural disasters, technological accidents, and military crises. The paper details the main components, such as sensors, info-communication networks, and robotic systems, that enable rapid response and effective decision-making.

Special attention is given to military systems such as drones and cyber defence technologies that increase precision and reduce human casualties. In addition to the benefits, challenges such as the threat of cyber-attacks and technological dependency will also be highlighted.

Future developments include further enhancing artificial intelligence, strengthening cybersecurity and increasing network redundancy to improve the effectiveness of CPS systems in crises.

References

1. Checco JC (2022) Cyber-Physical coordinated attacks: The emerging complexity of crisis management
2. Campioni L, Poltronieri F, Stefanelli C et al (2023) Enabling civil–military collaboration for disaster relief operations in smart city environments. Futur Gener Comput Syst 139:181–195. https://doi.org/10.1016/j.future.2022.09.020. https://www.sciencedirect.com/science/article/pii/S0167739X22003041. ISSN 0167-739X
3. Szatmáry R, Lazányi K (2024) Are Self-driving cars a safer solution?. In: Kovács TA, Nyikes Z, Berek T, Daruka N, Tóth L (eds) Critical infrastructure protection in the light of the armed conflicts HCC 2022. Advanced Sciences and Technologies for Security Applications, Springer, Cham. https://doi.org/10.1007/978-3-031-47990-8_39
4. Nardelli PH (2022) Cyber-physical systems: Theory, methodology, and applications. John Wiley. https://books.google.hu/books?id=6cp6EAAAQBAJ&lpg=PR11&ots=G0ZejtTUlS&dq=Military%20and%20Civilian%20Uses%20of%20Cyber-Physical%20Systems%20in%20Crisis%20Situations&lr&hl=hu&pg=PR9#v=onepage&q&f=false
5. Volkan G, Steffen P, Tony G, Frank V (2014) A survey on concepts, applications, and challenges in Cyber-Physical Systems. KSII Trans Internet Inf Syst. Korean Society for Internet Information (KSII). https://doi.org/10.3837/tiis.2014.12.001
6. Fraga-Lamas P (2017) Enabling technologies and cyber-physical systems for mission-critical scenarios. http://hdl.handle.net/2183/19143
7. Gomaa W (2021) Cyber physical systems: Prospects and challenges. In: 2021 International Mobile, Intelligent, and Ubiquitous Computing Conference (MIUCC). IEEE, Cairo, Egypt, pp 1–7. https://doi.org/10.1109/MIUCC52538.2021.9447595
8. Lazányi K (2024) Human dynamics in armed conflicts. In: Kovács TA, Nyikes Z, Berek T, Daruka N, Tóth L (eds) Critical infrastructure protection in the light of the armed conflicts HCC 2022. Advanced Sciences and Technologies for Security Applications, Springer, Cham. https://doi.org/10.1007/978-3-031-47990-8_45
9. Jin AS et al (2022) Resilience of Cyber-Physical systems: Role of AI, Digital Twins, and Edge Computing. IEEE Eng Manag Rev 50(2):195–203. https://doi.org/10.1109/EMR.2022.3172649
10. Simola J, Lehto M (2019) Effects of cyber domain in crisis management. In: Proceedings of the European conference on information warfare and security. Academic Conferences International. https://jyx.jyu.fi/handle/123456789/67101
11. Zhang D, Li C, Goh HH, Ahmad T et al (2022) A comprehensive overview of modeling approaches and optimal control strategies for cyber-physical resilience in power systems. Renew Energy 189:1383–1406. https://doi.org/10.1016/j.renene.2022.03.096. https://www.sciencedirect.com/science/article/pii/S0960148122003809. ISSN 0960-1481
12. Damm W, Hess D, Schweda M et al (2024) A reference architecture of human Cyber-Physical Systems-Part I: Fundamental concepts. ACM Trans Cyber-Phys Syst 8(1):1–32
13. Daruka N, Szalkai L (2024) Detection of explosives and explosive devices current issues. Németh Kornél. VIII. Turizmus és Biztonság Nemzetközi Tudományos Konferencia tanulmánykötet. ISBN: 9789633962862. pp 436–448
14. Kertész J, Kovács TA (2022) Importance of vehicle's unloaded weight optimization: The personal injury severity point of view. In: Kovács TA, Nyikes Z, Berek T, Daruka N, Tóth L (eds) Critical Infrastructure Protection in the Light of the Armed Conflicts HCC 2022. Advanced Sciences and Technologies for Security Applications. Springer, Cham. (2024). https://doi.org/10.1007/978-3-031-47990-8_27
15. Um JS (2019) Drones as cyber-physical systems. Springer 10:978–981. https://doi.org/10.1007/978-981-13-3741-3
16. Kertész J, Kovács TA (2022) Importance of vehicle's unloaded weight optimization: The personal injury severity point of view. In: Kovács TA, Nyikes Z, Berek T, Daruka N, Tóth L (eds) Critical Infrastructure Protection in the Light of the Armed Conflicts HCC 2022. Advanced

Sciences and Technologies for Security Applications. Springer, Cham. (2024). https://doi.org/10.1007/978-3-031-47990-8_27
17. Berek T (2022) Integrated physical protection of emergency water production facilities. In: Kovács TA, Nyikes Z, Berek T, Daruka N, Tóth L (eds) Critical Infrastructure Protection in the Light of the Armed Conflicts HCC 2022. Advanced Sciences and Technologies for Security Applications. Springer, Cham. https://doi.org/10.1007/978-3-031-47990-8_29
18. Gallagher PS, Bannan B, Blake-Plock S et al (2015) Embedding Cyber-Physical systems for assessing performance in training simulations. In: Proceedings of the I/ITSEC. https://d1wqtxts1xzle7.cloudfront.net/40961026/15263-libre.pdf?1451949213=&response-content-disposition
19. Daruka N (2024) The risks of insider attacks on critical infrastructure. Russian Aggression and European Security. ISBN: 9788382703375, pp 142–157
20. Aichaoui NEY, Kovács TA (2022) Risk assessment of industrial collaborative welding robots: A critical review of methodologies and limitations. In: Kovács TA, Nyikes Z, Berek T, Daruka N, Tóth L (eds) Critical Infrastructure Protection in the Light of the Armed Conflicts HCC 2022. Advanced Sciences and Technologies for Security Applications. Springer, Cham. https://doi.org/10.1007/978-3-031-47990-8_14
21. Halász C (2022) Current legal challenges related to social networking sites. In: Kovács TA, Nyikes Z, Berek T, Daruka N, Tóth L (eds) Critical Infrastructure Protection in the Light of the Armed Conflicts HCC 2022. Advanced Sciences and Technologies for Security Applications. Springer, Cham. https://doi.org/10.1007/978-3-031-47990-8_3
22. Négyesi I (2022) Social and ethical issues of the military use of artificial intelligence. In: Kovács TA, Nyikes Z, Berek T, Daruka N, Tóth L (eds) Critical Infrastructure Protection in the Light of the Armed Conflicts HCC 2022. Advanced Sciences and Technologies for Security Applications. Springer, Cham. https://doi.org/10.1007/978-3-031-47990-8_37
23. Toth A (2022) Industrial IoT and 5G in critical information infrastructures. In: Kovács TA, Nyikes Z, Berek T, Daruka N, Tóth L (eds) Critical Infrastructure Protection in the Light of the Armed Conflicts HCC 2022. Advanced Sciences and Technologies for Security Applications. Springer, Cham. https://doi.org/10.1007/978-3-031-47990-8_16
24. Durczy L, Lazányi K, Sziklai Z, Molnár A (2022) Legal aspects of network security. In: Kovács TA, Nyikes Z, Berek T, Daruka N, Tóth L (eds) Critical Infrastructure Protection in the Light of the Armed Conflicts HCC 2022. Advanced Sciences and Technologies for Security Applications. Springer, Cham. https://doi.org/10.1007/978-3-031-47990-8_35
25. Huszák C, Pinke P, Kovács TA (2022) Review of the maintenance strategies, and failures in the aspects of the reliability. In: Kovács TA, Nyikes Z, Berek T, Daruka N, Tóth L (eds) Critical Infrastructure Protection in the Light of the Armed Conflicts HCC 2022. Advanced Sciences and Technologies for Security Applications. Springer, Cham. https://doi.org/10.1007/978-3-031-47990-8_12
26. Barta J, Nyikes Z (2022) Cyber threats to consumers and other risks in the field of E-commerce security. In: Kovács TA, Nyikes Z, Berek T, Daruka N, Tóth L (eds) Critical Infrastructure Protection in the Light of the Armed Conflicts HCC 2022. Advanced Sciences and Technologies for Security Applications. Springer, Cham. https://doi.org/10.1007/978-3-031-47990-8_4

Chapter 18
Hydrogen in Mobility and Infrastructure

Mehmet Alican Polat and Tünde Anna Kovács

Abstract Mobility is key to our age. The number of vehicles is very high in Europe. Reducing fossil fuel-driven car CO_2 emissions is an essential goal of our continent. One solution is to use electric drive vehicles, but the accumulator's low efficiency and slow recharging limit the spread of these vehicles. The other solution is hydrogen-driven vehicles, which is a new challenge in the logistics of hydrogen. This paper discusses the logistics of hydrogen in different industries where it plays a critical role. It examines hydrogen's disadvantages, analyses the different logistic methods, and focuses on the infrastructure, design, installation, and daily operational phases of hydrogen refuelling stations.

Keywords Liquid hydrogen · Gaseous hydrogen · Hydrogen fuel cells · Risk assessment · Comparison

18.1 Introduction

The chemical industry is one of the most important sectors today, playing a crucial role in the functioning of modern society, industrial production, healthcare, agriculture, and daily life. The chemical industry is responsible for producing several basic and auxiliary materials that are essential for various sectors of the economy. Sulfuric acid, ethylene, sodium hydroxide, ammonia, chlorine, and methanol are a few of the most used chemicals in the world. Ammonia is no doubt the most used after sulfuric acid. 80% of ammonia is used in agriculture, and the rest is used in plastics, explosives, and purification of water. Methanol also has high consumption for plastics, paints, and construction materials. As you can see below, both methanol

M. A. Polat (✉)
Doctral School on Safety and Security Sciences, Obuda University, Budapest, Hungary
e-mail: mehmet.alicanpolat@uni-obuda.hu

T. A. Kovács
Bánki Donát Faculty of Mechanical and Safety Engineering, Óbuda University, Budapest, Hungary

T. A. Kovács and I. Fürstner (eds.), *Critical Infrastructure Protection: Advanced Technologies for Crisis Prevention and Response*, NATO Science for Peace and Security Series C: Environmental Security, https://doi.org/10.1007/978-94-024-2308-2_18

(1) and ammonia (2) production require hydrogen. This is the reason hydrogen is one of the top 5 reactants in the world used after water, oxygen, sulfur, and nitrogen.

$$CO_2 + 3H_2 = CH_3OH + H_2O \quad (18.1)$$

$$N_2 + 3H_2 \leftrightarrow 2NH_3 \quad (18.2)$$

18.2 Hydrogen in Industry

Hydrogen is a well-known material for industrial applications. Metal manufacturing used hydrogen many times ago for thermal cutting, welding and other processes. The transportation of hydrogen is always a problem because the hydrogen atom is the smallest one in our present knowledge. It can traverse through all materials, and this is the reason of the lost quantity of transportation and storage.

Hydrogen can be found in the industry for a wide range of applications, some of which are; Petrochemical Industry, Ammonia Production and Steel Manufacturing.

18.2.1 Petrochemical Industry

To be able to convert heavy oil products to lighter products such as diesel gasoline, naphtha hydrogen is needed for hydrocracking, and then for the purification of sulfur and nitrogen hydrogen is necessary for the desulphurisation, denitrogenating or hydrogen treating. Usually, the first reactor is for hydrogen treating, and then the second reactor is hydrocracking by saturating it with hydrogen [1]. Hydrogen in this industry is usually stored in high-pressure cylinders or large spherical tanks. It is transferred here in tubes or short pipelines if the distance is short. It is possible that the petrol industry generates its own hydrogen using naphtha, but it might come from outside as well.

18.2.2 Ammonia Production

As mentioned above, the fertiliser industry is one of the biggest industries, which needs hydrogen. Hydrogen can be produced at the site from NG and converted to ammonia, or small-scale companies can buy ammonia directly and use it in their process.

18.2.3 Steel Manufacturing

Conventional steelmaking requires carbon to be able to remove oxygen from iron ore with the following Eq. (18.3); this process is the reason for 11% of global emissions; 1 ton of steel emits 1.8 tons of carbon dioxide [2].

$$Fe_2O_3 + 3CO(g) \rightarrow 2Fe(s) + 3CO_2(g) \tag{18.3}$$

It needs to remark on the task of hydrogen in metallurgy, hydrogen as a reducing agent of iron oxides. Therefore, iron ore (Fe_2O_3) can undergo chemical reduction with hydrogen; Eq. (18.4) below shows the reaction [3]. This dramatically reduces the C print and increases the demand for hydrogen, but it is important not to forget that if hydrogen is not produced from electrolysis inhouse by the steel manufacturers, therefore there is no point; also, it is essential not to forget that hydrogen embrittlement is the huge holdup for this process to improve as embrittlement goes into steels crystal structure and increase the stress, low-temperature process, controlled hydrogen pressure, pretreatments, and more is essential to control embrittlement [4].

$$Fe_2O_3 + 3H_2 \rightarrow 2Fe + 3H_2O \tag{18.4}$$

18.2.4 Electronics

Hydrogen is needed in semiconductor manufacturing for annealing. Hydrogen is also needed for lithography; EUV lithography tools use around 600 L of hydrogen per minute [5]. Hydrogen recovery systems are playing a big role here. Hydrogen can be found in high-pressure cylinders in these facilities.

18.2.5 Other

Hydrogen can also be used in pharma such as hydrogenation in the glass industry as an oxygen controller. The amount will change depending on the scale of the industry. One other usage is la scala for experiments in HPLC, etc. Of course, this is less than the amount used in other usages. Hydrogen is usually transferred in cylinders for these purposes or pipelined or produced on the side.

18.2.6 Hydrogen in Mobility

Hydrogen in mobility shortly means $H_2(g)$, in the hydrogen storage tank, is used by fuel cell combining O_2 and this is going through the fuel cell, the protons are travelling through the electrolyte membrane, but electrons are introduced to another circuit therefore after thus electrochemical reaction Hydrogen is breaking down to protons and electrons travelling through external circuit producing heat and energy that is transferred to the electric engine, electrical engine changes this energy to mechanical energy and stores excess energy in the battery; also the braking or acceleration energy is stored in supercapacitors. This seems like a good way to reduce carbon, but there are some challenges. Once we talk about hydrogen in mobility, for now, we do talk about high-pressure hydrogen going up to 700 bars pressure or cryogenic hydrogen at around 253 °C. Hydrogen can be stored in many other ways, such as metal hybrids, carbon nanotubes, ammonia, etc. However, the most conventional and easy method is still high-pressure or cryogenic. From a transportation point of view, cryogenic hydrogen is very costly; one method left in our hands these days is high-pressure gas storage. This is the drawback of hydrogen. Another drawback of modern fuel stations is their complex structure due to hydrogen properties. One other drawback is also the production of hydrogen; if we produce it from natural gas or coal, all these carbon footprint points of view do not make sense; it is important to produce it from electrolysis or wind turbines using excess energy produced by alternative energy.

18.3 Hydrogen Cars and Hydrogen Fuel Station Infrastructure

Due to the high carbon release and environmental concerns, it became necessary to focus on alternative energies; for a long time, biomass was the main focus, but these days hydrogen has become more and more focused. Hydrogen is one of the most abundant elements in the world. Unfortunately, the energy density of the hydrogen is small, therefore more hydrogen is needed compared to other energy sources, meaning that the amount of hydrogen to run a car for 500 km is more than the amount of methane that would run that car for 500 km. In this case, we come across the drawback of the storage problem.

Hydrogen can be stored in different forms; physical storage is the still most common method of storing hydrogen. It includes gas compression, liquid cooling, and compression cooling. There is also material-based storage as well; this method basically relies on adsorption and absorption [6]. Carbon nanotubes are one way of adsorption-based storage, having a wide carbon surface and feeding hydrogen into it; once needed, the temperature difference will release the hydrogen. Absorption has the opportunity of metal hydrides, such as $LiBH_4$; the metals will absorb hydrogen in them for later usage. These are very bright ideas for the future of hydrogen for the drawbacks of storage. Thinking that having that small amount of hydrogen in a brick of hydrogen and changing this brick wherever it is needed seems very smart and easy. Unfortunately, the storage method we are using right now in mobility with hydrogen is gas and liquid hydrogen. Therefore, the next sections will focus on these two types of hydrogen cars and their challenges.

18.3.1 Cryogenic Hydrogen Cars

The energy density of hydrogen gas in 1 bar is around 10 MJ/m^3, and it can go up to only 2700 MJ/m^3, and by liquefying it, we can achieve 8700 MJ/m^3 [7]. This data shows that liquid hydrogen, in terms of energy density, is advantageous over gaseous hydrogen. Coming back to the disadvantages, we come across a very important phenomenon that will need to be focused on when it comes to talking about hydrogen cars: the boil-off effect. Sulfur oxide called boil-off of LH2 describes the cryogenic liquid in the tank starts evaporating in time and losing efficiency, not even efficiency but the H_2 (g) itself, and the prevention of this loss of LH2 in this low temperature is not possible 100% even though the vacuum super isolation [8, 9].

$$NER = \frac{Q}{\varphi * \rho l * V \tan k * \Upsilon}(5) \tag{18.5}$$

Equation (18.5) can be used to find out the boil-off loss per day in the dank. The geometry of the tank influences the boil-off effect. Spherical tanks tend to have lower surface area, therefore having better insulation performance. However, high-performance thermal insulation, such as multi-layer insulation, glass microspheres, and aerogels, is critical to reducing the boil-off effect.

Focusing on the liquid hydrogen fuel stations and the flow of hydrogen into the liquid hydrogen car. The fuel station needs infrastructure to cool it down below −253 C and the fuel nozzle has to be cryogenic then inside the car, also, the storage must be maintained to that degree. Also, as per the disadvantage of boil-off, a ventilation system in the car is needed to protect ignition; also, this boil-off has to be liquefied and reused again for the efficiency to be maintained.

One other point with the liquid hydrogen is that the hydrogen at −253 °C temperature will cause liquefaction of nitrogen and oxygen in the air (−196 °C and −183 °C respectively) and then they will freeze (−209 °C, −219 °C). This is going to cause problems in the components of all liquid fueling and also hydrogen car components as this frozen solid particle will create a multi-layer phase. The more complex issue here is the percentage of this solid-liquid phase, this depends on the volume of the liquid hydrogen, if hydrogen is more than air then the liquefied nitrogen and oxygen might mix into liquid hydrogen. Also, it is believed that 2% of hydrogen is lost in every 1 k increase [10]. The results are all dynamics dependent on many factors. Also, it is important to note that by temperature increase and decrease the density of air or hydrogen shows different values, it might be possible when the temperature is very low, the density of hydrogen might have a higher density than the air mixture for a while until the dilution is fixed or dispersed over the air.

18.3.2 Gaseous Hydrogen Cars

As discussed before, the hydrogen molecule in atmospheric pressure of 1 bar is around 0.01 mj/l, and this can be increased to 5.6 MJ/litre [11] only increasing its pressure up to 700–800 bars and this is still the half energy capacity of liquid hydrogen. This is the point where the challenge starts. As illustrated in Fig. 18.1, gas hydrogen fuel station infrastructure is a journey from production to logistics then compressing it to storage buffer tanks at 900 bars, then cooling it before dispensing to car.

This cooling process is important as hydrogen has a reverse Joule Thompson effect, once the inverse temperature is lower than the temperature of the gas in the atmospheric conditions then, that gas tends to increase the temperature with the pressure drop. This usually happens only with hydrogen and helium. Therefore, a pressure drop will be observed once the hydrogen gas at 900 bars in bulk cylinders is transferred from there to the dispenser unit. This process will cause hydrogen temperature to increase, the overheating temperature is taken as 85 °C in many safety standards, therefore chilling hydrogen is important but extra cost and efficiency loss are to be taken into account. Besides this, there is a phenomenon called embrittlement. Gaseous hydrogen atoms can come together and accumulate stress on the components or equipment of the process. This phenomenon can be in different forms: adsorption dislocation, hydrogen-enhanced decohesion, hydrogen-enhanced localised plasticity, hydrogen-induced cracking, and temperature attacks of hydrogen. These all referring different process temperatures and, pressure and process conditions.

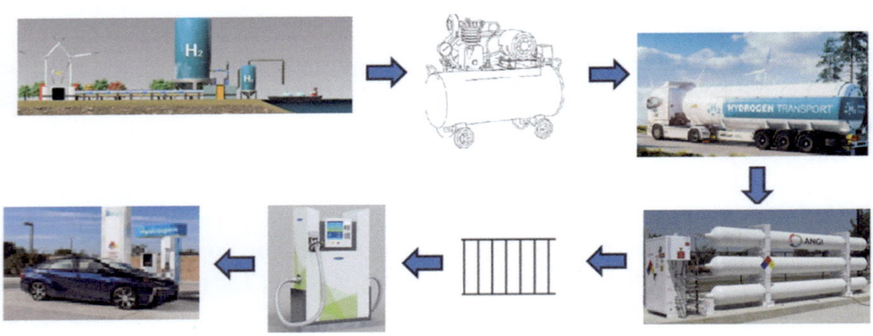

Fig. 18.1 Gas hydrogen from production to car

18.3.3 Short Risk Assessment Comparison of Liquid and Gaseous Fuel Station Scenarios

As per standards such as NFPA, IEC, Hydrogen is defined as one of the most dangerous gas groups because of its very low ignition energy and flammability. Of course, the gaseous and liquid phase has their characteristics in terms of danger. The methodology behind the two different phases of fuel stations is causing different infrastructures and a need for alternative approaches to safety systems. For example, once freeze burns are subject to liquid hydrogen, blast waves and overpressure are the topics of gaseous scenarios. As per the risk assessment of [11], catastrophic rupture of tube trailers and liquefied tanks are the worst-case accidents.

GHRS has been found to have a greater effect on surrounding people and obstructions, but a greater explosion effect was found for LHRS. The tube trailer of GHRS and the buffer tank of LHRF are the most risky parts. These risks are to be reduced with risk reduction methods such as applying Ex devices, safety barriers, more training of people, better management, safety devices, gas detectors, sensors, Ex dispensers, risk assessment for classification, correct equipment selection, etc.

18.3.4 Cost Comparison

To be able to structure the future of hydrogen technology, it is important to clearly see the cost effect of liquid or gaseous hydrogen, even if it is worth comparing it with other storage techniques. First of all, the research shows that comparing LH2 with ammonia, LH2 has higher energy efficiency, and ammonia has energy efficiency loss due to chemical processes. LH2 has lower CO_2 emissions, and LH2 costs around 5 EUR/kg; meanwhile, ammonia costs 5.9 Eur/kg in, for example, Rotterdam. LH2 cost target of Japan in 2030 is 2.5 Eur/kg [12].

What about LH2 and GH2, which one is more effective? Before this comparison, it is valuable to check Table 18.1 [13]; it can be seen that the production of hydrogen is significantly cheaper if it is produced from Natural gas using the steam reforming or gasification of coal. Still, the environmental methods are significantly expensive. As we are talking about green hydrogen, we will be focusing on the values taken from the electrolysis of hydrogen while comparing the LH2 and GH2. It is important to know these figures are changing with time, technology, location, etc.

Table 18.1 Hydrogen production cost

Production method	Cost (USD)
Natural gas	1.63–1.99
Coal	1.63–2.26
Water (Solar)	6.22–12.1
Water (Wind)	2.37–5.69

Table 18.2 Liquid hydrogen production to distribution cost

Production method	Production cost (USD)	Liquefaction cost (USD)	Transport cost (Road) (USD)	Distribution cost (USD)	Total (USD)
Green	2.55	0.53	1.07	2.3	6.45
Blue	1.22	0.53	1.23	2.3	5.28

As per the research from 2022, [14] below Table 18.2 can be obtained to see the approximate cost of green or blue hydrogen production, liquefaction cost, transport cost, distribution cost and final total amount.

Researches on gaseous hydrogen shows that the production, transportation and storage cost of gaseous hydrogen is lower compared to liquid hydrogen [15]. This is due to liquid hydrogen needing a cryogenic process, causing cost; the infrastructure of liquid hydrogen storage, thinking fuel stations, is requiring more cost as it needs extra units. It is also claimed that liquid hydrogen transportation is more beneficial if the distance is more than 300 km [15].

Gaseous hydrogen coming to the fuel station raises its price to around 35 USD; it is well known that a gaseous car such as Toyota Mirai consumes around 5 kg H2 for 600 km; therefore, the full tank is around 175 USD for 600 km, as per the comparison of Mirai with liquid hydrogen car examples, it is shown that the liquid hydrogen cars are consuming around double liquid hydrogen. Also, considering that liquid hydrogen production and refuelling are more expensive than gaseous hydrogen, gaseous hydrogen cars seem to be more advantageous than liquid ones.

One other study compared hydrogen fuel cell vehicles and hydrogen internal combustion vehicles, and it was found that hydrogen fuel cells are twice as efficient as hydrogen internal combustion engine vehicles; hydrogen fuel cell consumes 1 kg/H2, but hydrogen engines consume around 1.8 kg/H_2 [16]. It is also mentioned that if hydrogen is not produced as green hydrogen than causing more production of CO_2. Hydrogen engine produces more CO_2 than fuel cells.

18.4 Conclusion

It has already been mentioned that the production and storage methods of hydrogen mostly need high pressure or very low temperature. These extreme conditions bring up the question of safety conditions. The increase in hydrogen production also brings up the need for awareness of safety measures and even changing the law in some countries. Research on field safety, equipment selection, and personal training is growing. In case the hydrogen consumed in vehicles increases, the safety criteria of the filling stations should fall under this scope as well, and even the regulations and safety standards should be legalised.

Reminding the fact that hydrogen has reverse Joule Thompson effect, very low minimum ignition energy, hydrogen embrittlement, physical explosions such as

BLEVE or RPT, hydrogen damages such as HIC, HTHA, Hydrogen blistering, Hydrogen sulfide cracking etc. warning us to have extra safety criteria such as adding cooler for fueling stations to prevent temperature increase causing by rapid expansion. Doing hazardous area classification considering the low minimum ignition energy of hydrogen and having safety extensions. Material selection to prevent components from embrittlement. All these have to be part of liquid or gaseous fuel station infrastructure. SAE J2601-2020 standard as well as ATEX, IEC, NFPA, ISO 15916, ISO 16111:2018, and ISO TC 197 standards have to be considered.

It is also important to mention that if hydrogen is produced from coal or natural gas, then there is no point in discussion, it is valuable to improve electrolysis and utilise excess energy of wind or solar energy. Once green hydrogen is produced it is a point of discussion which technique to utilise for fuel stations and fuel cell technology. A summary, still there is a way to reduce the cost and improve safety measures.

References

1. Aitani AM Oil Refining and Products
2. Wikipedia (2024) Steel Making. In: https://en.wikipedia.org/wiki/Steelmaking
3. Liu W, Zuo H, Wang J et al (2021) The production and application of hydrogen in steel industry. Int J Hydrogen Energy 46:10548–10569. https://doi.org/10.1016/J.IJHYDENE.2020.12.123
4. Alican Polat M, Anna Kovács T Brittle Structures and Hydrogen Damage of Steels
5. Bailey C Hydrogen Recovery in EUV Lithography
6. Alican Polat M, Anna Kovács T Review of Hydrogen Properties, Production, Storage, Logistics, and Safety Criteria
7. Sorensen B (2005) Hydrogen and fuel cells. Emerg Technol Appl
8. Eberle U, Arnold G, Von Helmolt R (2006) Hydrogen storage in metal–hydrogen systems and their derivatives. J Power Sources 154:456–460. https://doi.org/10.1016/J.JPOWSOUR.2005.10.050
9. Wang H, Gao Y, Wang B et al (2024) Perspective for the Safe and High-Efficiency storage of liquid hydrogen: Thermal behaviors and insulation. Hydrogen 5:559–573. https://doi.org/10.3390/hydrogen5030031
10. Rong Y, Yuan W, Peng J et al (2024) An review of research on liquid hydrogen leakage: regarding China's hydrogen refueling stations. Front Energy Res 12
11. Yoo BH, Wilailak S, Bae SH et al (2021) Comparative risk assessment of liquefied and gaseous hydrogen refueling stations. Int J Hydrogen Energy 46:35511–35524. https://doi.org/10.1016/j.ijhydene.2021.08.073
12. Ishimoto Y, Voldsund M, Nekså P et al (2020) Large-scale production and transport of hydrogen from Norway to Europe and Japan: Value chain analysis and comparison of liquid hydrogen and ammonia as energy carriers. Int J Hydrogen Energy 45:32865–32883. https://doi.org/10.1016/j.ijhydene.2020.09.017
13. Al Ghafri SZ, Munro S, Cardella U et al (2022) Hydrogen liquefaction: A review of the fundamental physics, engineering practice and future opportunities. Energy Environ Sci 15:2690–2731
14. Suwidji P, Chung H, Ng YH (2021) Progress in practical hydrogen production and utilisation in East Asia. HKIE Trans 28:88–101. https://doi.org/10.33430/V28N2THIE-2020-0047

15. Collina G, Cantini A, Leoni L et al (2024) Hydrogen in glass sector: A comparison between Risk-Based maintenance and Time-Based maintenance approaches. IFAC-PapersOnLine 58:109–114. https://doi.org/10.1016/j.ifacol.2024.08.058
16. Durkin K, Khanafer A, Liseau P et al (2024) Hydrogen-Powered vehicles: Comparing the powertrain efficiency and sustainability of fuel cell versus internal combustion engine cars. Energies (Basel) 17. https://doi.org/10.3390/en17051085

Chapter 19
Humanitarian Response to the Basic Psychosocial Needs of Refugees

Brigitta Sáfár and Stelian Tampu

Abstract Displacement, resettlement, voluntary and forced displacement due to conflict, persecution, natural and humanitarian disasters and other crises often cause profound psychological and emotional challenges for individuals and communities alike. States are faced with the strain on their Social Infrastructure Systems every time a disaster occurs. Therefore, to mitigate the impact, humanitarian NGOs play a key role in providing psychosocial support services aimed at restoring a sense of normalcy, promoting mental well-being and facilitating the social integration of refugees and their return. This paper explores different aspects of effective humanitarian responses to basic psychosocial needs, including basic humanitarian needs, the provision of mental health services, community participation, cultural sensitivity and the creation of safe spaces. Drawing on an analysis of the activities of the Hungarian Charity Service of the Order of Malta and the Hungarian Red Cross, the research highlights and demonstrates the importance of cooperation between humanitarian organisations, host communities and refugees themselves in upholding and enforcing basic human security principles.

Keywords Refugees · Psychosocial needs · Humanitarian Aid · Social infrastructures

19.1 Introduction

Society in the twentieth and twenty-first centuries has created a very complex and dangerous world for itself, its environment and its descendants, both locally and globally. The picture that emerges for human security researchers is one of a vital space with disastrous and sometimes hopeless outcomes for the individual [1]. We

B. Sáfár · S. Tampu (✉)
Doctoral School on Safety and Security Sciences, Obuda University, Budapest, Hungary
e-mail: tampu.stelian@maltai.hu

B. Sáfár
e-mail: brigitta.safar@redcross.hu

T. A. Kovács and I. Fürstner (eds.), *Critical Infrastructure Protection: Advanced Technologies for Crisis Prevention and Response*, NATO Science for Peace and Security Series C: Environmental Security, https://doi.org/10.1007/978-94-024-2308-2_19

cannot choose a cardinal direction where we do not encounter natural disasters, displacement, war, traumatic events, severe humanitarian losses, suffering, human and non-human acts of destruction, disregard for human dignity and, in many cases, threats to children's lives [2]. In ontological terms, man is threatened by a multitude of global dangers, because the harmful and dangerous natural and humanitarian phenomena that have emerged threaten the very basis of man's existence. Human security experts now clearly see and understand that the man of our century no longer faces the same difficulties as the man of previous centuries, as the social structures of our times have changed, geopolitical strategies and interests have shifted, and the local approach to life has been replaced by a global mindset [3]. Today's disasters and refugees, voluntary and forced displacement do not take place without physical and psychological trauma and we know that the impact of all these causes serious harm to individuals and communities, especially children [4]. Disaster in itself causes severe psycho-social, but sometimes also severe ecological disturbance, which in most cases the affected individual or community is unable to cope with [5]. Wars, epidemics, famines, demographic disasters, lack of human rights, disregard for human dignity, child abuse, all can and do cause physical and psychological trauma, death and permanent loss. Considering the consequences of all this, it is possible to accurately diagnose the emergence of unwanted stressors, and therefore the specific nature of traumatic experiences and the difficulties of processing them in both children and adults requires intervention [6, 7]. It should also be noted that the encounter with different types of disasters and their specific effects can shape and determine the psychological trauma of children and adults [8]. Today, the factors of exposure, their intensity and duration are very significant, since direct contact with destruction, death and horror can be incorporated as a subjective experience in the development of the personality of both children and adults [9]. However, we must also draw attention to the dangers of exposure through the media, which can have the effect of making people immune to traumatic events, especially children. The global media often broadcast live the horrors of war and human suffering on a daily basis. Of course, it is also important to emphasise that traumatic experiences are also determined by individual factors, because age, gender and cognitive and maturational development are also important factors in the development and management of possible later psychological phenomena [10]. The determining role of community factors is also of great value. Education, the community support services that a society sets up, as well as the economic situation and the political culture and its structural existence, can be several protective factors. The existence of individual protective factors can be important, especially for children and adolescents, because the ability of an individual to recognise the potential for resistance in disadvantaged situations is a major determinant of a community's ability to survive [11]. Well-developed social skills, empathy, recognition of a person's strengths and limitations, a sense of inner control and a positive self-image, a sense of humour and religious affiliations are all necessary for resilience [12]. When a traumatic event occurs, if these skills are lacking or have faded, they need to be replaced by the psychosocial support professionals who carry out this responsible task, who enter the life of the traumatised person as a support person after the disaster has occurred [13, 14].

Observing and modelling all this helps us to understand the psychosocial impacts of disasters, which have very profound repercussions on both the human and political security structures of communities. It is clear that those fleeing disasters and poverty, those fleeing war and persecution, may be at excessive risk of psychosocial impacts, and thus a wide range of psychological impacts on individuals, families and communities can be identified [15].

19.2 The Psychological Impact of Disasters

In our human lives, there are many occasions when we have to deal with an inevitable variety of stressors. At stages of human development, such as birth and sibling birth, or perhaps the premature loss of parents, lack of love, family problems or separation of parents, can be a major stressor, but so can financial worries, an accident, illness or other unexpected unpleasant events. The same is true of natural disasters, which are usually accidental, forced or voluntary displacement, and man-made disasters, both deliberate and unintentional, which can leave each person affected with very serious, sometimes permanent, trauma [16].

War, as a humanitarian disaster, can trigger very strong fear reactions, for example an air raid lasting days or even months can be a huge shock, especially for children. Many factors can determine the psychological state of people in times of war, depending on the length of the war, the level of preparedness of the population, the social attitudes that permeate society, and the moral state of the society concerned [2]. We can see this in war-torn regions, or in the case of the Russia-Ukraine war, where a very large proportion of the population has left the country and those who are not allowed to leave the country and are even conscripted have a very high suicidal tendency. The encounter of NGOs with refugees requires first of all psycho-social assistance, when professionals listen to the traumatised, followed by the provision of other basic needs such as first aid, food, safe shelter [17].

The existence of terrorism is a constant threat, and it is more prevalent today than ever before, which means that we have to live with it. Adding to its stress is the feeling that it has no beginning and will never end. In certain geographical areas, anything can happen at any time, so fear is accompanied by a sense of uncertainty [18, 19].

Earthquakes are not stress-free phenomena either, they occur unexpectedly and cause fear shocks, they can recur and cannot be prepared for. Their consequences are often not only material but also cause many victims, as we have seen recently in the Turkey-Syria region, where many people have lost their lives, many have become homeless, and the psychological impact of all this will be felt for a very long time [20–22].

Floods are not an uncommon phenomenon in Hungary. It is a disaster that can perhaps be predicted, with the exception of sudden night-time flooding, which can cause fear of drowning [23, 24]. In such cases, the sense of hopelessness and hopelessness can reach a high level, and the possibility of recovery may be delayed, which is why intervention is very important.

Today, forest fires are also quite common and one of the most horrific deaths of mankind is being burnt alive, something we know well from history. Besides, summer forest fires are found on almost every continent, and although the number of victims is not high, the material damage, the destruction of homes and the natural damage are incalculable, a factor that can also cause frustration.

Nor can we ignore the psychological effects of chemical and nuclear disasters and epidemics, which have become common again today, because although they are not as frequent as the well-known phenomena mentioned above, when they occur, they also lead to displacement and flight [25].

Stress itself is a subjective experience known to all of us and is felt or experienced when we feel that we cannot cope with a task because we feel that the magnitude of it is beyond our capacity. The definition of stress was introduced into medicine by John Selye. He distinguishes three phases of stress: the first is when the body is confronted with stress, the second is when it copes or resists, and the third is the exhaustion phase [2].

In disasters, the first emotional and behavioural manifestations of flight are fright, terror, which triggers fear. And the consequence of persistent and intense fear is anxiety. This condition usually requires medical intervention or at least the help of a specialist, because anxiety can be followed by stress, which can turn into aggression due to the body's exhaustion, because apathy, depression and panic can lead to a state of complete hopelessness, despair and aimlessness, which almost certainly requires medical intervention. Therefore, understanding the psychological effects of disasters is a very complex and multifaceted topic in psychology, which has been and is still widely studied by researchers and professionals [26]. Disasters, whether natural or man-made, can cause significant physical, emotional and social disturbance to individuals and communities. The psychological impact of a disaster can vary depending on a number of factors, such as the type, severity and duration of the disaster, the extent of exposure and loss, the availability of social support and resources, the individual's pre-existing mental health and coping skills, and cultural factors in the affected community [27]. Therefore, it is important to assess the needs and preferences of each refugee person in order to provide appropriate and tailored interventions to promote recovery and resilience [28]. Evidence-based interventions that have been shown to be effective in reducing the psychological impact of a disaster include psychological first aid, which is a brief and practical approach, but which aims to provide immediate emotional and practical support, reduce distress and enhance coping [29]. Cognitive-behavioural therapy, a form of psychotherapy, helps people identify and challenge negative thoughts and beliefs, and learn new skills and strategies to manage emotions and behaviours. Exposure therapy is also a type of CBT

(Cognitive Behavioural Therapy), in which you gradually confront fearful or evoked stimuli, such as memories, images or situations, in a safe and controlled way to reduce anxiety. Group therapy is also a form of psychotherapy, involving meetings where people are listened to, trust is developed, and new relationships are possible [2].

19.3 Refugee and Refugee Issues Today

Two basic categories of refugees, displaced persons or disaster refugees in the twentieth century and in the twenty-first century can be distinguished according to their place of origin, and in both categories, we can speak of two types of refugees: forced and voluntary.

(1) Internally displaced persons (IDPs), disaster refugees are internal refugees of a disaster within a state, whose accommodation, basic care and supply infrastructure are most often provided jointly by the security services of the state and its accredited, contracted and qualified NGOs. If the State and its agencies cannot provide for the care of disaster refugees on their own, it may request assistance from international security agencies and organisations with which it has contractual or membership relations.

(2) We talk about external disaster refugees when a disaster in another state forces populations of people to leave their homes and seek safety in a neighbouring state, as we have seen in the case of groups of refugees from Syria to Turkey, from Ukraine to Hungary [30, 31]. However, the choice of a third state is also common, in which case the decision to seek, receive and settle in a third country is often motivated by socio-political reasons [32].

    (a) Forced displacement occurs when someone is forced by some external factor to leave their current place of residence and choose another safe space. This can be an unforeseen natural disaster such as a flood, earthquake or volcanic eruption, but it can also be a man-made disaster, whether deliberate or accidental, such as war or industrial disaster. Today, these events are becoming more and more frequent, just think of the earthquakes in Turkey-Syria and Morocco, the Russo-Ukrainian war, or the events in Syria, Iraq, Afghanistan and Africa. Millions are displaced by force, temporarily or permanently [28].

    (b) The case of voluntary refugees is becoming one of the most serious global problems of our time, mostly for economic and demographic reasons, since the hope of a better life can motivate very large numbers of people to move. Hunger, poverty and unemployment in a particular area can be described as a natural disaster, but the hope that the welcoming attitude of developed regions will change all this is a motivating factor for voluntary departure. Developing countries, as emitting regions, are also mostly the losers of voluntary emigration, or at least suffer a loss of human resources. And the host developed regions often put their social infrastructure systems under

> strain, because they also risk the arrival of masses of people who do not want to work with cheap labour, which can be interpreted as a burden [33].

The security strategy of the European Union states and the coordination of disaster-crisis situations are essentially determined by the policy framework regulated by the European Commission, the European Council and the European Parliament, and the Council Regulations on humanitarian aid set out the framework that can be used, which is not only earmarked for disaster relief in the European Union, but also for the relief of all global disasters. The EU Member States are also all members of sub-organisations run by the UN, such as OCHA, the UN Office for the Coordination of Humanitarian Affairs, through which it aids disaster victims. However, most EU Member States are also members of NATO, of which Hungary has been a member since 1999. Civil Contingency Planning (CEP) is the Alliance's (NATO) military programme that oversees the activities of the North Atlantic Disaster Response Coordination Centre (EADRCC). NATO's military alliance is increasingly linked to the security of civil society problems, which, while new challenges, in addition to operational coordination of disaster response, also include the protection of critical infrastructure and partners in the fight against climate change, and NATO has developed a broad spectrum of links with its strategic partners [34].

The UN, NATO and other security organisations around the world are constantly monitoring the world's population, keeping track of and supporting the millions of people displaced by natural and humanitarian disasters, whatever the cause: epidemics, extreme natural phenomena, war, political persecution, famine, economic hardship. At present, it is estimated that nearly 70 million people are displaced from their homes, living somewhere in temporary camps or reception facilities. According to UNHCR, in 2018, 25 people were forced to flee their homes every minute [35, 36] Africa and the Middle East have the largest refugee camps in the world, but there are also refugee reception camps all over the world [27]. Refugees arriving in Europe come from different countries and cultural backgrounds, which makes it very difficult for them to integrate, because the cultural attitudes they bring do not allow them to adopt, accept or even assimilate into another culture, so it is very important to provide psychosocial support for the new arrivals, psychological help to strengthen the factors that facilitate integration, reintegration, acceptance and openness [37].

19.4 General Definition of Basic Humanitarian Needs

In order to understand and analyse the basic nature of the humanitarian activities of NGOs, it is important to define the ideals and objectives that determine their activities [38]. Most charitable civil society humanitarian organisations subscribe to the foundations of the Universal Declaration of Human Rights, as formulated and promulgated by the United Nations, because basic human needs can be, or most often are, also basic rights. Of course, when defining basic human needs, almost everyone thinks of the well-known Maslow's pyramid scheme, because a basic need is access to food, but it can also be the desire for self-fulfilment, as it is also a human right, as is also stated in the UN Charter of Human Rights. Refugees, voluntary and forced displaced persons are also covered by the Charter of Human Rights, because all human beings are entitled to all human rights and all human beings need the most basic needs to live safely, especially children [39]. In the event of a natural disaster (Turkish-Syrian earthquake, Moroccan earthquake) or a war (Russian-Ukrainian war, Israeli-Hamas war), in most cases more children than adults are affected. For these children, traumatic events become a complex part of their personality development and may therefore have an impact on their adult life [40, 41].

Humans, regardless of age and gender, need to satisfy the most basic physiological needs, such as clean air, access to food, access to clean water, because without these, a sense of danger to life develops, which leads to neurosis, i.e. anxiety and despair, and then they try to control and prevent trouble by compulsive actions [42]. The issue of security in today's globally insecure world is not a marginal issue for societies that are often hit by natural or man-made disasters, because most phenomena such as earthquakes, floods, wars, result in the loss of human lives, or at least can cause very serious physical and psychological damage, especially to children [43]. If the basic needs listed above are lacking, other basic needs, such as self-esteem, the possibility of self-fulfilment for adults and children alike, become impossible, and it is therefore very important to provide the basic needs of refugees, the possibility of survival, physical and psychological security for children and adults alike, so that as little permanent and unavoidable trauma as possible is built into their developing personalities, in the hope that integration and resilience will be successful [28]. In addition to the state social infrastructures (law enforcement and rescue state organisations), charitable humanitarian organisations such as the Hungarian Charity Service of the Order of Malta and the Hungarian Red Cross also play an extremely important role, as their first aid activities, mobile medical clinics, food provision and safe neighbourhoods have been key psychosocial activities for decades [44, 45].

In the past, little documentation was available on the psychosocial impact of disasters, but from the 1990s onwards, psychosocial support programmes received increasing attention. In 1991, the Red Cross launched a unified programme, which was later institutionalised with the establishment of the IFRC (International Federation of Red Cross and Red Crescent Societies) Psychosocial Support Centre. This institution laid the foundations for the modern psychosocial support methodology, which was further developed jointly by the IFRC and the WHO. In 2003, the IFRC and the WHO jointly laid down the Psychosocial Support Policy, which defines the psychosocial aspects of humanitarian assistance. Based on the joint policy, an "intervention pyramid" was developed to increase the effectiveness of psychosocial support for both the affected population and the aid workers [46]. The effectiveness of support programmes depends on the participation of well-trained professionals. The Red Cross Psychosocial Support Centre, for example, develops training, surveys and new programmes that are adapted by national societies worldwide. These principles have also been applied in Hungary during various disasters and emergencies.

Experience has shown that mass acute psychiatric problems do not usually occur in disasters. Most of those affected need accurate information, practical help and social and psychological support. General support is often sufficient to help them overcome their difficulties. However, the assistance should always be tailored to the situation.

Psychosocial support always requires an assessment of the circumstances. For psychosocial support to be effective, it is necessary to consider local resources, capacity building, voluntary networks and opportunities for cooperation with other organisations. To be effective, programmes must be tailored to the needs and specificities of the community. It is the role of professionals to adapt psychosocial support programmes to local circumstances, thereby increasing the coping capacity of individuals and communities.

The long-term objectives of the programmes are:

- Support community organising.
- Establish support groups and mental health support systems.
- Facilitating support groups and support networks to create support groups and support networks for people in need.
- To increase the effectiveness of humanitarian aid.
- Addressing the mental health problems of those affected and protecting the mental health of those helping.

During the preparation period, it should be remembered that relief workers are also exposed to the effects of disasters and therefore need specific training. These trainings will help them to better respond to the needs of those in need, while also protecting their own mental health. Psychosocial support is a key element in dealing with disasters and crisis situations. Cooperation between local and international organisations and continuous professional development ensure that humanitarian assistance is effective and sustainable for both the affected population and the aid workers [47].

19.5 Humanitarian NGOs' Response to the Psychosocial Needs of Refugees

a. Response of the Hungarian Charity Service of the Order of Malta

The response of the Hungarian Charity Service of the Order of Malta to the psychosocial needs of refugees is now very broad. The history of the Order of Malta, which founded the organisation, goes back more than 900 years, and its ideas and activities have shaped European and often global developments for centuries. Over the centuries, its solidarity activities for one another have created a global human security value system, as the life-protecting activities of its institutions have never been localised or ceased but have continuously followed the course of history and human development worldwide [48, 49].

The Hungarian Charity Service of the Order of Malta has been playing a prominent role in providing psycho-social support since its establishment. The organisation has been active in the field of human security for many years in Hungary, Europe and beyond, as well as in Africa, South America and the Middle East. It makes serious efforts to support forced and voluntary refugees and vulnerable groups who face psychological and social challenges [50, 51].

The Organization's mental health services are delivered by its employed professionals and volunteers, in addition to other physiological needs such as counselling, therapy and crisis intervention. These services help refugees cope with trauma, stress and other psychological difficulties. In relation to the events of 1989, when citizens of the GDR (People's Democratic Republic of the former East Germany) fled through Hungary, the Hungarian Charity Service of the Order of Malta and the Hungarian Red Cross, alongside the state law enforcement agencies, provided the bulk of the human security tasks, both nationally and internationally. The two organisations took it upon themselves to set up and run refugee camps, with the Hungarian Charity Service of the Order of Malta setting up a refugee camp in the courtyard of the Zugligeti Parish Church and the Hungarian Red Cross taking care of the residents of the refugee camp set up in the Csillebérc orphanage. The camps were of course designed to meet basic human physiological needs, such as shelter, food and health care. But it is difficult to suppose that the social activities of the volunteers were free of spiritual talks, encouragement and counselling, as families spent their days in the camps in a very tense atmosphere, taking serious risks for themselves and their children because of the political conditions of the time. This activity is also confirmed by the after-action reports of the priests and pastors who worked in the camps. We therefore know from personal accounts that the East German citizens who accepted asylum were afraid, sometimes even terrified, and therefore found it very difficult to open up. They made every effort to leave the camps as soon as possible and to cross Austria into West Germany. The volunteer activists of the 1989 events were not qualified psychologists, but the refugees who checked in and out of the camps said in the recollective interviews that the staff were welcoming, understanding, helpful and communicative, so their fears were reduced, they found the camp life bearable, and they trusted the helpers. This is very important information because, especially because of the large number of minors, special attention was needed [51].

At the time of the war in the South, many young Hungarians, mainly of enlistment age, left the then disintegrating Yugoslavia and sought refuge in Hungary. These young people were fleeing from fear and terror, and the social workers who received them had to provide these deeply traumatised young people with the most basic humanitarian needs, such as food, shelter, health care, confidential conversations, attitudes of understanding and compassion, as well as the opportunity to talk to them in a confidential way. The organisation also contributed with ambulances and buses to the evacuation of the population of Vukovar, where the bombing had in fact caused a fear of death. The intervention staff carefully coordinated both the refugees and the evacuated residents [51].

The first conscious intervention of the Hungarian Charity Service of the Order of Malta with psychologists began with the 2010 flood in Felsőzsolca and since then, in every natural and humanitarian disaster intervention, the organisation's Emergency Response Team has consciously, prepared and professionally coordinated a team of psychologist volunteers to assist both the victims of the disaster and the intervention staff [51].

Group therapies and sessions are essential. The organisation organises several group therapy sessions where participants can share experiences and support each other. The industrial disaster in Devecser, which flooded the inner and outer areas of Kolontár and Devecser with red sludge on 4 October 2010, resulted in fatalities, many houses becoming uninhabitable and many personal injuries. The event was so sudden that the losses could not be dealt with overnight. It was important to contact the victims quickly and effectively, to follow up closely and to provide basic humanitarian needs to stay alive. Group discussions and sessions eased the anger resulting from hopelessness, but also reduced the tendency to aggression. These sessions helped the victims to form new friendships and learn new skills to start afresh [51].

In the case of the 2015 refugee wave, when hundreds of thousands of refugees and migrants from Hungary's southern border practically marched across the country, all charitable NGOs tried to provide humanitarian aid to people who were tired from the long journey, including many children and women. In this case, too, the first aid is material support to provide the necessities of life, such as food, medicines, clothing and rest. First aid was provided by the mobile medical clinics of the Hungarian Charity Service of the Order of Malta, while social volunteers and psychologists looked for opportunities to encourage and listen. Volunteer psychologists and specialist therapists were among the staff both entering and leaving the country, constantly monitoring the needs of the crowd and individuals, the exhausted looks on people's faces, and trying to make contact with refugees as much as possible [50].

In addition to political persecution, wars, industrial disasters, mass epidemics are once again appearing on a global level, and these epidemics are the triggers of fear, and this fear has been increased in people by mass media, so it was very important to set up the Hungarian Charity Service of the Order of Malta's screening stations, and the reassurance during the screenings resulted in a reduction of people's fear. And care for the elderly, not only in an institutionalised form, but also for pensioners trapped in their homes, was undertaken, was carried out and is still carried out in many municipalities [17].

Because of the Russ-Ukrainian war, Hungary has allowed and continues to allow in a large number of refugees and provides for their resettlement and care. When refugees, mostly children and women, flee a war, we have to face the fact that all basic human needs are needed. Crossing the border can bring peace of mind because their own lives are safe, but the lives of family members left at home remain precarious, the survival of young people who have been driven to the front is minimal, and the constant anxiety, stress and worry is a burden. The Hungarian Charity Service of the Order of Malta continuously monitors the lives of the families after they have crossed the border, and in addition to basic services, it organises accommodation, work, schooling and psychosocial support to make it easier for them to integrate, accept and cope with the situation. These activities contribute to the social integration of refugees [17]. Through these activities, refugees have the opportunity to integrate into new communities and adapt to a new life. They also organise regular educational programmes to raise awareness and break down taboos about mental health issues among refugees. Awareness and education is therefore one of its most important programmes. Its social integration programmes thus contribute to the social inclusion of refugees [31]. This will enable refugees to integrate into new communities and adapt to a new life. The organisation's children and youth programmes are also very important because children and young people are particularly vulnerable to psychosocial difficulties. The Hungarian Charity Service of the Order of Malta ensures that it organises programmes to help young people to deal with trauma and to develop positive emotions. These show that the organisation is committed to improving the mental and emotional well-being of refugees and vulnerable groups through its psychosocial activities. The support provided by the organisation helps refugees to cope with difficulties and to start a new life in a new environment [52–55].

b. **Response of the Hungarian Red Cross**

Psychosocial activities of the Hungarian Red Cross during the Ukrainian crisis.

The humanitarian crisis resulting from the Ukrainian-Russian conflict posed a huge challenge to the Hungarian Red Cross, which responded quickly and effectively to the psychosocial support needs of the people affected. The organisation was involved at several levels to ensure the mental and emotional stability of refugees and affected communities, and to manage the crisis. For refugees from Ukraine, the stress, trauma and insecurity caused by the war have resulted in serious psychosocial problems. The Hungarian Red Cross supported the victims through the following activities:

During the first encounters, volunteers and professionals of the organisation provided psychological first aid to help the refugees to stabilize emotionally. This included active listening, teaching stress management techniques and creating a calm and safe atmosphere. In transit points and refugee shelters, supportive community spaces were set up where refugees could interact with other stakeholders and receive social and psychological support. Safe zones were created for children, where playful and creative activities helped them to process trauma and regain a sense of normality. Hungarian Red Cross volunteers and staff have been trained to deal with war trauma and the refugee crisis. The trainings included professional skills that proved to be indispensable in the provision of assistance:

- Crisis management techniques: recognising the impact of conflict and trauma, basics of mental first aid.
- Developing cultural sensitivity: to understand the specific needs and background of Ukrainian refugees.
- Self-help and stress management: to maintain the mental health of volunteers and professionals.

The Hungarian Red Cross has implemented long-term psychosocial programmes tailored to the needs of the people concerned, with an emphasis on family reunification. Support was provided for the contact of refugees with their families, including information and reunification services. Where necessary, psychologists and social workers provided individual therapy sessions. Group discussions and support programmes helped participants to re-establish community links and process emotions. Psychosocial support became a key pillar of refugee care. To this end, long-term programmes are in place to promote the integration of refugees, including expanding access to mental health services. The commitment and training of volunteer professionals will enable effective and people-centred assistance to those in need during the Ukrainian crisis, setting an example in the management of humanitarian crises [12].

The Hungarian Red Cross has also worked closely with other international and national organisations, including the Hungarian Maltese Relief Service, to provide refugee assistance. This has enabled the effective coordination of resources and the expansion of services and accessibility for refugees.

19.6 Summary

This study examined the foundations and approaches for effective humanitarian responses to address the basic psychosocial needs of refugees. It is well known that the Social Welfare Systems of States, which are now considered critical infrastructures in terms of social and human security, are often under strain, and the depletion of staff, lack of resources and supplies in the welfare system justifies the intervention of civil society voluntary organisations, the extensive use and employment of volunteers in disaster situations, as they alleviate the burden on state systems, which can reallocate significant resources to other areas and tasks. NGOs generally do not use central budgets, relieve the burden on critical public infrastructure, can mobilise highly skilled professionals, are politically independent, may sometimes belong to a religious denomination, and have an unquestioned commitment to human security, so that organisations with international networks (Malteser International, International Federation of Red Cross and Red Crescent Societies) can be deployed in disasters anywhere in the world [56]. Therefore, timely counselling, therapy and psychological support can help individuals and groups to manage the symptoms of trauma, depression and anxiety. In addition, safe spaces created in refugee camps or host communities promote a sense of safety and facilitate access to support, which is essential in the healing process. Voluntary or forced migration often results in traumatic experiences, loss, uncertainty and disrupted social structures, which bring with them a range of mental health challenges [41]. Through their multifaceted response, humanitarian organisations provide psychosocial support to refugees to help them process trauma, rebuild social relationships and develop resilience. This paper examines the foundations and approaches for effective humanitarian responses to address the basic psychosocial needs of refugees.

References

1. János B (2017) Fences and border protection: the question of establishing technical barriers in Europe. Acad Appl Res Military Public Manag Sci 16(1):77–87. https://doi.org/10.32565/aarms.2017.1.7
2. Zellei G (1993) Katasztrófapszichológia. Credit Kiadó
3. Besenyő J, Buzan's B (2019) Securitization theory and the case of iraqi kurdish military action against ISIS in 2014. J Secur Sustain Issues 8(3):295–306, https://doi.org/10.9770/jssi.2019.8.3(1)
4. Wolmer L (2018) Preschool children facing mass trauma: disasters, war and terrorism. J Mental Health Clin Psychol 2(1):1–5
5. Tol WA, Song S, Jordans MJ (2013) Annual research review: resilience and mental health in children and adolescents living in areas of armed conflict–a systematic review of findings in low- and middle-income countries. J Child Psychol Psychiatry 54(4):445–460
6. Sim T, Garai J (2019) Natural disaster: from individual-focused to community-based psychosocial work. In: Mental health and social work, pp 1–19
7. Sharma G (2010) The vulnerability of older adults and natural disaster. In: Disaster Advances. Besenyő, J. Saharawi refugees in Algeria, AARMS 9(1):67–78, https://ssrn.com/abstract=3610819
8. McNeill I, Ronan K (2017) Children in disasters: the role of household preparedness. Natural Hazards, 89
9. Boer F et al (2009) Impact of a technological disaster on young children: a five-year postdisaster multiinformant study. J Trauma Stress 22:516–524
10. Magyar Máltai Szeretetszolgálat. Gyermekprogramokkal is segít a Máltai Szeretetszolgálat a szabadkai Befogadó Állomáson. 2017 November 23, 2023]; Available from: https://maltai.hu/cikk/hir/2643
11. Rosenkoetter MM et al (2007) Perceptions of older adults regarding evacuation in the event of a natural disaster. Public Health Nurs 24(2):160–168
12. Sáfár B (2018) A humanitárius segítségnyújtás elméleti és gyakorlati kérdései a Vöröskereszt nemzetközi tevékenységének tükrében PhD értekezés, in Egyetem Katónai Műszaki Doktori Iskola. Nemzeti Közszolgálati Egyetem Katónai Műszaki Doktori Iskola
13. Preparecenter (n.d.) Mental health and psychosocial support for migrants. November 8, 2023]; Available from: https://preparecenter.org/toolkit/humanitarian-service-points-toolkit/implementation-programmes-and-activities/mental-health-and-psychosocial-support/
14. Pfefferbaum B, Newman E, Nelson SD (2014) Mental health interventions for children exposed to disasters and terrorism. J Child Adolesc Psychopharmacol 24(1):24–31
15. Patel PP et al (2011) Transitioning mental health & psychosocial support: from short-term emergency to sustainable post-disaster development. Humanitarian Action Summit 2011. Prehosp Disaster Med 26(6):470–81
16. Najarian LM et al (2001) The effect of relocation after a natural disaster. J Trauma Stress 14(3):511–526
17. Magyar Máltai Szeretetszolgálat. Menekülteket és Migránsokat segítő szolgálat. n.d. November 23, 2023]; Available from: https://maltai.hu/menekultszolgalat
18. Fremont WP (2004) Childhood reactions to terrorism-induced trauma: a review of the past 10 years. J Am Acad Child Adolesc Psychiatry 43(4):381–392
19. Belga CR (2008) Gestión de las necesidades en caso de crisis—EURESTE Handbook, Proyecto de acción preparatoria para víctimas de terrorismo-2006. Comisión Europea, Dirección General de Justicia, Libertad y Seguridad
20. Zhang Y, Ho SM (2011) Risk factors of posttraumatic stress disorder among survivors after the 512 Wenchuan earthquake in China. PLoS ONE 6(7):e22371

21. World Health Organozation (2023) Dealing with the psychological aftershocks of the Türkiye earthquakes: why mental health and psychological support are so desperately needed. 2023 Oktober 23, 2023]; Available from: https://who.int/europe/news/item/09.03.2023-dealig-with-the-psycological-aftershoks-of-the-türkiye-earthquakes-why-mental-health-and-psychological-support-are-so-desperately-needed
22. United Nations International Children's Emergency Fund (UNICEF). For every child: Recovering from earthquakes in Syria and Türkiye. n.d. Oktober 18, 2023]; Available from: https://unicef.org/emergencies/Syria-Turkiye-earthquakes.
23. Muhoray Á, Muhoray R (2022) Áldozati mentalitás és félelmek katasztrófahelyzetekben, in Katasztrófák, következmények, segítők és áldozatok, T. Hábermayer, I. Varga, and Z. Ackermann, Editors. 2022, Tolna Megyei Katasztrófavédelmi Igazgatóság. p 10, 93–102, 145
24. Komlósi A et al (2010) Katasztrófa utáni komplex pszichológiai támogatási rendszer elvei. Javaslatok a 2010 tavaszi, hazai árvíz tapasztalat alapján. Alkalmazott Pszichológia 12(3–4):125–138
25. Math SB et al (2015) Disaster management: mental health perspective. Indian J Psichol Med 37(3):261–271
26. Bognár J, Szekeres G (2009) Katasztrófa és kríziskommunikáció lélektani alapjai. 2009, Zrinyi Miklós Nemzetvédelmi Egyetem Kossuth Lajos Hadtudományi Kar
27. Mutunga JN, Odhiambo EO, Iyaya MW (2022) Psychological First Aid Practices Applied By Humanitarian Responders during Disasters in Kakamega County, Kenya. Int J Soc Sci Humanities Invention 9(09):7230–7238
28. Robila M (2023) Refugees and Social Integration in Europa. United Nations department of economic and social affairs (UNDESA). Division for social policy and development 2018 December 8, 2023]; Available from: https://www.annalindhfoundation.org/resources/publications/refugees-and-social-integration-europe
29. Office for Health Improvement and Disparities (2023) Guidance: mental health: migrant health guide. Advice and guidance ont he health needs of migrant patients for healthcare practitioners. 2017 November 9, 2023]; Available from: https://www.gov.uk/guidance/mental-health-migrant-health-guide
30. Medecins Sans Frontieres. After the dust settles: The mental health needs in Türkiye. 2023 September 4, 2023]; Available from: https://www.msf.org/mental-health-support-people-affected-earthquakes-t%C3%BCrkiye.
31. Magyar Máltai Szeretetszolgálat. Integrációs program szociálisan rászoruló harmadik országbeli állampolgárok részére. n.d. June 27, 2023]; Available from: https://maltai.hu/mmiaprojekt
32. Frivaldszky J (2020) A menekültek alapvető jogainak védelme és a befogadó társadalom önvédelme a tömeges bevándorlás és az asylum shopping korában, in Máltai Tanulmányok, pp 161–181
33. Comfort LK et al (2001) Complex systems in crisis: anticipation and resilience in dynamic environments. J Contingencies Crisis Manag 9(3):144–158
34. NATO (n.d.) January 2, 2023]; Available from: https://katasztrofavedelem.hu/159/nato
35. McClelland M, How to Build a Perfect Refugee Camp. 2014 November 30, 2022]; Available from: https://nytimes.com/2014/02/16/magazine/how-to-build-a-perfect-refugee-camp.html
36. Aburamadan R, Trillo C, Makore BCN (2020) Designing refugees' camps: temporary emergency solutions, or contemporary paradigms of incomplete urban citizenship? Insights from Al Za'atari. City, Territory Archit 7(1):12
37. Magyar Máltai Szeretetszolgálat. A hovatartozás lemoshatatlan jele. 2023 November 23, 2023]; Available from: https://maltai.hu/cikk/hir/4659
38. Magyar Máltai Szeretetszolgálat. Nem álláspontunk, küldetésünk van—Szakmai fórumot szervezett a Szeretetszolgálat a menekültek világnapján. 2017 November 23, 2023]; Available from: https://maltai.hu/cikk/hir/2728
39. Danese A et al (2020) Child and adolescent mental health amidst emergencies and disasters. Br J Psychiatry 216(3):159–162

40. Kövecsi-Oláh P (2017) Mi történik jelenleg Törökországban a menekültek tömegeivel. Migrációkutató Intézet
41. De Mendonca Lima CA (2023) Older adults' mental health during humanitarian crisis. Eur Psychiatry 66(S1):S539–S539
42. Marks MR et al (2017) REACT: A paraprofessional training program for first responders-a pilot study. Bull Menninger Clin 81(2):150–166
43. Kar N (2009) Psychological impact of disasters on children: review of assessment and interventions. World J Pediatr 5(1):5–11
44. Magyar Máltai Szeretetszolgálat. Aláírták az ENSZ és a Magyar Máltai Szeretetszolgálat együttműködését az ukrán menekültekért. 2022 November 23, 2023]; Available from: https://maltai.hu/cikk/hir/4574
45. International Federation of Red Cross and Red Crescent Societies Psychological Support Centre, Strengthening resilience: a global selection of psychosocial interventions. 2014, International Federation of Red Cross and Red Crescent Societies. Psychosocial Centre
46. International Federation of Red Cross and Red Crescent Societies (IFRC), Caring for volunteers: A Psychosocial Support Toolkit. 2013, International Federation of Red Cross and Red Crescent Societies. Psychosocial Centre
47. International Federation of Red Cross and Red Crescent Societies (IFRC). Mental health and psychosocial Support. n.d. October 24, 2023]; Available from: https://ifrc.org/our-work/health-and-care/community-health/mental-health-and-psychosicial-support
48. Török J, Legeza L (2009) Máltaiak. Szerzetesrendek a Kárpát- Medencében. Mikes Kiadó
49. Hunyadi Z (2019) Szent János Ispotályos testvérei. A johanniták Magyarországon a 12. század közepétől a konstanzi zsinatig, in Máltai Tanulmányok, pp 1–114
50. Solymári D, Győri-Dani L (2016) Gondoskodó kíséret. Magyar Máltai Szeretetszolgálat
51. Romhányi T, Solymári D, Tihanyi G (2021) Szolgáló Szeretet: A Magyar Máltai Szeretetszolgálat Története. Magyar Máltai Szeretetszolgálat
52. Magyar Máltai Szeretetszolgálat. Így segít a Máltai Szeretetszolgálat Szíriában | VIDEÓK. 2023 May 27, 2023]; Available from: https://maltai.hu/cikk/hir/4628
53. Magyar Máltai Szeretetszolgálat. Pillepalackokból tankolják a Máltai Szeretetszolgálat mozgó orvosi rendelőit Szíriában. 2023 May 29, 2023]; Available from: https://maltai.hu/cikk/hir/4627
54. Magyar Máltai Szeretetszolgálat. Osztoztunk tűzön és hidegen—Történetek a Törökországi felmérő útról. 2023 October 24, 2023]; Available from: https://maltai.hu/cikk/hir/4638
55. Magyar Máltai Szeretetszolgálat. Gyógyszerellátmányt küldött Törökországba a Máltai Szeretetszolgálat. 2023 November 9, 2023]; Available from: https://maltai.hu/cikk/hir/4643
56. Magyar Máltai Szeretetszolgálat. Humanitárius világtalálkozón a Máltai Szeretetszolgálat. 2017 November 23, 2023]; Available from: https://maltai.hu/cikk/hir/2908

Chapter 20
Critical Infrastructure Protection of the Warehouse Logistics 4.0 Based on AGV Systems

János Simon

Abstract As warehouses continue to evolve toward Industry 4.0 paradigms, the adoption of Automated Guided Vehicle (AGV) systems plays a crucial role in revolutionizing logistics operations. AGVs enable increased efficiency, reduce human error, and contribute to enhanced safety by automating material handling tasks, from inventory management to transportation. However, this rapid digital transformation introduces new vulnerabilities, particularly in the realm of critical infrastructure protection (CIP). Cybersecurity threats, physical security risks, and operational disruptions represent significant challenges in maintaining secure and reliable AGV operations. This paper explores the key issues involved in protecting warehouse logistics 4.0 infrastructures that rely on AGV systems, providing insights into the strategies required to ensure their security and efficiency. The primary goal of this research is to develop a comprehensive understanding of the security, efficiency, and reliability aspects of AGV systems within the context of critical infrastructure. Achieving these goals will help ensure that AGV-based warehouse logistics are not only optimized for performance but also resilient against potential threats.

Keywords AGV · Logistics 4.0 · IoT · WMS · ERP

20.1 Introduction

Warehouse logistics have undergone a dramatic transformation with the advent of Industry 4.0 technologies. AGVs, part of this technological revolution, have become a fundamental tool in modern warehouses, facilitating the automation of tasks such as material handling, loading, and inventory movement. By operating without the

J. Simon (✉)
Department of Mechatronics and Automation, Faculty of Engineering, University of Szeged, Szeged, Hungary
e-mail: simon@mk.u-szeged.hu

Subotica Technology—College of Applied Sciences, Subotica, Serbia

T. A. Kovács and I. Fürstner (eds.), *Critical Infrastructure Protection: Advanced Technologies for Crisis Prevention and Response*, NATO Science for Peace and Security Series C: Environmental Security, https://doi.org/10.1007/978-94-024-2308-2_20

need for direct human intervention, AGVs improve efficiency, reduce labor costs, and enhance operational safety. However, the integration of these autonomous systems within critical infrastructures introduces significant security challenges. The growing connectivity of AGVs through wireless networks, cloud platforms, and data management systems creates a larger attack surface for cybersecurity threats, while the physical nature of AGV systems exposes them to sabotage and operational risks.

20.2 Security Threats to AGV Systems

Warehouse logistics infrastructures relying on AGVs are susceptible to multiple layers of threat. Cybersecurity concerns are among the most pressing, as AGV systems are often connected to warehouse management systems (WMS), enterprise resource planning (ERP) platforms, and the broader supply chain network. These connections make AGVs vulnerable to a range of cyber attacks such as ransomware, denial-of-service (DoS) attacks, and data breaches [1]. These attacks can lead to disruptions in warehouse operations, compromise sensitive data, and even cause physical harm to the facility and personnel. Physical security is another crucial consideration. Unauthorized access to warehouse environments can result in theft, vandalism, or sabotage of AGV systems, potentially damaging infrastructure and halting operations. Additionally, operational risks, including system malfunctions and human errors, pose internal threats that can result in costly downtime and inefficiencies. Without proper maintenance, training, and operational oversight, AGV systems may experience failures that compromise the security and productivity of the entire logistics operation.

20.3 Critical Infrastructure Protection Strategies

To mitigate these risks, robust CIP strategies are essential. Cybersecurity measures, such as firewalls, intrusion detection systems (IDS), and data encryption, must be implemented to safeguard AGV systems from external threats. Regular software updates and patch management are crucial to prevent vulnerabilities from being exploited by malicious actors. In addition, employing strong access control protocols and multi-factor authentication (MFA) for networked devices can further secure AGV operations [2]. On the physical security side, securing the warehouse perimeter with surveillance systems, access control gates, and real-time monitoring tools is key to preventing unauthorized access to AGVs. Effective physical barriers, such as locking mechanisms and alarms on AGV charging stations, can deter theft or tampering. Furthermore, operational best practices—such as scheduled maintenance, staff training on cybersecurity awareness, and regular system audits—help maintain the long-term reliability of AGV systems. The rapid advancement of warehouse logistics under the Industry 4.0 framework has led to the widespread adoption of AGV systems. These systems have revolutionized material handling, inventory management, and transportation, becoming integral to warehouse operations. However, alongside these technological advancements, there are emerging challenges concerning the protection of critical infrastructure in logistics environments that depend on AGV systems. This research aims to explore the CIP aspects of AGV systems, focusing on securing their operations and optimizing their performance [3].

20.3.1 AGV Systems in Warehouse Logistics

AGV systems are autonomous vehicles used in various industries, primarily for tasks like transporting materials across facilities without human intervention. Their role in modern warehousing is pivotal for achieving high levels of operational efficiency. These systems are designed to interact seamlessly with WMS, ERP systems, and other smart logistics technologies. While AGVs offer numerous benefits, such as reducing human labor, improving safety, and increasing productivity, their integration into critical infrastructure poses unique security challenges [4]. As more warehouses adopt AGV technologies, it is essential to ensure these systems are robustly protected against a variety of threats. AGV systems can be vulnerable to cyber attacks, physical disruptions, and operational failures, all of which can negatively impact warehouse logistics.

20.3.2 Ensuring Robust Security Measures

A fundamental concern in critical infrastructure protection is safeguarding AGV systems from various security threats. These threats range from cybersecurity breaches to physical intrusions. Cyber attacks on AGV systems can result in disruptions to the entire warehouse operation, as AGVs are often connected to cloud platforms, databases, and other IT systems. Security measures such as firewalls, intrusion detection systems, and data encryption are necessary to prevent these attacks. Physical security also plays a vital role in protecting AGVs from sabotage or theft. Effective access control measures, surveillance systems, and security personnel are essential components of a robust protection strategy [5]. This research will explore best practices in both cyber and physical security to ensure that AGV systems are resilient to external and internal threats.

20.3.3 Maximizing Operational Efficiency

Operational efficiency is another key objective in optimizing AGV systems. While AGVs are designed to enhance productivity by automating repetitive tasks, their efficiency can be compromised if proper protection and maintenance measures are not in place. The research will investigate how maintaining security directly contributes to the operational efficiency of AGV systems. By preventing system failures due to external threats or operational mishaps, AGVs can operate at full capacity, reducing downtime and increasing output. Optimizing efficiency also involves integrating AGV systems seamlessly with other technologies within the warehouse. Effective communication between AGVs, WMS, and other smart systems is critical to avoid bottlenecks and improve overall workflow [6]. This research will provide insights into optimizing AGV operations through efficient system integration and process automation.

20.3.4 Maintaining System Reliability

Reliability is essential for AGV systems to support continuous warehouse operations. A single point of failure can lead to significant delays and losses, especially in environments where logistics are time-sensitive [7]. To maintain high reliability, it is crucial to have regular system updates, real-time monitoring, and contingency plans in place. System reliability also depends on proper maintenance and human oversight. While AGVs operate autonomously, trained personnel are still necessary to monitor their performance and address issues as they arise. This research will examine strategies for improving system reliability, including predictive maintenance, real-time diagnostics, and staff training.

20.4 Components of an AGV Based Warehouse Logistics 4.0 System

AGVs are robotic units designed to autonomously handle material transport tasks within industrial and warehouse settings. These vehicles are essential components in modern logistics and manufacturing environments, enabling efficient and streamlined operations.

20.4.1 AGV Unit (Automated Guided Vehicle)

The AGV unit itself is composed of several core elements: motors, sensors, onboard computers, and navigation systems. Together, these components allow the AGV to move autonomously, perceive its environment, and make decisions in real time. Additionally, AGVs are integrated with broader control systems or ERP platforms, enabling task assignments, performance monitoring, and seamless communication within the larger operational framework [8]. Motors are the primary drivers of movement in an AGV, allowing it to transport materials from one location to another. Most AGVs use electric motors, which are avoured for their energy efficiency, reliability, and minimal maintenance requirements. The drive mechanism is typically linked to the AGV's wheels, providing precise control over acceleration, speed, and direction. This control is essential for navigating complex environments, especially where the AGV must share space with human workers or other vehicles. The motors are regulated by control algorithms that adjust their function based on data from onboard sensors and the navigation system, ensuring smooth and accurate movement. Sensors are integral to the AGV's ability to detect and respond to its surroundings. Different types of sensors are utilized, including laser scanners, ultrasonic sensors, and proximity sensors. These sensors provide real-time data about the environment, allowing the AGV to detect obstacles, navigate safely, and avoid collisions. For example, laser scanners enable high-precision mapping and localization, which is critical for moving through predefined paths without error. Proximity and ultrasonic sensors further enhance safety by detecting objects within close range, enabling the AGV to slow down or stop when necessary. Together, these sensors contribute to the AGV's awareness and adaptability in dynamic environments [9]. The AGV's onboard computer functions as its brain, processing data from sensors, motors, and navigation systems to execute tasks autonomously. This computer is equipped with a real-time operating system that can handle various inputs, make calculations, and issue commands to different components of the AGV. Advanced AGVs may incorporate machine learning algorithms that allow them to improve their performance over time by learning from operational data. The onboard computer also interfaces with the central control system or ERP platform, receiving instructions, updating status, and transmitting operational data. This interconnectivity allows for coordinated actions and enhances the AGV's ability to adapt to changing tasks or schedules.

20.4.2 Navigation System

Navigation systems are critical for guiding the AGV along specified routes or paths. Different AGVs may use varied navigation technologies, such as laser-based navigation, magnetic strips, or even camera-based visual guidance. Laser navigation is one of the most precise methods, allowing the AGV to identify its exact location within the warehouse by referencing markers or reflectors [10]. For simpler applications, magnetic or painted strip navigation can guide the AGV along predetermined paths. Some AGVs combine multiple navigation methods to improve accuracy and reliability in complex environments. The navigation system ensures that the AGV can reach its destination precisely and efficiently, even if the path involves turns, intersections, or dynamic obstacles. Common navigation techniques include magnetic strips, laser guidance, or vision-based navigation. GPS (for outdoor environments) or QR codes (for indoor).

20.4.3 Warehouse Management System (WMS)

AGVs are not isolated units; they are integrated with central control systems or ERP platforms for task management and operational synchronization. Through wireless connectivity (often via Wi-Fi or other industrial communication standards), AGVs receive task assignments, such as where to pick up and drop off materials. The control system can monitor each AGV's status, location, and performance in real time, optimizing fleet management and ensuring that tasks are distributed effectively. Integration with ERP systems enables a higher level of planning and resource management, as AGVs can be assigned tasks based on inventory levels, production schedules, or other operational priorities [11]. This connectivity supports an adaptive, agile workflow, where AGVs can be dynamically reassigned to different tasks as needed. The system that sends tasks and monitors AGV performance. AGVs communicate with the ERP or WMS using wireless or wired connections.

20.4.4 Control and Communication System

The control and communication system of AGVs is critical to their efficient and coordinated operation within logistics and industrial environments. Communication is often achieved via Wi-Fi or other wireless networks, enabling seamless data exchange between AGVs and the central control system. This wireless connection allows AGVs to receive task assignments, report their status, and communicate with other AGVs for collision avoidance or path adjustments. The robust communication system ensures flexibility and adaptability in dynamic environments, where AGVs may need to alter routes or tasks in real time [12]. A Supervisory Control and Data Acquisition (SCADA) system often supports this communication framework, providing a real-time monitoring and control layer. SCADA gathers data from AGVs and other connected devices, allowing operators to oversee operations, track performance metrics, and quickly respond to any anomalies. With SCADA, operators have a centralized view of AGV statuses, battery levels, and potential issues, enhancing decision-making and ensuring continuous operation. This integration of

Wi-Fi and SCADA enables efficient fleet management, allowing AGVs to operate cohesively, adaptively, and securely, which is essential for maintaining productivity in high-demand, automated environments.

20.4.5 Safety Systems

Safety systems in AGVs are crucial to protecting both the vehicle and its surrounding environment. These systems incorporate emergency stop buttons, bump sensors, and laser-based object detection to prevent accidents. Emergency stop buttons provide a quick, manual way to halt the AGV, allowing operators to respond immediately if an issue arises. Bump sensors enable the AGV to detect physical contact with objects or obstacles, causing it to stop instantly to prevent potential damage. Laser-based object detection offers more advanced, real-time scanning of the AGV's path, identifying obstacles at a distance and enabling the vehicle to adjust its course or stop before contact occurs. In addition to these hardware features, AGVs are equipped with integrated safety protocols that dictate how they respond to various scenarios [13]. For example, if an obstacle is detected in the AGV's path, it may stop, slow down, or automatically reroute to avoid it. These protocols work in harmony with safety sensors, creating a multi-layered safety net. By integrating these systems, AGVs can operate autonomously with minimal risk, ensuring safe interactions within dynamic environments, including warehouses and industrial settings where both human workers and machines operate.

20.4.6 Charging Infrastructure

Charging infrastructure is essential for maintaining uninterrupted operations in AGV systems. AGVs typically rely on two main charging methods: inductive (wireless) charging stations and designated areas for manual recharging. Inductive charging allows AGVs to recharge without physical connections, usually at designated spots where they can pause momentarily to receive power. This method is highly efficient, as it enables vehicles to recharge in brief, idle moments during operation, reducing downtime and enhancing productivity. Alternatively, manual charging stations require AGVs to dock and connect physically, often during scheduled breaks or low-demand periods. Both methods are integrated with advanced battery management systems to optimize energy use and ensure efficient power distribution across the fleet [14]. Battery management monitors each AGV's charge levels, schedules charging sessions based on usage, and prevents overcharging, thus extending battery life and reducing operational costs. Effective charging infrastructure, coupled with intelligent battery management, is vital to ensuring AGV energy efficiency. This infrastructure supports continuous operation by minimizing interruptions, allowing AGVs to function around the clock in high-demand environments like warehouses and manufacturing facilities, ultimately boosting overall system productivity and reliability.

20.5 Conceptual Framework of Logistics 4.0

Logistics 4.0 is an advanced framework that applies the principles of Industry 4.0 to logistics, integrating intelligent systems, data-driven technologies, and automation to optimize supply chain management. It revolves around the utilization of digitalization, smart devices, and real-time data to improve efficiency, reduce costs, and enhance flexibility in logistics operations.

The conceptual framework for Logistics 4.0 is built on several key components, including IoT, big data analytics, cloud computing, artificial intelligence (AI), and autonomous systems like AGVs (see Fig. 20.1). These components are interwoven to form a highly interconnected and automated logistics system [15].

20.5.1 Integration of IoT in Logistics

The Internet of Things (IoT) plays a critical role in the Logistics 4.0 framework by enabling real-time tracking, monitoring, and control of logistics assets such as vehicles, inventory, and equipment. IoT sensors and devices are embedded across various

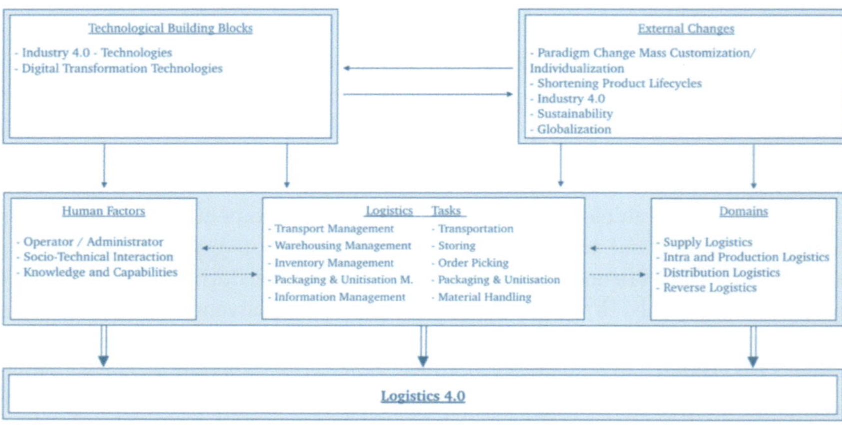

Fig. 20.1 Conceptual framework of Logistics 4.0

points in the supply chain, generating vast amounts of data related to location, environmental conditions, and performance metrics. This data is sent to central systems that use advanced algorithms and machine learning to provide real-time insights and make informed decisions [16]. In this framework, IoT ensures that every process in the logistics chain, from transportation to warehousing, is interconnected, enabling seamless communication between systems.

20.5.2 Big Data Analytics for Decision-Making

Big data analytics is the backbone of the decision-making process in Logistics 4.0. With the integration of IoT devices, companies collect massive volumes of data that can be ehaviou to gain insights into operational efficiencies, demand forecasting, and customer ehaviour. Advanced analytics tools allow for the identification of patterns and trends that may not be visible in traditional systems [17]. This capability enables predictive maintenance, route optimization, and even anticipatory logistics, where goods are moved based on future demands rather than current orders. Logistics 4.0 thus shifts from reactive to proactive operations, driven by data intelligence.

20.5.3 Cloud Computing for Scalability and Flexibility

The Logistics 4.0 framework depends on cloud computing to provide scalable, flexible infrastructure for data storage, processing, and sharing. With cloud technology, logistics firms can connect and synchronize various platforms across multiple locations, ensuring that information is always available and up-to-date. Cloud platforms also enable collaboration between supply chain partners, allowing different companies to share data, resources, and applications in real time. Moreover, the scalability offered by cloud computing means that logistics operations can quickly adapt to fluctuations in demand, ensuring continuity without disruptions.

20.5.4 Artificial Intelligence and Automation

AI is a fundamental element of Logistics 4.0, enabling automation in decision-making and operational processes. AI algorithms power various aspects of logistics, from demand forecasting and inventory management to automated route planning and customer service chatbots. AI-driven systems are capable of learning from historical data and improving their accuracy over time [18]. Combined with automation technologies like robotics and AGVs, AI ensures that warehouses, distribution centers, and transport fleets operate with minimal human intervention. Automation not only improves speed and efficiency but also minimizes errors and enhances safety.

20.5.5 Autonomous Systems and AGVs

AGVs and other robotic systems are crucial to achieving full automation in the Logistics 4.0 framework. AGVs operate in warehouses and distribution centers to transport goods, load and unload cargo, and organize inventory with high precision. These systems interact seamlessly with WMS and ERP platforms to execute tasks autonomously [19]. The ability of AGVs to operate continuously, without breaks, significantly enhances throughput and productivity in logistics operations. Moreover, they are equipped with advanced sensors and navigation systems, allowing them to avoid obstacles and ensure safety in dynamic environments.

20.5.6 Cybersecurity and Data Protection

Given the high level of interconnectivity in Logistics 4.0, cybersecurity becomes a critical concern. The use of cloud platforms, IoT devices, and data analytics makes logistics systems more vulnerable to cyberattacks. Therefore, robust cybersecurity measures, such as encryption, multi-factor authentication, and network monitoring, are integral to ensuring the protection of sensitive data and maintaining the integrity of logistics operations. Furthermore, compliance with data privacy regulations, such as GDPR, is essential in safeguarding customer information in the digital era.

The conceptual framework of Logistics 4.0 integrates IoT, big data analytics, cloud computing, AI, and autonomous systems to create a smart, efficient, and flexible logistics system [17]. By leveraging these technologies, Logistics 4.0 enables real-time decision-making, automation, and optimization across the entire supply chain. The result is a logistics network that is more responsive to market demands, capable of predictive operations, and resilient to disruptions. As the digital transformation of logistics continues, companies that adopt the Logistics 4.0 framework will be well-positioned to lead in an increasingly competitive global market.

20.6 Protection Strategies for AGV Systems

AGV systems have become an integral part of modern logistics and warehouse operations under the Industry 4.0 framework. These systems enhance efficiency, reduce operational costs, and improve safety by automating material handling tasks. However, the increased reliance on AGVs introduces vulnerabilities that must be addressed to ensure the continuous and secure operation of logistics systems. Protection strategies for AGV systems encompass both cybersecurity and physical security measures to safeguard them from a wide range of potential threats [15]. These strategies aim to maintain operational integrity, prevent disruptions, and protect critical infrastructure.

20.6.1 Cybersecurity Measures

AGV systems are typically integrated with broader networks, including WMS, cloud platforms, and ERP systems. This interconnected nature exposes AGVs to potential cybersecurity threats such as data breaches, unauthorized access, and cyberattacks [11]. Therefore, implementing robust cybersecurity measures is crucial to protect the data and operations associated with AGVs.

Network Security: One of the primary cybersecurity strategies involves securing the communication networks that AGVs rely on. Firewalls, virtual private networks (VPNs), and secure communication protocols should be used to prevent unauthorized access to AGV systems. Ensuring that all data transmitted between AGVs and control systems is encrypted can further protect against data interception and manipulation.

Access Control: Restricting access to AGV systems is another essential protection measure. Only authorized personnel should have access to critical AGV operations. Implementing MFA and role-based access control (RBAC) helps ensure that only the right individuals can interact with the system, reducing the risk of unauthorized actions.

Intrusion Detection and Prevention Systems (IDPS): AGVs should be equipped with intrusion detection and prevention systems that monitor for unusual ehaviour or unauthorized attempts to access the system. These systems can alert operators to potential security breaches in real time, allowing for quick responses to mitigate risks.

Software Updates and Patches: Regular software updates are critical to addressing security vulnerabilities. Ensuring that AGV systems are equipped with the latest firmware and security patches helps protect them from known threats. Automated update systems can be used to ensure timely implementation of patches across all connected AGV units.

20.6.2 Physical Security Measures

In addition to cybersecurity, physical security is essential to protect AGV systems from sabotage, theft, or accidental damage [9]. These vehicles often operate in environments with high traffic and valuable materials, making them potential targets for malicious activities.

Securing AGV Pathways: One strategy to enhance physical security is to restrict access to AGV operating areas. Unauthorized personnel should not be allowed in areas where AGVs are operating. This can be achieved through physical barriers, surveillance cameras, and access control systems, reducing the risk of tampering or interference.

AGV Safeguards and Fail-Safe Mechanisms: AGVs should be equipped with safety features such as emergency stop buttons, bump sensors, and collision avoidance systems. These features help prevent accidents or intentional disruptions by ensuring that AGVs can detect obstacles or dangerous situations and stop immediately if necessary.

Tamper-Resistant Hardware: To prevent physical sabotage, AGVs should use tamper-resistant hardware. This can include protective casings around critical components, locked access points, and secured charging stations. Tamper-resistant hardware makes it more difficult for malicious actors to gain access to sensitive AGV components.

20.6.3 System Redundancy and Resilience

Another key protection strategy involves designing AGV systems with redundancy and resilience in mind. Ensuring that there are backup systems in place in the event of a failure can help maintain continuous operations, even in the face of a cyberattack or physical disruption [4].

Redundant Communication Channels: AGVs should have multiple communication channels that allow them to stay connected to central control systems in case one network fails. This redundancy ensures that AGVs remain operational, even if one part of the communication infrastructure is compromised.

Backup Power and Charging Solutions: To prevent downtime due to power outages or battery failures, AGV systems should have backup power solutions and reliable charging infrastructure. This can include backup batteries or alternative charging methods that allow AGVs to continue operating during power disruptions.

Regular Maintenance and Monitoring: Continuous monitoring and regular maintenance are crucial for detecting potential vulnerabilities early. Predictive maintenance systems can identify issues with AGVs before they lead to system failures, reducing the likelihood of unplanned downtime.

Protecting AGV systems is vital to ensuring the efficient and secure operation of modern warehouse and logistics operations. By implementing a combination of cybersecurity and physical security measures, organizations can safeguard AGVs from a wide range of potential threats [12]. Incorporating system redundancy, regular maintenance, and continuous monitoring further enhances the resilience of AGV systems. As AGV systems become increasingly integrated into logistics operations, developing robust protection strategies will be essential to maintaining the safety, security, and reliability of critical infrastructure.

20.7 Conclusion

AGV units are sophisticated machines that combine mechanical, electronic, and digital technologies to operate autonomously in industrial environments. Motors provide the power for movement, sensors enable environmental awareness, onboard computers process information and execute tasks, and navigation systems guide the AGV's path. The integration of AGVs with control systems or ERP platforms enhances their functionality, enabling efficient and synchronized operations. As technology advances, AGVs continue to evolve, becoming increasingly capable and indispensable in the logistics and manufacturing sectors. Their integration within smart, automated systems is central to the development of Industry 4.0, transforming traditional operations into highly efficient, data-driven processes. As AGV systems become increasingly integral to the future of warehouse logistics 4.0, protecting these infrastructures from evolving threats is vital for ensuring both security and operational efficiency. By implementing comprehensive CIP strategies that encompass cybersecurity, physical security, and operational excellence, organizations can safeguard their AGV systems against a wide range of potential risks. This proactive approach will not only protect critical logistics infrastructure but also enhance the overall resilience, productivity, and profitability of modern warehouses.

References

1. Arifin R, Egbelu P (2000) Determination of vehicle requirements in automated guided vehicle systems: A statistical approach. Prod Plan & Control 11(3):258–270
2. Chavez Z, Hauge JB, Bellgran M (2022) Industry 4.0, transition or addition in SMEs? A systematic literature review on digitalization for deviation management. Int J Adv Manuf Technol, 119 (1–2), 57–76
3. Chen G, Wang P, Feng B, Li Y, Liu D (2020) The framework design of smart factory in discrete manufacturing industry based on cyber-physical system. Int J Comput Integr Manuf 33(1):79–101
4. Elia V, Gnoni MG, Tornese F, Guglielmi A, Pellicci M, De Merich D, Campo G (2022) Applications of smart technologies for automatic near miss detection in the industrial safety. Procedia Comput Sci 200:1282–1287
5. Gnoni MG, Elia V, Bragatto P (2016) An IOT-based system to prevent injuries in assembly line production systems. In: IEEE international conference on industrial engineering and engineering management (IEEM), 1889–1892
6. Grau A, Indri M, Lo Bello L, Sauter T (2021) Robots in industry: The past, present, and future of a growing collaboration with humans. IEEE Ind Electron Mag 15(1):50–61
7. Hofmann E, Rüsch M (2017) Industry 4.0 and the current status as well as future prospects on logistics. Comput Ind 89:23–34
8. José DL (2022) Industry 4.0 and the small business. SerbN J Manag, 17(1), 161–178
9. Kesen SE, Baykoç OF (2007) Simulation of automated guided vehicle (AGV) systems based on just-in-time (JIT) philosophy in a job-shop environment. Simul Model Pract Theory 15(3):272–284
10. Kiel D, Müller JM, Arnold C, Voigt KI (2020) Sustainable industrial value creation: benefits and challenges of industry 4.0. Int J Innov Manag, 21(8), 1–34

11. Oztemel E, Gursev S (2020) Literature review of industry 4.0 and related technologies. J Intell Manuf, 31(1), 127–182
12. Piccarozzi M, Aquilani B, Gatti C (2018) Industry 4.0 in management studies: a systematic literature review. Sustain, 10(10), 3821
13. Sabattini L et al (2018) The PAN-Robots Project: Advanced automated guided vehicle systems for industrial logistics. IEEE Robot Autom Mag 25(1):55–64
14. Tamang D, Pozzebon A, Parri L, Fort A, Abrardo A (2022) Designing a reliable and low-latency LoRaWAN solution for environmental monitoring in factories at major accident risk. Sensors 22(6):2372
15. Um I, Cheon H, Lee H (2009) The simulation design and analysis of a flexible manufacturing system with automated guided vehicle system. J Manuf Syst 28(4):115–122
16. Sarcevic P, Kincses Z, Pletl S (2014) Wireless sensor network based movement classification using wrist-mounted 9DOF sensor boards. In: 2014 IEEE 15th international symposium on computational intelligence and informatics (CINTI), Budapest, Hungary, pp 85-90, https://doi.org/10.1109/CINTI.2014.7028654
17. Ballai G, Sőrés MA, Vásárhelyi L, Szenti I, Kun R, Hartmann B, Sebők D, Farkas F, Zahoor A, Mao G, Sápi A, Kukovecz Á, Kónya Z (2023) Exploration of Li-Ion batteries during a Long-Term heat endurance test using 3D temporal microcomputed tomography investigation. Energy Technol 11:2300207. https://doi.org/10.1002/ente.202300207
18. Bús, Nikolett Katalin, Nyikes, Zoltán (2021) Cyber security challenges in construction. In Critical Infrastructure Protection in the Light of the Armed Conflicts, Springer Nature Switzerland. https://doi.org/10.1007/978-3-031-47990-8_44
19. Mészáros A, Sárosi J (2022) Soft robotics: state of art and outlook. Ancta Tech Szeged, 16 (1): 8–13. https://doi.org/10.14232/analecta.2022.1.8-13

Chapter 21
Nonverbal Communication and the Psychology of Lying for a Security Environment

Darjana Sredić and Ljubica Janjetović

Abstract Non-verbal communication is a way of communicating and conveying messages that is achieved without the use of words, i.e. non-verbally. The most frequently studied channels of nonverbal communication, through which people transmit their messages and receive messages from others, are facial expression, eye contact and gaze, gestures, body posture, touch, and vocal behavior. Apart from the various ways in which this form of communication is realized, non-verbal communication also has numerous functions such as expressing emotions, establishing relationships, establishing identity, regulating verbal messages and others. However, a particular feature that arouses interest in research in this area is the assumption that a large part of non-verbal behavior often takes place unconsciously or automatically. In this regard, there is a widespread opinion that lying, as an often-indispensable part of human interaction, can be recognized by analyzing the non-verbal behavior of a person who is suspected of lying. A lie is considered a deliberate and conscious act of deceiving another person, and the types of lies, with different motives behind them, vary from well-intentioned ones to unacceptable ones that, according to some, can have serious consequences. Understanding subtle cues in body language, detecting shifts in facial expressions, and recognizing patterns of behavior are paramount in assessing potential threats and de-escalating tense situations. In a security environment, Non-Verbal Communication plays a critical role in conveying and interpreting information, often supplementing or even replacing verbal exchanges in certain situations. One key aspect is the use of body language and gestures to signal and interpret security-related cues.

Keywords Non-verbal communication · Psychology of lying · Security environment · Lying detection

---

D. Sredić (✉) · L. Janjetović
PIM University, Faculty of Phylosophy, Banja Luka, Bosnia and Herzegovina
e-mail: darjana.sredic@univerzitetpim.edu.ba

T. A. Kovács and I. Fürstner (eds.), *Critical Infrastructure Protection: Advanced Technologies for Crisis Prevention and Response*, NATO Science for Peace and Security Series C: Environmental Security, https://doi.org/10.1007/978-94-024-2308-2_21

21.1 Introduction

While the word, as a means of verbal communication, is defined by itself, every person can experience and interpret an unspoken message in their own way [5]. Accordingly, the study of non-verbal communication is demanding [5], but also very useful because the skill of non-verbal communication is an integral aspect of human social competence [2, 4]. "The skill of non-verbal communication implies individual differences in the ability to transmit or encode, as well as receive or decode non-verbal signals" [4]. As Aronson et al. state, "encoding is the display or expression of nonverbal behavior, and decoding is the interpretation of the meaning of nonverbal behavior expressed by other people" [1].

Scientific study of non-verbal communication was carried out mainly after the Second world war. However, this does not mean that significant earlier contributions cannot be found throughout history. For example, Quintilian's work Institutio Oratoria, which in translation means the speaker's education, is an important source of data on gestures, and dates to the first century BC. The study of nonverbal cues has never been the domain of one particular discipline so we can find it in fields such as zoology, anthropology, dance art, linguistics, psychology, psychiatry, public speaking and so on. In the second half of the nineteenth century Delsarte tried to systematically list and describe the rules of regulation, speech culture and movement/gesture. In 1872, one of the most influential works before the twentieth century was written, namely Darwin's Expression of Emotions in Man and Animals. This work is the initiator of the modern study of facial expressions, and later his observations and thoughts were confirmed by other researchers. In the first half of the twentieth century, there were individual studies of speech, physical appearance and clothing of that person. However, an unsystematic review of publications indicates that it is the study of nalysing nalysing, the environment and body movements received even less attention, while the least attention was paid to the study of touch and sight. During that period, there are two different but significant events. The first event involved controversial teaching and scandal, while the second event was a work that had an extraordinary influence in the study of non-verbal nalysing. In 1925, Kretschmer wrote the book Body structure and character, and fifteen years later Sheldon published the book Variations in the Human Body. The listed works are based on the assumption that precise measurement and analysis of the human body can be found about the intelligence, temperament, moral qualities and future achievements of the individual. However, none of the authors proved the validity of such perceptions. It was published in 1941 Efron's book Gestures and surroundings, which soon became a classic. It's innovative and exhaustive methods of studying gestures and body language, together with a systematic classification of non-verbal behaviors left an impact on generations of authors. During the 1950s, there was the rise of the study of nonverbal nalysing as the 1960s brought the nuclear explosion. Individual parts of the body, vocal expression of emotions, pupil dilation, personal space and arrangement in space, pauses and hesitations in spontaneous speech, etc. In the 1970s for years, non-verbal communication was studied through newspaper reports, and after the success story there is an appearance of many books that want to make non-verbal communication comprehensible to the audience. In the 1980s, the focus was on identifying the ways in which different non-verbal signals work together to achieve general communication goals. In the following years from the study of situations in which there is no interaction, the focus is placed on interactive situations, both participants in the interaction, changes over time, complex nalysing, signal perception during the interaction, the existence of multiple goals and meanings, the way certain nalysing appears, etc. [15].

One of the characteristics of non-verbal communication compared to verbal communication is that sending and receiving non-verbal signals takes place simultaneously. More precisely, when a person transmits information through his appearance and non-verbal behavior, he simultaneously receives information about the appearance and behavior of another person. Furthermore, sending and receiving a large part of non-verbal messages happens automatically, without control or unconsciously. It also indicates the cognitive effectiveness of non-verbal communication since, by exchanging non-verbal messages while interacting with others, a person can focus his attention on other things. Ultimately, nonverbal behavior largely conveys information about current experiences and face-to-face interactions. Therefore, although language is useful for describing and analyzing past and even future events, it lacks the immediacy, effectiveness and power of non-verbal communication [19].

21.1.1 Functions of Non-verbal Communication

Non-verbal communication has many functions, and different authors single out the most important ones according to their opinion. Aronson et al. note that "nonverbal ehaviour is primarily used to express emotions, reveal attitudes, reflect personality traits, and regulate verbal communication" [1]. Remland on the other hand, explains that the functions of non-verbal communication are key communication activities that are necessary for the survival of the species [21]. One of the common misconceptions is that non-verbal ehaviour is only used to convey emotional messages [14]. However, as Remland states, there are four functions of non-verbal communication: identification function, relationship function, emotion expression function and delivery function [21]. These functions are universal, all animals have developed unique signal systems that allow relatives within the species to identify each other, establish relationships with others, express their inner state and provide important information about their environment. Therefore, it is understandable that humans have developed the same abilities [21].

In the context of the identification function, the animal must be able to signal its identity to others. In other words, in the animal world, failure to recognize identity can have threatening consequences, such as the inability to recognize predators, attempts to reproduce with individuals of other species, prohibition of joining other members of one's group, and the like. Accordingly, identification signals contribute to the survival of the species, and include various sounds, smells and physical characteristics. In a similar way, people use their appearance and ehaviour to signal their age, race, gender, ethnic origin, social status, occupation, values, and other characteristics. But different non-verbal signs such as facial features, body shape and voice distinguish people more concretely and emphasize the uniqueness of everyone [21].

Furthermore, the relationship function implies that both animals and humans communicate non-verbally to make friends and partners. The importance of these non-verbal signs is evident in the fact that it is almost impossible to interact face-to-face without showing a certain level of liking or disliking and a certain level of dominance or subordination to another person. Different physical features that people are born with attract the opposite sex, but apart from looks, different non-verbal behaviors have evolved with the purpose of improving a person's relationships with other people in his environment [21]. One of the primary functions of non-verbal communication is the transmission of symbolic messages, which implies that most species use non-verbal signs to exchange information about the outside world. As in human speech, the symbols used by animals refer to certain things. Thus, it was observed that chimpanzees often use gestures that have different meanings in different contexts. For example, during a fight, a chimpanzee can extend its hand to another chimpanzee asking for help, but with the same gesture it can show that it is looking for food if it sees another chimpanzee eating. Although some animals use signs in ways that mimic the properties of language, there is no evidence that they can use symbols that will refer to things distant in space and time, which is one of the main characteristics of non-verbal communication. On the other hand, people are born with the capacity to speak, but in conveying the verbal content of the message, they simultaneously use body movements that can change the meaning and impact of the words themselves [21].

21.2 Lying and Types of Lies

In defining a lie, one should consider both the liar and his victim. In a lie, the victim does not ask to be deceived, nor does the liar give any warning of his intention to deceive. It would be bizarre to call actors liars. Their audience agrees to be fooled for a while because that's what they're there for. Unlike people who abuse other people's trust, actors approach the performance only after announcing that it is an act. If a broker told a buyer that he was giving him reliable but incorrect information, the buyer would not intentionally follow his advice. It wouldn't be a lie either if the psychiatric patient Mary confided to her doctor that she would invoke feelings she doesn't have, or if Hitler told Chamberlain not to rely on his promises [9].

There are two basic ways of lying: concealment and falsification. In concealment, the liar withholds some information and does not actually say anything untrue. In the case of counterfeiting, additional measures are taken. Not only does a liar withhold true information, but he presents false information as true. It is often necessary to combine concealment and forgery to commit fraud, but sometimes the liar can get away with concealment alone. There are people who do not think that concealment is lying; some people save this word only for the more reckless act of forgery.

Fear is harder to hide than worry, just as anger is harder to hide than resentment. The stronger the emotion, the more likely some sign of the emotion will "leak", despite the liar's best attempts to hide it. "Putting on" another emotion, one that is not experienced, can make it possible to mask the experienced emotion that is being hidden. Faking an emotion can prevent the hidden emotion from "leaking" [10].

A mask is the best way to hide strong emotions. Covering the face or part of it with a hand, or turning the head away from the person speaking, usually indicates lying. The best mask is a false emotion. It not only deceives, but it is the best camouflage. It is terribly difficult to keep a straight face or inactive hands when the emotion experienced is strong. It is most difficult to maintain a cool, unemotional or neutral appearance when experiencing emotions. It is much easier to strike a pose, stop, or counter actions that express the emotions experienced with another set of actions.

Poker is a situation where masking cannot be used to hide emotions. When a player gets excited about the prospect of winning a lot because he has great cards, he must hide any sign of his excitement so that other players don't give up. Masking real excitement with signs of any other unexperienced emotion can be dangerous. If a player tries to hide his excitement by pretending to be disappointed or angry, others will think he has been dealt bad cards and expect him to fold rather than stay in the game. He must look poker-faced. He can use a mask if he chooses to hide his disappointment or anger at bad cards by bluffing, trying to force others to fold. By faking happiness or excitement, he could hide his disappointment and create the impression that the card wanted him. It won't be convincing to other players unless they consider him a beginner. An experienced poker player is assumed to have mastered the art of not showing any emotion when the cards are dealt [13].

Faking any emotion can make it easier to hide another emotion. A smile is the most used mask. It serves as the opposite of all negative emotions—fear, anger, pain, disgust and so on. He is often chosen because performing many tricks requires showing some variation of luck. Another reason for the frequent use of smiling as a mask is that smiling is part of standard greetings and that polite greetings often require smiling.

21.2.1 Lie Detection

Understanding subtle cues in body language, detecting shifts in facial expressions, and recognizing patterns of behavior are paramount in assessing potential threats and de-escalating tense situations. In a security environment, Non-Verbal Communication plays a critical role in conveying and interpreting information, often supplementing or even replacing verbal exchanges in certain situations. One key aspect is the use of body language and gestures to signal and interpret security-related cues.

Lying detection is an extremely complex task for those who carry it out. In addition to the already well-known method of polygraph testing, police officers also use non-verbal, verbal and vocal signs of lying to detect lies. This implies knowledge about the emotional and cognitive signs of lying, considering that lying requires additional mental effort, and can generate different emotions in people (fear, excitement, guilt, etc.) Because lies are predominantly an emotional construct, thus the issue of emotions appears to be extremely important for the detection of lying [7]. Namely, the scientific explanations of non-verbal signs of lying are based on the claim that during lying, people are under the influence of three psychological processes, which are designated as emotional, cognitive and behavioral control processes. Since there is a connection between lying and non-verbal behavior, it is believed that the key to recognizing a person's body language while speaking lies in the ability to understand their current emotional state and to perceive the circumstances under which a person is speaking [8, 11]. For this reason, the greatest discoveries in the field of lie detection were made when the scientific focus of research was extended to include emotions.

The foundations for discovering the connection between emotions and facial expression were laid by the French neurologist Guillaume-Benjamin-Amand Duchenne with his research on the physiology of facial expressions in the nineteenth century [28]. But the greatest scientific contribution to the study of emotions and expression is attributed to the American psychologists Sylvan Tomkins and Paul Ekman. Namely, Tomkinson's ideas (1963) about the relationship and influence of emotions and expression on human experience were realized in the works of Paul Ekman [3]. So, Ekman, together with Wallace Freisen, conducted one of the more respectable research projects in New Guinea with members of the for tribe, who, despite living isolated and far from civilization, interpreted photos of different facial expressions like any resident of the Western world. Based on this finding, it was concluded that human facial expressions are biologically innate, not culturally determined.

Expanding his studies on non-verbal behavior, from body movements to the face, Ekman singled out 43 different facial muscles and their movements and calculated that their combinations can result in over ten thousand possible facial expressions, of which 3000 are unmistakable indicators of emotion. Those 3000 expressions can be reduced and categorized into six basic emotions: fear, sadness, disgust, happiness, surprise and anger; and which are equally manifested in all cultures around the world. It should be mentioned here that happiness belongs to the group of positive emotions, unlike fear, anger, disgust and sadness, which are considered negative emotions. Surprise is an emotion that can be both positive and negative. On the other hand, Driver and Van Aalst state that there are seven primary emotional states: happiness, sadness, fear, anger, surprise, contempt and disgust, along with the seventh, contempt [6]. It is believed that introverted people are better at hiding their emotions condition since their facial expressions are calmer and less expressive. On the other hand, extroverted people show their current mood and any changes in it on their face (Fig. 21.1). However, you should know that the list of emotions that have a universal facial expression is much shorter than the number of emotions, which is understandable if you bear in mind that some positive emotions such as, for example, amusement, relief and elation have only one facial expression, which is a smile [7].

Fig. 21.1 Seven primary emotions

On the other hand, there is no reliable knowledge as to whether some other emotions besides the above have universal facial expressions.

In research in which facial expression was studied without the inclusion of body movements, it was found that evaluators are only to a certain extent capable of accurately identifying emotions. Therefore, it is important to point out that the dynamics of movement contain additional information about the strength of the experienced emotion, as well as whether it is sincere. Movement of the head down, backwards, forwards or sideways can be involved in expressions of sadness, fear, interest or disgust. However, you should know that these head movements alone do not convey these emotions, while the hands can also provide additional information. For example, a clenched fist is a movement that is consistent with the facial expression of anger, while crossed arms can be part of an expression of enjoyment, but also of shyness [3]. If this dynamic movement is supported by additional vocalization, it is certain that the accuracy in identifying emotions largely increases.

21.3 Detecting Deception in Words, Voice or Body

The notion that lies are transparent and can be detected through nonverbal behavior dates back a long time. As early as 900 BC, it was claimed that liars shiver and engage in fidgeting behaviors [23]. In 1908, Münsterberg pointed to the utility of observing posture, eye movements, and knee jerks for lie detection purposes [24].

Popular culture reflects the enduring belief that liars give themselves away through nonverbal ehaviour. For example, there are numerous books aimed at a popular audience recommending ways to decipher deception based on demeanor [16, 18]. Also, the basic mythology of the leaky liar is integrated into American police interrogation manuals [27]. The belief in the detectability of lies is also built into many legal presumptions—for example, jurors in criminal cases are often asked to pay attention to the defendant's nonverbal ehaviour to assess their truthfulness [22, 27].

The relationship between nonverbal behavior and deception intrigues people and represents big business. In this article, we review what is currently known about this topic. We discuss theories about why liars and truth tellers display different behaviors and then review research on actual nonverbal cues to deception. As we describe, research consistently shows that attempting to read truth and deception results in very poor accuracy rates, most likely because the behavioral traces of deception faint.

21.3.1 Words

Linguistic lapses—first identified by Freud as a linguistic lapse. In The Psychopathology of Everyday Life, Freud showed that mistakes in everyday life, such as language mistakes, forgetting the names of close people and errors in reading and writing, are not random but significant events that reveal inner psychological conflicts. Lapses expresses, he said, what someone did not want to say: it becomes a mode of self-betrayal". Freud did not specifically deal with fraud, but one of his examples shows how a mistake betrays a lie.

Tirades are the third way liars can betray themselves through words. A tirade is different from a blunder. Speaking carelessness betrays a liar more than a word or two. Information does not come out of the speaker's mouth—it simply flows. The liar is carried away by emotion and therefore loses self-control, realizing too late the consequences of what he himself discovered. If a liar manages to keep his cool, he often does not reveal information that could harm him. Overwhelming emotions, such as rage, horror, fear, discomfort, pressure and cause the liar to reveal the information [9].

21.3.2 A Voice

The voice includes everything that is included in the speech, except for the words themselves. The most common vocal signs of cheating are pauses. Breaks may be too long or too frequent. Hesitation at the beginning of someone's speech, especially if it occurs when someone is answering a question, can arouse suspicion. Suspicion can also be caused by short pauses during speech if they occur often enough. Speech errors can also indicate fraud. By him a are meant "non-words", such as "ah", "aaa", and "uhh"; repetitions such as, for example: "I, I, I think, I really ..."; and parts of words, as in the example: "I really like that". The best documented vocal cue of emotion is pitch. In almost 70% of the subjects studied, the tone of voice p remained higher when he was upset. This is probably most true for a state of anxiety provoked by feelings of anger or fear. There is evidence that the pitch decreases with sadness or grief, but this is not entirely certain. Scientists have not yet learned whether the pitch changes with excitement, pain, disgust or contempt. Other signs of emotion, which are not as well established but promising, are: louder, faster speech in anger or fear, and softer, slower speech in sadness. Progress in this area of study is likely to depend on measuring other aspects of the voice, such as voice quality, voice type, the energy spectrum of the voice in different frequency groups and measuring changes in the voice associated with breathing. Changes in voice caused by emotion are not easy to hide. Especially—if one tries to hide the emotions experienced at the time of lying—there is a good chance that "things will leak". If the purpose of lying was to hide fear or anger, the voice should sound higher and louder, and the speed of speech should be faster. Completely opposite changes in the voice can "miss" the feeling of sadness that the liar is trying to hide. Raising your voice is not a sign of fraud. It is a sign of fear or anger, maybe even excitement [9].

21.3.3 The Body

The shrug and the middle finger are two examples of actions that are called emblems, to make a difference in relation to all other gestures that people show. Emblems have a very precise meaning, which everyone who lives within a culture knows (Fig. 21.2). Everyone knows what the middle finger means, and that shrugging means "I don't know", "I'm helpless" or "what does it matter?". Most other gestures do not have such a precise definition, and their meaning is vague. If not accompanied by words, most gestures do not mean much. Not so with emblems—they can be used instead of words or when words cannot be used. Today there are about sixty emblems in public use in the United States of America. Emblems are a nod for "yes", a shake of the head for "no", a wave of the hand "come here", a wave of the hand for "hello/see you", finger to finger for "shame", "hand behind the ear" as a request to something is said louder, hitchhiker's thumb, and so on. Emblems are almost always performed on purpose [12]. The person performing one emblem knows what he is doing. She chose to communicate a message. But there are exceptions. Just as there are lapses in speech, there are also lapses in body movement—emblems that leak the information a person is trying to hide. There are two instances in which an emblem can be said to be an oversight that reveals hidden information rather than an intentionally sent message. One is when only a part of the emblem is performed, and not the whole action. Shrugging can be performed by raising both shoulders, or by turning the palms upwards, or by facial movements that include raising the eyebrows and lowering the upper eyelid and setting the mouth into a horseshoe shape, or a combination of all these actions and, sometimes, by throwing the head to the side. When an emblem leaks hidden information, only one element will be shown and even that will not be complete. It may be raised alone on one shoulder, and not very high, or only the lower lip may be thrust upwards, or the palms may be turned upwards only slightly. The emblem "middle finger" entails not only a special grouping of five fingers, but the hand is thrown forward, up, often in several swings. Another warning sign that the emblem is an involuntary rather than a deliberate act is that it is performed outside the usual place of display. Most of the emblems are performed right in front of the person, in the area between the waist and the neck. The emblem must be noticed when displayed in a common place. An emblem that leaks hidden information is never performed in a normal position. Although not every liar shows the emblem of a slip, when these slips occur, they are completely reliable. An emblematic slip can be taken as a sign to be trusted, an actual hint of a message that a person does not want to reveal. Compared to the interpretation of most other fraud signs, their interpretation is less susceptible to Brokov's risk or Othello's error. Some people always speak in a roundabout way, and a small number of people regularly make emblematic blunders.

One of the important elements of communication is body language. A communication in which the body language cannot be resolved remains incomplete. Our physical behavior also plays major role in presenting our speech effectively. Our views, expressions, stances and movements have a great effect on the burden of meaning and emotion of our sentences and words [25]. Body language has important functions in understanding and explaining emotions and thoughts. It plays a role in the correct expression and interpretation of messages in the communication process. It also supports the oral communication and creates integrity in communication [17]. 55% of how we impress others while speaking is based on our body language, 7% are the words we use and 38% are our discourse. To be a successful speaker, it is important to focus on this nuance. The audience mainly observes body language, because

Fig. 21.2 Body emblems

a person quickly registers it in the subconscious. Even if the spoken language is not understood by the audience, the body language helps them effectively understand the message [20]. Body language may vary according to culture. However, eye expressions have common features in many cultures. Speech is a typical feature of people and we look at each other for a long time while we speak. Thus, the signals indicating the direction of view become very important [25, 26]. Amy Cuddy, a social psychologist and Harvard professor has researched about the importance of body language. She mentions the consequences of posture and much more and notes that, poses can help us feel better and be more successful (Fig. 21.3). Cuddy emphasizes how having a powerful posture can increase testosterone levels in people and lower cortisol levels [20].

Speech errors can be a sign of various stresses, but not necessarily the stresses associated with lying. Emblem blunders are usually not that vague, since emblems have a very specific message, which is very similar to words. If a person gets the message "screw you" or "I'm crazy" or "I don't mean it" or "I'll wait for you across the street"—all of which can be shown through emblems—then it shouldn't be too difficult to interpret on what was meant.

Illustrators are another type of body movements that can be used to depict deception. Illustrators often get confused with emblems, so it's important to distinguish between them because these two types of body movements can change in opposite

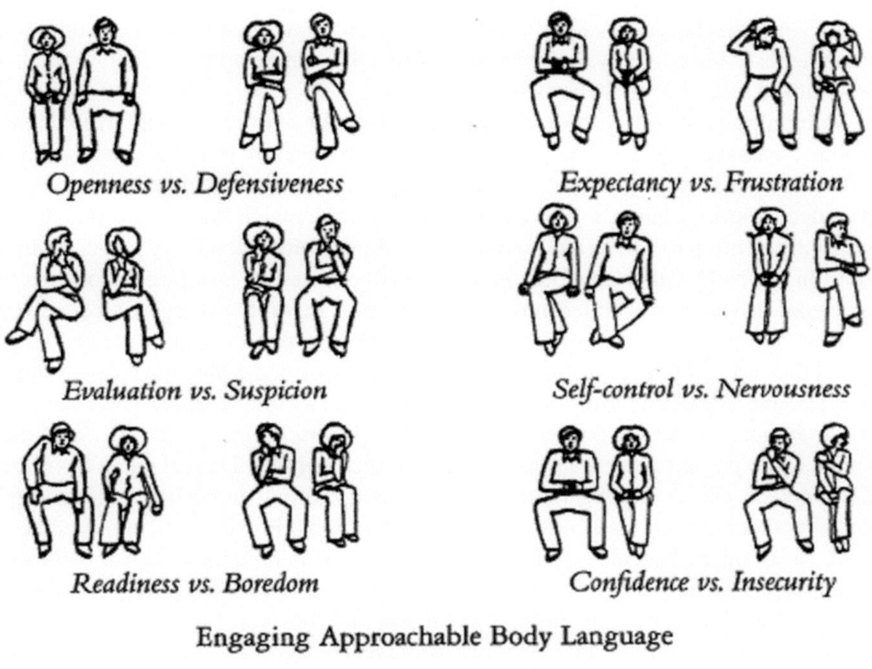

Fig. 21.3 Body language

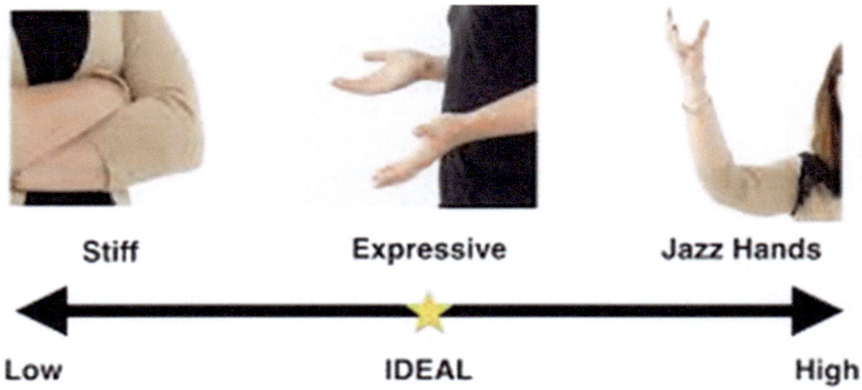

Fig. 21.4 Illustrators

ways when people lie (Fig. 21.4). While emblematic blunders are on the rise, illustrators will usually be on the decline. Illustrators are called by that name because they parallel illustrate the speech as it unfolds. There are many ways to do this: emphasis can be achieved with a word or phrase, very similar to an accent mark or like underlining; the flow of thought can be outlined in the air, as if the speaker is following his speech; the hands can draw a picture in space or show the act of repeating or amplifying what has been said. It is usually the hands that illustrate speech, although movements of the eyebrow or upper eyelid often illustrate emphasis, which can also be achieved with the whole body or upper torso.

Manipulators include almost all those movements by which one part of the body grooms, massages, rubs, holds, pinches, nibbles, scratches, or otherwise manipulates another part of the body. Manipulators can last very short, or their duration can be extended to several minutes. With some manipulators, which last a short time, there is a clear intention to fix the hair, remove something from the auditory canal, scratch a part of the body. Other manipulatives, especially those that last longer, don't seem to be goal-driven: hair is rolled and unrolled, fingers are rubbed, feet are tapped. The hand is a typical manipulator. The hand can also be the recipient, just like any other part of the body. Manipulating actions can also be performed within the face—tongue in cheek and slight lip biting—and leg over leg. Accessories can also become part of a manipulative action—a match, a pencil, a paper clip or a cigarette. Why are people more prone to one manipulative action than another. Does it mean anything if the person chose rubbing rather than squeezing, biting rather than scratching? And can a message be read considering the part of the body that is being scratched: hand, ear or nose? Part of the answer is idiosyncrasy. People have their favorites, a certain type of manipulator that is their signature. For one person it may be turning a ring, for another it may be biting the cuticles, and for a third it may be twirling a mustache. There is some subtle evidence to suggest that certain manipulative actions reveal more than just discomfort. We found manipulative acts of biting in psychiatric patients who did not express anger. Covering the eyes was common for shy patients.

Scientists have pretty much confirmed the lay belief that people fidget and that their movements give off restlessness when they are uncomfortable or nervous. Any form of discomfort prompts a person to engage in manipulative actions such as body scratching, pinching, biting, mouth cleaning, and grooming.

Behavioral signs that may miss hidden information, that show that the person has not prepared his strategy, or that may reveal an emotion that does not fit into the chosen strategy. Linguistic slips, emblem slips, and tirades can miss hidden information of any kind—emotions, past endeavors, plans, intentions, fantasies, ideas, etc. The environment in the speech, pauses, speech errors and reduced use of illustrators can show that the speaker cares a lot about what is being said, but does not have a prepared strategy. They are signs of any negative emotion. The reduction of illustrators, too, goes with boredom. A raised tone and louder, faster speech go with fear, anger, and perhaps excitement. The voice changes in the opposite direction with sadness, and perhaps guilt. Changes in breathing or increased sweating, more frequent swallowing and very dry mouth are signs of strong emotions. Perhaps in the future, it will be possible to determine which emotion is involved, based on the structure of the described changes.

21.3.4 Facial Signs of Cheating

Some of the facial expressions of emotions—at least those that show happiness, fear, anger, disgust, sadness and anxiety, and maybe other emotions—are universal, ie. the same for all people regardless of age, gender, race or culture. These facial expressions are the richest source of information about emotions because they reveal subtle nuances of current feelings [12]. A face can reveal the details of an emotional experience that only a poet can put into words. The face can show:

- what emotion was experienced—anger, fear, sadness, disgust, anxiety, happiness, pleasure, excitement, surprise and contempt can be conveyed by special expressions.
- are two emotions mixed—often two emotions are experienced and the person registers elements of both.
- strength of the experienced emotion—each emotion can vary in intensity, from annoyance to anger, from fear to horror, etc.

21.3.5 A Smile

Smiles are probably the most underrated facial expressions, much more complicated than most people realize. There are many smiles that differ from each other in appearance and the message they convey. There are many positive emotions that are expressed by smiling—enjoyment, physical or sensory enjoyment, satisfaction and cheerfulness. People also smile when they feel miserable [4]. These smiles are not the same as fake smiles used to convince others that positive feelings are being experienced when they are not, often masking the expression of a negative emotion.

Eighteen different kinds of smiles—not one of them is a smile of deception.

An actual smile (as an expression of an experienced emotion). The experienced smile lasts longer and is more intense when the positive feelings are more extreme.

The smile of fear has nothing to do with positive emotions but is sometimes misunderstood. In the facial expression of fear, the rectangular shaped mouth (with or without the corner of the lip turned up) will be accompanied by the specific position of eyebrows and eyes.

A scornful smile is another misnomer, as this expression has little to do with positive feelings, although it is often interpreted as such. The version of contempt involves tightening of the muscles at the corners of the lips, the creation of a muscular bulge in and around the corners, often a dimple, and a slight upturning of the corners of the lips.' In this case too, the upward slanting of the corners of the lips is an action that causes confusion, because it is also characteristic of an experienced smile (a smile that shows positive emotions). Another common element is the dimple, which sometimes appears with an experienced smile. The main difference between a contemptuous smile and an experienced smile is the tightening of the corners of the lips, which is present in contempt and not present in an experienced smile.

With a suppressed smile, the person really feels positive emotions but tries to show that these feelings are less strong than they really are. The goal is to dampen (but not stifle) the expression of positive emotions, keeping both the expression and perhaps the emotional experience within permissible limits. The lips may be pressed, the corners of the lips tightened, the lower lip pushed up, the corners of the lips pulled down, or any combination of these actions can be combined with a simple smile. A suppressed smile with three suppressed muscle actions fused with the muscle action of the experienced smile.

The patient's smile confirms that negative emotions have been experienced. It is not an attempt to hide something, but a facial illustration of making someone feel miserable. A patient's smile also usually means that the person showing it will not, at least at that moment, protest much about their suffering. If we accept that the patient's smile is an attempt to control the expression of fear, anger or pain, it can look very similar to a suppressed smile. Pressed lips, the lower lip pushed up due to the contraction of the chin muscles and the corners of the lips tightened or turned down—can serve to control the onset of some of these negative feelings.

In a coquettish smile, a coquette shows an experienced smile as she turns and looks away from the person she is interested in and then, for a moment, sneaks a look at that person, long enough to be noticed, before looking away again. One of the elements that makes the painting of the Mona Lisa so unusual is that Leonardo painted her by capturing her in the middle of such a flirtatious smile as she looks to one side, but out of the corner of her eye looking askance at the object of her interest. In actual circumstances, along with looking away, this action only takes a moment. In a confused smile, the gaze is directed downwards or to the side, so that the confused person does not meet other people's eyes.

Chaplin's smile is unusual, produced by a muscle that most people cannot voluntarily move. This smile, in which the lips are slanted upwards much more sharply than in the experienced smile, could be abelled as Charlie Chaplin's smile. It's a haughty smile. There are numerous signs by which fake smiles differ from real ones, which they pretend to resemble. Fake smiles are more asymmetric than real smiles. A fake smile will not follow the engagement of the muscles around the eyes, so in a fake smile of weak or moderate strength there will be no raised cheeks, baggy wrinkled skin under the eyes, wrinkles in the outer corners of the eyes or a slight lowering of the eyebrows that will appear in an experienced smile of weak or moderate strength [9]. Several different signs of deception can be found on the face: micro expressions, suppressed expressions, permeability of reliable facial muscles, blinking, dilated pupils, tears, redness and pallor, asymmetry, temporal inconsistency, mislocation and false smiles. Some of these signs allow for leaks, revealing hidden information; others provide signs of deception that show something is being hidden—but not what; and still others indicate that the expression is false. These facial signs of deception, as well as the signs of deception in words, voice, and body, which were described in the previous chapter, differ in the precision of the information they convey. Some signs of cheating reveal exactly what kind of emotion the liar is really experiencing, even if he tries to hide that feeling. Other signs of deception only reveal whether the hidden emotion is positive or negative, and do not reveal exactly which negative or positive emotion the liar is feeling. However, there are signs that are even less differentiated, which reveal only that the liar feels some emotion, but do not reveal whether the hidden feeling is positive or negative. That may be enough. Knowing that an emotion has been experienced can sometimes indicate that a person is lying if the situation is such that no emotion should be experienced in it, unless it is a matter of lying. Other times, the lie will not be revealed unless there is more precise information about what emotion is being hidden. It depends on the lie, the chosen strategy of the person who is supposed to be lying, the situation and the available alternative explanations, regardless of lying, searching for why an emotion could be experienced but also hidden.

21.4 Conclusion

The ability to decode non-verbal cues is a particularly desirable skill when dealing with the psychological phenomenon of lying. A popular assumption that people can easily recognize the presence of lies in communication, if they direct attention to certain non-verbal signs, understandable is due to the fact that sending a large number of non-verbal messages takes place unconsciously. Besides, lying it can often be a cognitively demanding task, and the main "enemy" of a person trying someone emotions are mixed. The increased level of arousal caused by emotions and the fact that expression emotion, almost the most important function of non-verbal communication, support the point of view towards which it is it is easier to recognize a lie by observing non-verbal signs than to lie. On the other hand, lying it implies a conscious and deliberate act, which is why a possible attempt at control must not be ignored non-verbal behavior because some channels are more susceptible to conscious control than others. The success of control depends, among other things, on the type of lie, that is, on how much a person is motivated to tell present your lie as the truth.

In general, the results of various research on non-verbal signs of lying, which are the most common analyzed as part of facial expressions, eye contact and gaze, gestures and vocal behavior, they are often contradictory. Although there are a few researchers who agree that certain nonverbal cues can be good indicators of lying, what they all particularly emphasize is the uniqueness of each communication situations. Considering individual differences, each person may exhibit a specific set non-verbal sign, so ignorance of a person's usual non-verbal behavior can additionally make it difficult to see the signs that distinguish a lie from the truth. In this sense, research in this area may be limited, as well as because non-verbal behavior is rarely observed in relation to verbal content of communication. Despite the opinion of some that the representation of non-verbal communication in of the total communication process is 93%, it seems that making conclusions about lying impossible without simultaneous analysis of verbal and non-verbal signs. Understanding subtle cues in body language, detecting shifts in facial expressions, and recognizing patterns of behavior are paramount in assessing potential threats and de-escalating tense situations. In a security environment, Non-Verbal Communication plays a critical role in conveying and interpreting information, often supplementing or even replacing verbal exchanges in certain situations. One key aspect is the use of body language and gestures to signal and interpret security-related cues.

References

1. Aronson E, Wilson TD, Akert RM, Sommers SR (2015) Social psychology (9. izd.). Pearson Educ
2. Aronson E, Wilson TD, Akert RM (2005) Socijalna psihologija. Zagreb: Mate
3. Baić V, Areh I, Radovanović D, Umek P, Mitrović D (20115) Detekcija laganja. Sinapsa edicije
4. Burgoon JK, Bacue AE (2003) Nonverbal communication skills. U Greene J. O. i Burleson B. R. (Ur.), Handbook of communication and social interaction skills (str. 179–219). Lawrence Erlbaum Assoc, Publishers
5. Burić Moskaljov M (2014) Poruke bez riječi: Umijeće neverbalnog komuniciranja. TIM press, Zagreb
6. Driver J Van, Aalst M (2012) Odajete se više no što mislite. Škorpion. Zagreb
7. Ekman P (1992) An argument for basic emotions. Cogn Emot 6(3–4):169–200
8. Ekman P (1989) Why lies fail and what behaviors betray a lie. U Yuille, J. C. (Ur.), Credibility assessment (str. 71–81). Kluwer Academic Publishers
9. Ekman P (2009) Lie catching and microexpressions. U Martin C. W. (Ur.), The philosophy of deception (str. 118–133). Oxford University Press
10. Ekman P, Friesen WV (1971) Constants across cultures in the face and emotion. J Pers Soc Psychol 17(2):124–129
11. Ekman P (1976) Movements with precise meanings. J Commun 26(3):14–26
12. Ekman P, Friesen WV, Scherer KR (1976) Body movement and voice pitch in deceptive interaction. Semiotica 16(1):23–27. https://doi.org/10.1515/semi.1976.16.1.23

13. Gozalova MR, Gazilov MG, Kobeleva OV, Seredina MI, Loseva ES (2016) Non-verbal communication in the modern world. Mediterr J Soc Sci 7(4):553–558
14. Hall JA, Knapp ML (2010) Neverbalna komunikacija u ljudskoj interakciji. Jastrebarsko: Naklada Slap
15. Hall JA, Horgan TG, Murphy NA (2019) Nonverbal communication. Annu Rev Psychol 70(1):271–294
16. Houston P, Floyd M, Carnicero S (2012) Spy the Lie New York: St. Martin's Press
17. İpek O (2017) Yüz İfadelerinin Tefsiri HakkındaTecrübi Bir Tektik. Ana Dili Eğitimi Dergisi 5(3):485–492
18. Meyer P (2010) Liespotting: Proven techniques to detect deception. St. Martin's Press. https://scholar.google.com/scholar_lookup?title=Lie+Spotting%3A+Proven+Techniques+to+Detect+Deception&author=P+Meyer&publication_year=2010&
19. Patterson ML (2011) More than words: The power of nonverbal communication. Aresta
20. Rao M (2017) Tools and techniques to boost the eEloquence of your body language in public speaking. Ind Commer Train 49(2):75–79
21. Remland MS (2016) Nonverbal communication in everyday life. Sage Publications
22. Sporer SL, Schwandt B (2006) Paraverbal indicators of deception: a meta-analytic synthesis. Appl Cogn Psychol. 20:421–46. https://scholar.google.com/scholar_lookup?title=Paraverbal+indicators+of+deception%3A+a+meta-analytic+synthesis&author=SL+Sporer&author=B+Schwandt&journal=Appl.+Cogn.+Psychol.&volume=20&pages=421-46&publication_year=2006&
23. Trovillo PV (1939) A history of lie detection. I J Crim Law Criminol. 29:848–881
24. Trovillo PV (1939) A history of lie detection. II J Crim Law Criminol. 30:104–119
25. Uçar I (2012) Body language in kutadgu bilig. J Turk Stud 7(7):3045–3058
26. Vrij A, Granhag PA (2007) Interviewing to detect deception. Offenders' Memories of Violent Crimes SA Christianson 279–304 Chichester, UK: Wiley. https://onlinelibrary.wiley.com/doi/book/10.1002/9780470713082#page=291
27. Vrij A, Turgeon J (2018) Evaluating credibility of witnesses: Are we instructing jurors on invalid factors. Weiner Law J In press. https://scholar.google.com/scholar_lookup?title=Weiner+Law+J&author=A+Vrij&author=J+Turgeon&publication_year=2018&
28. Yiu EM, Kornberg AJ (2008) Duchenne muscular dystrophy. Neurol India 56(3):236–247

Chapter 22
Ballistic Response of High–Strength Steel Against Small–Arm Projectiles—Numerical Modelling

Jovan Trajkovski

Abstract Critical infrastructure, as well as military and civil equipment, buildings, and Light Armored Vehicles (LAV), may encounter blast or ballistic impacts during their operational lifespan. These extreme forces induce significant deformations and stresses within fractions of a second. Addressing this challenge requires rigorous assessment and enhancement of structural response against such intense, transient loads. For this purpose, numerical simulations are often used as an alternative solution to the analytical or experimental approaches. However, the accuracy of numerical simulations relies on the precise characterization of material properties under varying conditions. To reliably evaluate the structural response to this kind of loads, careful validation of the numerical model represents an essential step. This paper presents the numerical aspects in modeling ballistic response of high strength steel against small-arms projectiles as 7.62 mm armor piercing ammunition.

Keywords Ballistic response · Critical infrastructure · Projectile impact · Numerical modelling

22.1 Introduction

Selecting suitable armor materials is crucial for enhancing system mobility while ensuring safety. Armor design requires identifying materials with the lowest areal density that can resist predefined threats. Steel remains a key candidate due to its excellent mechanical properties, extensive technological database, ease of production, and relatively low cost [1]. Although steel's high density is a drawback, it can be mitigated by increasing hardness through different heat treatments. High-strength aluminum alloys also show potential for armor use due to their higher specific strength

---

J. Trajkovski (✉)
Faculty of Mechanical Engineering, University of Ljubljana, Ljubljana, Slovenia
e-mail: jovan.trajkovski@fs.uni-lj.si

T. A. Kovács and I. Fürstner (eds.), *Critical Infrastructure Protection: Advanced Technologies for Crisis Prevention and Response*, NATO Science for Peace and Security Series C: Environmental Security, https://doi.org/10.1007/978-94-024-2308-2_22

[1–3], while titanium alloys, though promising, are limited by their high cost. Laminated composites [4–7], featuring a ceramic front layer and a metallic or composite backing layer, offer significant weight reduction compared to armor steels [1]. The ceramic layer, with its high hardness, erodes and fractures projectiles, while the backing layer absorbs residual energy [5]. Harder ceramic layers further improve ballistic protection. However, laminated composites are costly, and spalling under ballistic impacts reduces their multi-hit capability [1].

Numerical modeling of the ballistic response of materials is among the most challenging areas of numerical analysis, as it requires an accurate representation of material behavior under high strain rates, up to and including failure. This encompasses the entire nonlinear part of plastic deformation as well as the initiation and progression of fracture. Due to the relative simplicity of conducting experimental ballistic tests, they often provide a more cost-effective research method compared to numerical simulations. Therefore, unlike blast response analysis, ballistic experiments are highly popular among industrial manufacturers. Conversely, numerical modeling of ballistic response remains a vital research tool in the scientific community [8–10]. Beyond analytical approaches, it is frequently employed to validate material parameters and material models under extreme conditions involving high strain rates or to verify newly developed numerical methods.

Several numerical methods have been developed to address the challenges encountered in numerical modeling of the Taylor test and ballistic impact using the Lagrangian approach. Eulerian methods, commonly used for fluid simulations, have been applied to handle large deformations and material flow. The Arbitrary Lagrangian–Eulerian (ALE) method [11, 12] was introduced to combine the strengths of both Lagrangian and Eulerian approaches, allowing for better management of mesh distortions. Mesh-free methods, such as Smooth Particle Hydrodynamics (SPH) [13, 14] and the Element-Free Galerkin (EFG) method [15], have been explored to overcome issues with mesh entanglement and enhance computational accuracy. Additionally, the Material Point Method (MPM) and Peridynamics have emerged as promising alternatives, particularly for problems involving fractures, discontinuities, and high strain rates. These advanced methods offer improved robustness and flexibility in simulating complex ballistic and impact phenomena.

This paper presents the perforation of high-strength 20 mm PROTAC 500 steel plates using a Nammo AP8 projectile according to Level 11 of the VPAM standard, with a focus on numerical modeling. The model was validated and subsequently applied to further investigate the behavior of thinner plates (15 mm and 10 mm). Additionally, guidelines for effective numerical modeling of projectile impacts, such as contact parameter settings mesh characteristics and element formulations are also provided.

22.2 Material Models

22.2.1 Plate PROTAC 500

For the numerical representation of the plate made of PROTAC 500, the Johnson–Cook (J–C) model was utilized. The J–C model is based on von Mises plasticity theory, where the yield stress is adjusted based on the equivalent plastic strain, strain rate, and temperature. The model is expressed as:

$$\bar{\sigma} = \left(A + B\left(\bar{\varepsilon}_{pl}\right)^{n}\right)\left(1 + Cln\dot{\bar{\varepsilon}}^{*}_{pl}\right)\left(1 - T^{*m}\right), \tag{22.1}$$

In which:

- A, B, n, C and m are material constants
- $\bar{\varepsilon}_{pl}$ is the equivalent plastic strain,
- $\dot{\bar{\varepsilon}}^{*}_{pl} = \dot{\bar{\varepsilon}}_{p}/\dot{\varepsilon}_{0}$ dimensionless strain rate, with $\dot{\bar{\varepsilon}}_{p}$ as the equivalent plastic strain rate and $\dot{\varepsilon}_{0}$ as the reference strain rate.
- T^{*}, the dimensionless temperature is defined as $T^{*} = (T - T_{r})/(T_{m} - T_{r})$, where T is current temperature, T_r is reference temperature and T_m is the material's melting temperature.

he corresponding damage model is described by:

$$\bar{\varepsilon}^{f}_{p} = \left(D_1 + D_2 \exp\left(D_3\sigma^{*}\right)\right)\left(1 + D_4 ln\dot{\bar{\varepsilon}}^{*}_{pl}\right)\left(1 + D_5T^{*}\right), \tag{22.2}$$

where:

- D_1–D_5 are material constants determined from experiments,
- $\sigma^{*} = \sigma_H / \bar{\sigma}$ is a stress triaxiality ratio, with σ_H representing the hydrostatic pressure.

PROTAC 500 is a low-carbon, high-strength steel alloyed with Si (1.01 wt.%), Cr (0.69 wt.%), Mo (0.33 wt.%), and microalloyed with Ti (0.027 wt.%) and B (0.002 wt.%) [16, 17]. The material parameters for the J–C strength and fracture models for PROTAC 500, as determined by Trajkovski et al. [10], are presented in Table 22.1.

Table 22.1 J–C model parameters for PROTAC 500 [10]

J–C strength parameters	A (MPa)	B (MPa)	n (-)	C (–)	m (–)
	1380	948	0.2351	0.0035	1.087
J–C fracture parameters	D_1 (–)	D_2 (–)	D_3 (–)	D_4 (–)	D_5 (–)
	0.0001	1.586	-1.718	0.00695	3.247

22.2.2 Experiments

Preliminary ballistic tests conducted at the NATO-accredited Beschussamt Institute in Ulm, Germany, demonstrated that PROTAC 500 is suitable for military and civilian applications requiring high protection [16, 17]. Ballistic testing was conducted using 7.62 mm armor-piercing projectiles manufactured by the Swedish company Nammo, conforming to the German VPAM Level 11 and the American STANAG 4569 Level 3 standards (Table 22.1). These tests were performed to investigate the interaction between the projectiles and a steel plate. The armor-piercing projectiles feature a rigid core, typically made of high-strength steel, which facilitates the transfer of the projectile's total kinetic energy into the deformation of the target. A notable feature of these projectiles is their tungsten carbide (WC–Co) core, which enhances their ability to penetrate armored materials (Table 22.2). The results of the experimental tests are presented in Table 22.3. In interpreting the results of the ballistic tests, the key factor is whether the projectile penetrates the target or not. In the experiments, all projectiles were stopped by the plate. None of the shots resulted in perforation of the panel which can be also notices in Fig. 22.1.

Table 22.2 Projectile characteristics [1, 16]

Bullet mass [g]	8.3	
Core mass [g]	5.9	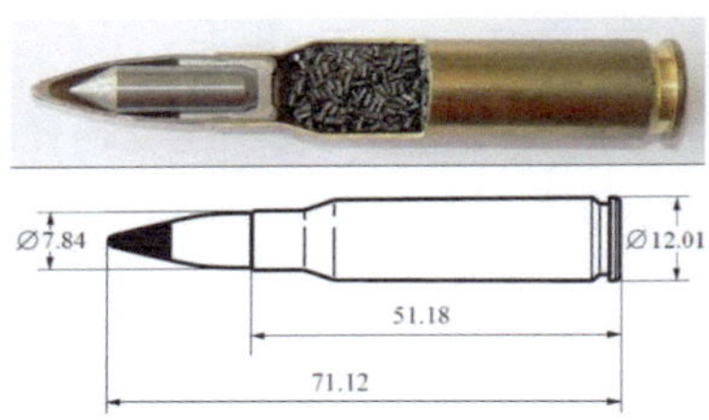
Core diameter [mm]	5.5	
Core point angle [°]	58	
Bullit cover material	Steel and brass	
Bullit core material	WC–Co	
Core hardness [HV]	1450	

Table 22.3 Parameters of the ballistic tests and results description from Ref. [17]

Parameter	Units	Sample 1	Sample 2	Sample 3
Distance	[m]	10	10	10
Shot angle	[°]	90	90	90
Impact velocity	[m/s]	929	931	937
Bullet energy	[J]	3642.77	3640.40	3687.47
Break trough	/	No	No	No

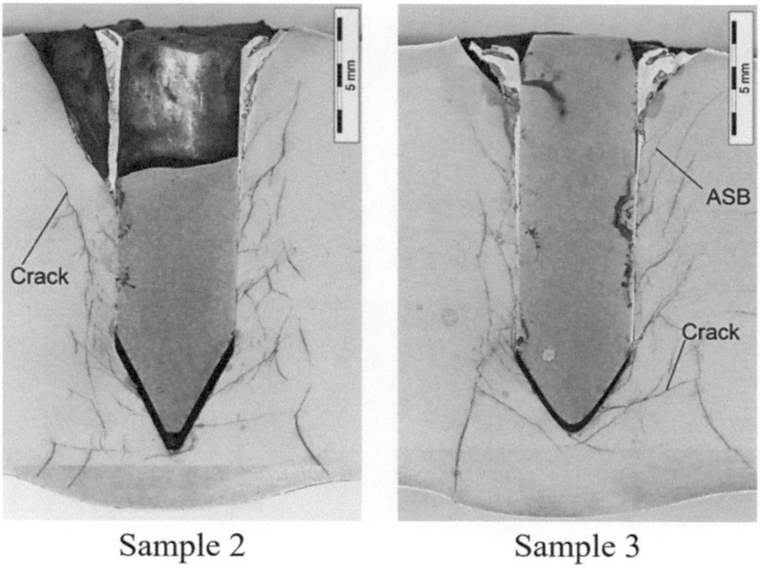

Fig. 22.1 Test samples from the ballistic experiments from Ref. [17]

22.3 Numerical Model

The Finite element (FE) model of the projectile was represented with 95,600 solid elements having an approximate element size of 0.12 mm, representing only one-quarter of the model due to symmetry (Fig. 22.2). The plate, with a side length of 120 mm and thickness of 20.8 mm, was modeled using 1,640,000 under-integrated solid elements with default element formulation with one integration point. Hourglass energy was controlled by using Type 5 hourglass control, incorporating Flanagan-Belytschko stiffness formulation with exact volume integration for the solid elements, utilizing the default parameters. The initial velocity of the projectile was set to 930 m/s, and the material properties of the projectile were defined as rigid. To accurately model the interaction between the projectile and the plate, ERODING_SURFACE_TO_SURFACE contact was used with a soft option, utilizing a pinball segment-based penalty formulation to efficiently represent the transfer of energy between the two bodies.

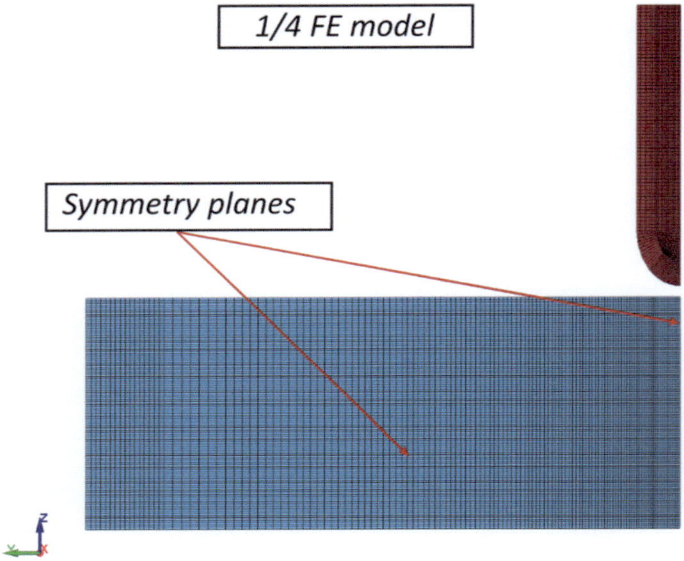

Fig. 22.2 Representation of the FE model

22.4 Validation Study

Since the projectile impact problem can be classified as an impulsive loading regime, the entire kinetic energy is transformed into the deformation of the plate (into internal energy). Figure 22.3 illustrates the energy distribution during the numerical simulation of a projectile impacting a 20 mm thick PROTAC 500 plate, with energy components plotted in time. The kinetic energy of the projectile decreases rapidly in the initial phase, reflecting energy transfer to the plate as the projectile decelerates upon impact. Simultaneously, the internal energy of the plate rises sharply, indicating absorption of energy through deformation, and eventually stabilizes as equilibrium is reached. The total energy remains constant throughout the simulation, demonstrating conservation of energy within the system. Hourglass and sliding energies remain negligible, indicating minimal numerical artifacts and frictional effects, respectively.

The results of this simulation are also relevant for validation of the numerical model. The average kinetic energy from three experimental tests is 3656.88 J Fig. 22.4, which aligns well with the numerical simulation, suggesting the model accurately captures the physics of the impact scenario.

Figure 22.5 illustrates the progression of effective stress (von Mises) in a 20 mm thick plate during a projectile impact simulation, captured at six distinct time intervals: t = 0.0 μs, 5.0 μs, 10.0 μs, 20.0 μs, 30.0 μs, and 45.0 μs. It can be notices at time t = 5.0 μs that a high-stress concentration (red region) appears at the impact zone as the projectile begins to interact with the plate and stress wave propagates

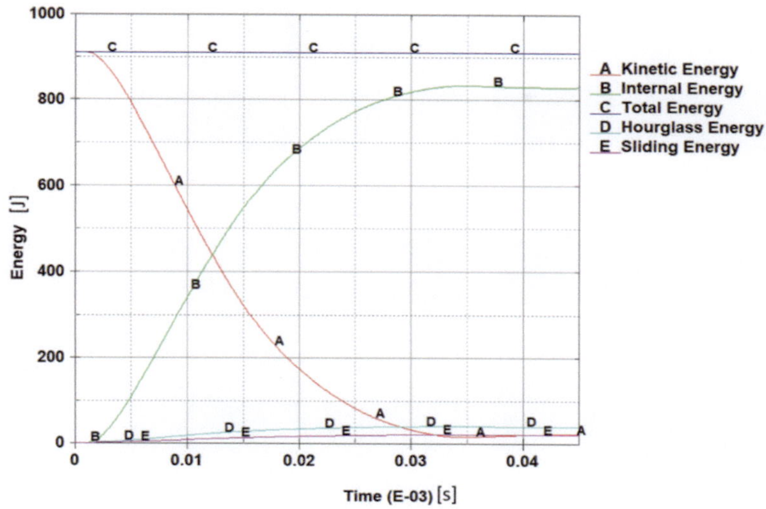

Fig. 22.3 Display of energies for 7.62 mm projectile impact (930 m/s) into 20 mm thick plate

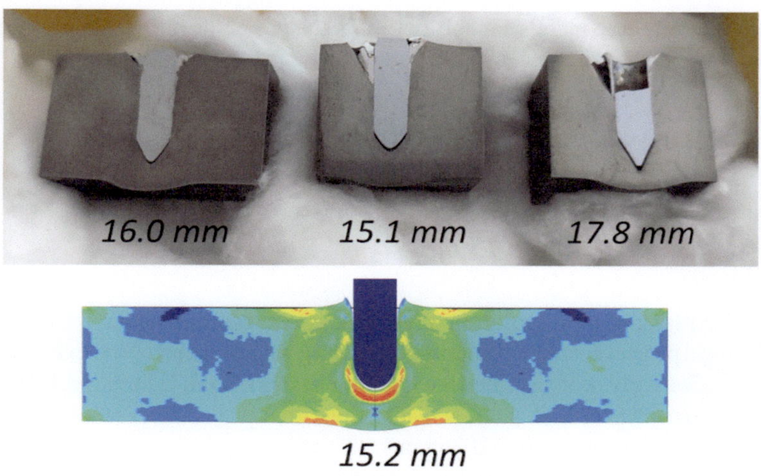

Fig. 22.4 Comparison between experimental and numerical final penetration

away from the contact zone by t = 10.0 μs. After t = 20.0 μs, stress intensities are highest near the contact point, indicating further material yielding and deformation. Finally, at t = 45.0 μs the projectile is stopped and slightly rebounds backward, as observed in both the simulation and experiments (see Fig. 22.5). The permanent depth of penetration into the plate is 15.2 mm, closely aligning with experimental results (referenced in Fig. 22.5), validating the accuracy of the numerical simulation.

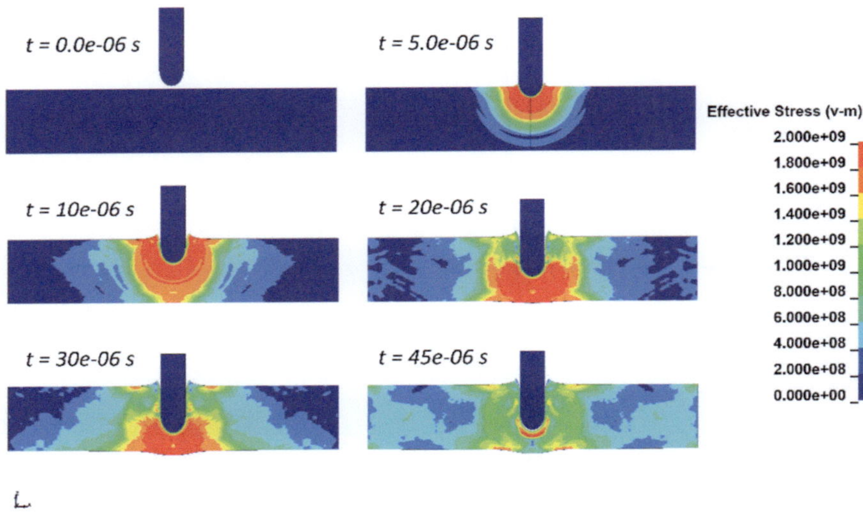

Fig. 22.5 Visual presentation of the projectile impact into 20 mm plate thickness

This demonstrates the plate's capability to stop the projectile and its good agreement with experimental findings.

22.5 Results and Discussion

22.5.1 Plate Thickness 15 mm

Figure 22.6 illustrates the evolution of effective stress (von Mises) in a 15 mm thick plate during a projectile impact simulation, captured at six distinct time intervals. This sequence effectively demonstrates the dynamic response and failure mechanisms of the plate under projectile impact. It is also clear that projectile perforates the plate having a residual velocity around 265 m/s.

Figure 22.7 presents the evolution of various energies during the numerical simulation of a projectile impacting a 15 mm thick plate. The projectile's kinetic energy undergoes a rapid decline in the simulation as it transfers energy to the plate, causing the projectile to decelerate. Simultaneously, the plate's internal energy experiences an increase due to deformation and gradually stabilizes as the system approaches equilibrium. Similarly, the total energy remains almost constant throughout the simulation, confirming energy conservation within the system. Both hourglass energy and sliding energy remain minimal, indicating negligible numerical artifacts and frictional losses.

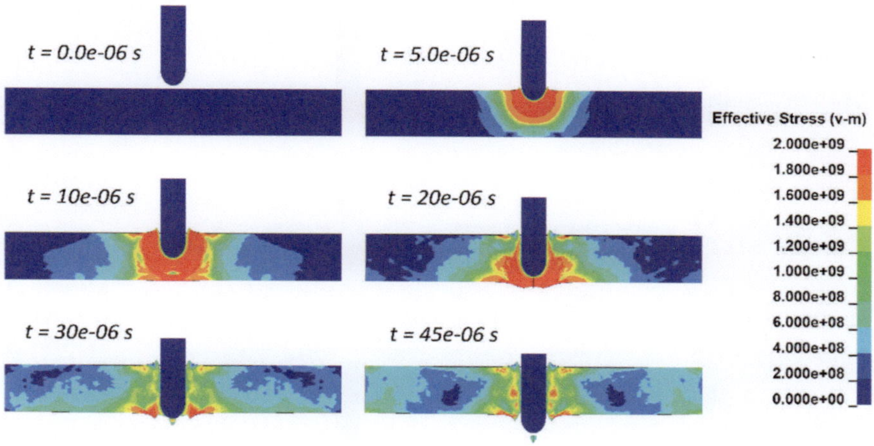

Fig. 22.6 Visual presentation of the projectile impact into 15 mm plate thickness

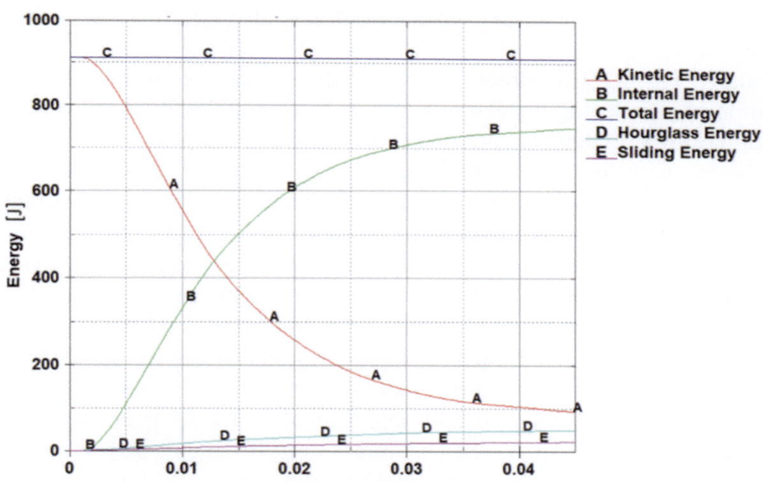

Fig. 22.7 Display of energies for 7.62 mm projectile impact (930 m/s) into 15 mm thick plate

22.5.2 Plate Thickness 10 mm

Figure 22.8 illustrates the evolution of effective stress (von Mises) in a 10 mm thick plate during a projectile impact simulation, captured at six distinct time intervals. This sequence effectively demonstrates the dynamic response and failure mechanisms of

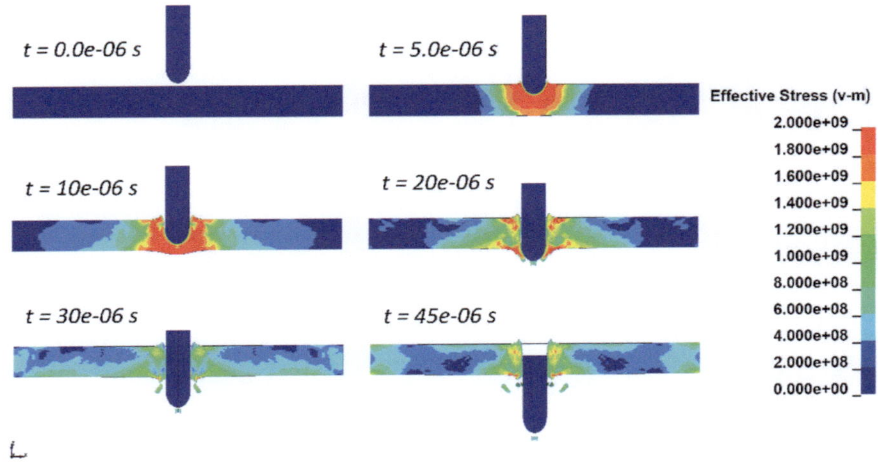

Fig. 22.8 Visual presentation of the projectile impact into 10 mm plate thickness

the 10 mm thick plate under projectile impact. It is also clear that projectile easily perforate the plate.

Similarly, Fig. 22.9 illustrates the evolution of energies during the numerical simulation of a projectile impacting a 10 mm thick plate, with energies plotted as a function of time. It demonstrates the energy transfer and conservation during the impact event, with the projectile's kinetic energy primarily transforming into the plate's internal energy. The kinetic energy decreases rapidly at the beginning of the simulation as the projectile transfers its energy to the plate, decelerating upon impact. Concurrently, the internal energy rises sharply as the plate absorbs energy through deformation and stabilizes over time as equilibrium is reached. The total energy remains nearly constant throughout the simulation, indicating conservation of energy within the system. Both hourglass energy and sliding energy stay minimal, reflecting negligible numerical artifacts and frictional losses, respectively.

The Fig. 22.10 shows the projectile velocities over time for plates with different thicknesses: 20 mm (curve A, red), 15 mm (curve B, green), and 10 mm (curve C, blue). The results indicate that the thickest plate (20 mm) effectively stops the projectile, as its velocity drops to zero within the simulation time. In contrast, the thinner plates (15 mm and 10 mm) fail to completely stop the projectile. For the 10 mm plate, the residual velocity is approximately 610 m/s, while for the 15 mm plate, it is around 265 m/s. This illustrates that the plate thickness significantly influences the projectile's ability to be stopped, with the 20 mm plate providing the most effective resistance.

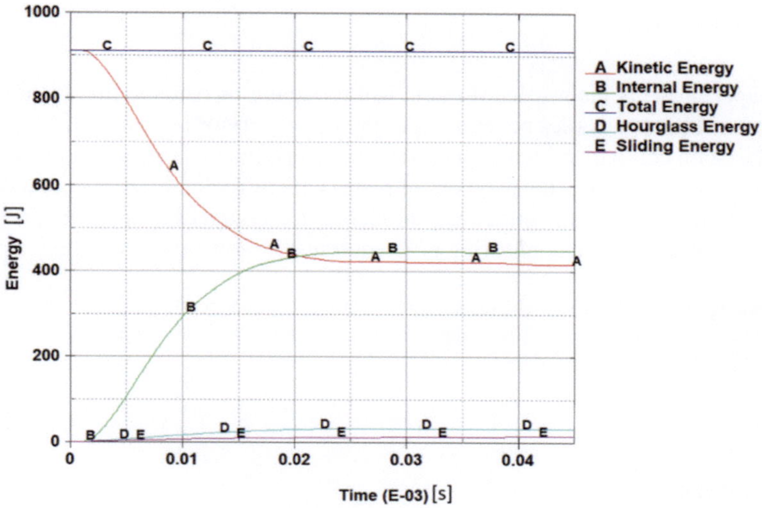

Fig. 22.9 Display of energies for 7.62 mm projectile impact (930 m/s) into 10 mm thick plate

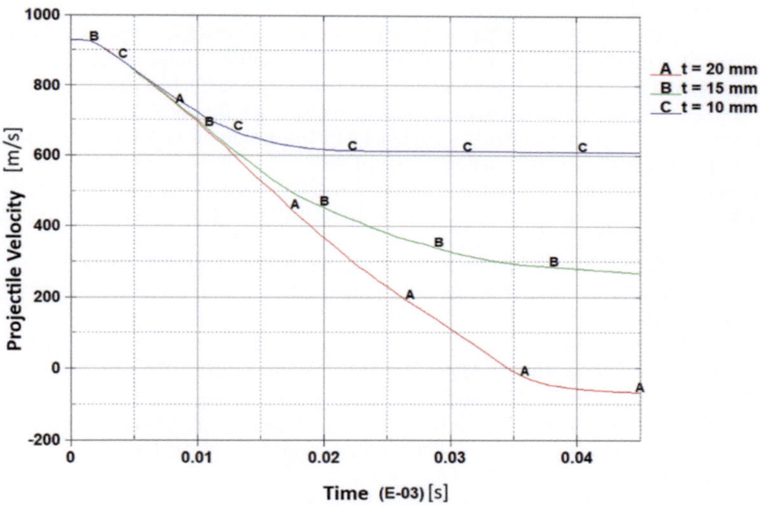

Fig. 22.10 Projectile velocities into plates of different velocities

22.6 Conclusion

This paper presents the results of numerical modeling of ballistic investigations of 7.62 mm caliber projectiles penetrating high-strength armor plate made of PROTAC 500. Based on comparison with experimental results, the numerical model was validated and used to analyze the response of plates with different thicknesses. Additionally, recommendations for effectively modeling the contact between the projectile and the plate are provided, ensuring efficient transfer of energy to the plate.

Acknowledgements The research presented in this article has been partly funded by the Slovenian Research Agency as part of the "Modelling in technics and medicine" (code P2-0109 (C)) research programme.

References

1. Demir T, Übeyli M, Yildirim RO (2008) Investigation on the ballistic impact behavior of various alloys against 7.62 mm armor piercing projectile, Mater Des 29. https://doi.org/10.1016/j.matdes.2008.04.010
2. Jung J, Cho YJ, Kim SH, Lee YS, Kim HJ, Lim CY, Park YH (2020) Mi-crostructural and mechanical responses of various aluminum alloys to ballistic im-pacts by armor piercing projectile, Mater Charact 159. https://doi.org/10.1016/j.matchar.2019.110033
3. Kristoffersen M, Costas M, Koenis T, Brøtan V, Paulsen CO, Børvik T (2020) On the ballistic perforation resistance of additive manufactured AlSi10Mg alumini-um plates, Int J Impact Eng 137. https://doi.org/10.1016/j.ijimpeng.2019.103476
4. Jinnapat A, Doungkom P, Somton K, Dateraksa K (2023) Ballistic performance of composite armor impacted by 7.62 mm armor projectile J Metals, Mate-Rials Miner 33. https://doi.org/10.55713/jmmm.v33i2.1698.
5. Peng L, Tan MT, Zhang X, Han G, Xiong W, Al Teneiji M, Guan ZW (2022) Investigations of the ballistic response of hybrid composite laminated structures, Compos Struct 282. https://doi.org/10.1016/j.compstruct.2021.115019
6. Übeyli M, Yildirim RO, Ögel B (2007) On the comparison of the ballistic per-formance of steel and laminated composite armors, Mater Des 28. https://doi.org/10.1016/j.matdes.2005.12.005
7. Wei Zhang R, Long Chao Z, Tao Jiang L, Peng Chen S, Wei Wang Z, Qi Du S, Luo T, Min Han H, Zhuo Han B, Qin Chen G, Mei Y, Hui Wu G (2023) Ballistic performance and mechanism of B4C/Al laminated targets with different configura-tions. J Mater Res Technol 27. https://doi.org/10.1016/j.jmrt.2023.11.227
8. Trajkovski J (02017) MM-ALE modelling technique for blast response analysisof light armoured Vehicles (LAV) According to AEP-55 Standard: Pros and Cons, 2017. https://doi.org/10.4028/www.scientific.net/KEM.755.159
9. Trajkovski J, Kunc R, Prebil I, Blast response of centrally and eccentrical-ly loaded flat-, U-, and V-shaped armored plates: comparative study, Shock Waves 27. https://doi.org/10.1007/s00193-016-0704-6
10. Trajkovski J, Kunc R, Pepel V, Prebil I (2015) Flow and fracture behavior of high-strength armor steel PROTAC 500, Mater Des 66. https://doi.org/10.1016/j.matdes.2014.10.030
11. Sun WH, Ju GL, Yang BQ, Wang LJ, Pan JW (2014) Ballistic resistance of AA5083-H116 aluminum plates against conical-nose steel projectiles impact. Adv Mat Res. https://doi.org/10.4028/www.scientific.net/AMR.941-944.1428

12. Zhai YX, Wu H, Fang Q (2021) Impact resistance of armor steel/ceramic/UHPC layered composite targets against 30CrMnSiNi2A steel projectiles, Int J Impact Eng 154. https://doi.org/10.1016/j.ijimpeng.2021.103888
13. Baranowski P, Kucewicz M, Małachowski J, Sielicki PW (2021) Failure behav-ior of a concrete slab perforated by a deformable bullet, Eng Struct 245. https://doi.org/10.1016/j.engstruct.2021.112832
14. Kubit A, Trzepieciński T, Kiciński R, Jurczak K (2022) Three-Dimensional smooth particle hydrodynamics modeling and experimental analysis of the ballistic performance of steel-based FML targets, Materials 15. https://doi.org/10.3390/ma15103711
15. Cheng YH, Wu H, Jiang PF, Fang Q (2023) Ballistic resistance of high-strength armor steel against ogive-nosed projectile impact. Thin-Walled Structures 183. https://doi.org/10.1016/j.tws.2022.110350
16. Bernetič J, Kosec B, Kosec G, Gojić M, Burzić Z, Foder J, Vrsalović L, Alec SK, Karpe B, Bizjak M, Nagode A (2019) New generation of armored steel. In: Mech Technol Struct Mater
17. Bernetič J, Kosec B, Kosec G, Burzić Z, Podlipec B, Nagode A, Karpe B, Kanalec S, Vodopivec F, Kosec L (2016) Phenomena in penetrating piercing bul-lets in armored steel plate, Metalurgija 55

Chapter 23
Android Knowledge Application as an Additional Advanced Technology for Use in Critical Infrastructure Protection

Darko Trifunovic and Branislav Todorovic

Abstract Critical infrastructure systems are vital to the functioning of a society and economy. Furthermore, disruptions of critical infrastructure systems can have cascading effects, impacting economic stability, national security, and public safety. As the basis for analysis the paper provides overview of key technological advancements in crises prevention and response, with a focus on critical infrastructure protection and resilience. The proposed Android Knowledge Application (AKA) is given in the context of the latest developments in CIP and resilience in EU and NATO, as a part of improvement process of crisis response strategies. Practical example of AKA is presented in a form of a TOC-s tool, a dynamic data base which offers comprehensive information on global terrorist network and help researchers, analysts, students and other stakeholders working to prevent terrorism. TOC-s database sources, application and perspectives have been discussed in the results section. The main conclusion could be that the continuous evolution of all mentioned technologies and systems underscores the importance of investment in research and development, training, and collaborative efforts among stakeholders.

Keywords Android Knowledge Application (AKA) · Critical Infrastructure Protection (CIP) · Crisis response strategies

D. Trifunovic (✉) · B. Todorovic
Institute for National and International Security (INIS), Belgrade, Serbia
e-mail: darko@intelligence-security.rs

B. Todorovic
e-mail: bane@intelligence-security.rs

T. A. Kovács and I. Fürstner (eds.), *Critical Infrastructure Protection: Advanced Technologies for Crisis Prevention and Response*, NATO Science for Peace and Security Series C: Environmental Security, https://doi.org/10.1007/978-94-024-2308-2_23

23.1 Introduction

In today's interconnected world, critical infrastructure (CI), including sectors like energy systems, healthcare, transportation networks, and communication platforms, plays an essential role in societal functioning and for maintaining societal stability and economic growth. With the increasing frequency and sophistication of threats, including natural disasters, cyberattacks, and terrorism, the need for advanced technologies in critical infrastructure protection (CIP), which is also related to crisis prevention and response, has never been more crucial.

European Union members, United States, China and other countries globally are putting a lot of efforts to enhance national resilience and security. CIP and critical infrastructure resilience (CIR) play an important role in these activities due to wide ranging of issues and unknown threats to area of operation of CI and its influence to related human activity (e.g. energy, transportation, public health, water networks, cybersecurity, and many other interconnected segments). Resilience offers the capability to better review how systems may continually adjust to changing information, relationships, goals, threats, and other factors in order to adapt in the face of change and uncertainty—particularly those potential changes that could yield negative outcomes [1].

The merciless war and the planners of that war designate water sources, water supply, power transmission lines, hydropower plants, thermal power plants, heating plants, and nuclear power plants as the main targets. These objects were not only the target of kinetic force projection but were also attacked from the cyber zone. The role of the intelligence services in the protection of critical infrastructure facilities is irreplaceable considering the mechanisms and information that the intelligence services have at their disposal. It is also important to point out that the process of educating members of the security services and intelligence services that deal with the protection of critical infrastructure is extremely important because through this process new threats are detected that come from different sources with the application of different enforcement techniques [2].

Besides various tools and systems currently used for CIP, due to the nature of most of the critical infrastructures, it might be considered as an advantage to utilize android knowledge applications since they are mobile, versatile and handy. One example of such application is a powerful educational and practical tool—Terrorist and Organized Criminal Search Data Base (TOC-s), which is in use for quite some time and has proven its efficiency.

TOC is a dynamic data base which offers comprehensive information on global terrorist network and help researchers, analysts, students and other stakeholders working to prevent terrorism. It is result of a common project realized by the Faculty of Security Studies and Faculty of Mathematics, University of Belgrade, with the support of the George C. Marshall European Center for Security Studies [3].

23.1.1 Overview and Objectives

Critical infrastructure refers to the physical and virtual systems that are vital to the functioning of a society and economy. Key sectors often classified as critical include:

- Energy: Power generation and distribution systems.
- Transportation: Roads, bridges, railways, airports, and shipping lanes.
- Water Supply: Systems for water treatment, distribution, and waste management.
- Communication: Telecommunications networks, internet infrastructure, and satellite systems.
- Public Health: Hospitals and emergency response facilities.

Disruptions of critical infrastructure systems can have cascading effects, impacting economic stability, national security, and public safety.

As threats evolve, from natural disasters to cyberattacks, the need for advanced technologies to enhance crisis prevention and response capabilities has become paramount. Following is the overview of key technological advancements in crises prevention and response, with a focus on critical infrastructure protection and resilience.

Key Technological Advancements

Artificial Intelligence (AI) and Machine Learning (ML)

Predictive Analytics: AI/ML algorithms can analyze vast datasets to identify potential threats and vulnerabilities in critical infrastructure systems. This includes predicting natural disasters, identifying anomalies in cyber traffic, and forecasting potential supply chain disruptions.

Anomaly Detection: AI/ML models can detect unusual patterns in sensor data, network traffic, and operational parameters, alerting operators to potential cyberattacks, equipment failures, or other threats.

Decision Support Systems: AI-powered tools can assist decision-makers in real-time during crises, providing recommendations for optimal resource allocation, evacuation routes, and response strategies.

Internet of Things (IoT):

Enhanced Situational Awareness: IoT sensors deployed across critical infrastructure networks can provide real-time data on system performance, environmental conditions, and potential threats. This data can be used to improve situational awareness and enable faster response times.

Remote Monitoring and Control: IoT enables remote monitoring and control of critical infrastructure assets, allowing operators to identify and address issues proactively.

Predictive Maintenance: IoT sensors can monitor the health of equipment and predict potential failures, enabling preventative maintenance and minimizing downtime.

Blockchain Technology:

Supply Chain Resilience: Blockchain can enhance the transparency and traceability of critical supply chains, improving resilience against disruptions and ensuring the authenticity of critical components.

Cybersecurity: Blockchain can be used to enhance cybersecurity by providing secure and tamper-proof records of system activity and access.

Data Integrity: Blockchain can ensure the integrity and authenticity of critical data, which is crucial for effective crisis response and decision-making.

Cybersecurity Technologies:

Advanced Threat Detection: Technologies like intrusion detection systems (IDS), intrusion prevention systems (IPS), and security information and event management (SIEM) systems can detect and prevent cyberattacks on critical infrastructure.

Threat Intelligence Sharing: Platforms for sharing threat intelligence among critical infrastructure operators, government agencies, and cybersecurity researchers can improve collective defense capabilities.

Cybersecurity Automation: Automating cybersecurity tasks, such as vulnerability scanning and patch management, can improve efficiency and reduce the risk of human error.

Robotics and Automation:

Disaster Response: Robots can be deployed for search and rescue operations, hazardous material handling, and infrastructure inspection in disaster zones.

Autonomous Systems: Autonomous vehicles and drones can be used for logistics, transportation, and surveillance during crises.

Automation of Critical Functions: Automating critical functions in infrastructure systems can improve resilience and reduce the risk of human error during emergencies.

Geospatial Technologies:

Remote Sensing: Satellite imagery and other remote sensing technologies can be used to monitor critical infrastructure, assess damage from natural disasters, and track the movement of people and resources.

Geographic Information Systems (GIS): GIS can be used to visualize and analyze spatial data related to critical infrastructure, such as location, vulnerabilities, and interdependencies.

Applications in Critical Infrastructure Protection and Resilience

Energy Sector:

Smart Grids: AI-powered smart grids can optimize energy distribution, improve grid stability, and enhance resilience against cyberattacks and physical disruptions.

Renewable Energy Integration: AI/ML can be used to optimize the integration of renewable energy sources into the grid, improving reliability and reducing reliance on fossil fuels.

Cybersecurity: Advanced cybersecurity technologies are crucial for protecting critical energy infrastructure from cyberattacks, which can have devastating consequences.

Transportation Sector:

Autonomous Vehicles: Autonomous vehicles can improve traffic flow, reduce congestion, and enhance transportation resilience in the face of disruptions.

Predictive Maintenance: IoT sensors can monitor the health of transportation infrastructure, such as bridges and tunnels, enabling predictive maintenance and preventing catastrophic failures.

Disaster Response: Drones and robots can be used for search and rescue operations, traffic management, and logistics during natural disasters.

Healthcare Sector:

Telemedicine: Telemedicine technologies can improve access to healthcare, especially in remote areas, and enhance the resilience of healthcare systems during emergencies.

Predictive Analytics: AI/ML can be used to predict disease outbreaks, optimize resource allocation, and improve patient outcomes.

Cybersecurity: Strong cybersecurity measures are essential to protect sensitive patient data and ensure the integrity of healthcare systems.

Water Sector:

Smart Water Grids: Smart water grids can optimize water distribution, detect leaks, and improve water quality monitoring.

Flood Prediction and Early Warning Systems: AI/ML can be used to predict and forecast floods, enabling early warnings and reducing the impact of these events.

Drought Mitigation: Technologies such as desalination and water recycling can help mitigate the impacts of drought and ensure water security.

Challenges and Considerations

Data Privacy and Security: The collection and use of vast amounts of data raise concerns about privacy and security. Robust data protection measures must be implemented to safeguard sensitive information.

Ethical Considerations: The use of AI and other advanced technologies raises ethical considerations, such as the potential for bias, job displacement, and the misuse of these technologies.

Interoperability and Standardization: Ensuring interoperability between different technologies and systems is crucial for effective crisis response.

Cybersecurity Risks: The increasing reliance on technology also increases the risk of cyberattacks. Robust cybersecurity measures must be implemented to protect critical infrastructure from cyber threats.

Funding and Investment: Significant investment is required to develop, deploy, and maintain advanced technologies for crisis prevention and response.

23.2 Methods

23.2.1 The Latest Developments in CIP and Resilience in EU

The European Union has been proactive in enhancing critical infrastructure protection and resilience, particularly in response to emerging threats and challenges. Following are some examples of the latest developments and activities:

- *EU Cybersecurity Strategy*—The EU has updated its cybersecurity strategy, focusing on strengthening the resilience of critical infrastructures against cyber threats. This includes the establishment of the European Cybersecurity Agency (ENISA) as a key player in promoting best practices and collaboration among member states.
- *Directive on Security of Network and Information Systems (NIS2)*—The NIS2 Directive aims to enhance the overall level of cybersecurity in the EU by expanding the scope of the original NIS Directive. It imposes stricter security requirements on essential and important entities across various sectors, including energy, transport, and healthcare, fostering a more resilient infrastructure landscape.
- *European Critical Infrastructure (ECI) Regulation*—The EU is revising regulations concerning the protection of European critical infrastructures. This includes increased scrutiny and standards for risk assessments, security measures, and incident reporting, ensuring that critical sectors are better prepared for potential threats.
- *Investment in Resilience and Innovation*—The EU is funding initiatives aimed at enhancing resilience through research and innovation. Programs like Horizon Europe focus on developing new technologies and methodologies for protecting critical infrastructure against both physical and digital threats.
- *Public–Private Partnerships (PPP)*—The EU promotes collaboration between public and private sectors to improve critical infrastructure resilience. By fostering partnerships, member states can leverage private sector expertise and resources, leading to more comprehensive protection strategies.
- *Focus on Climate Resilience*—Recognizing the impact of climate change on critical infrastructure, the EU is investing in adaptation strategies. This includes developing guidelines for climate-resilient infrastructure, encouraging member states to assess vulnerabilities to extreme weather events and natural disasters.
- *Joint Cyber Units*—The establishment of Joint Cyber Units among EU member states aims to enhance information sharing and coordination in response to cyber threats. These units facilitate collaborative efforts to protect critical infrastructures across borders.

Through these activities, the EU is making significant efforts in enhancing the security and resilience of its critical infrastructures, addressing both current challenges and future risks effectively.

23.2.2 Recent Developments in Critical Infrastructure Protection and Resilience Within NATO

Following incidents like the Nord Stream pipeline sabotage, besides CIP efforts similar to EU, NATO has significantly increased its focus on protecting critical undersea infrastructure. Following are some examples of the latest developments in CIP and resilience within NATO. As the risk and threat area continues to evolve, NATO is committed to adapting its approach to ensure the security and resilience of CI in its domain.

- Establishment of the Undersea Infrastructure Coordination Cell: This cell aims to improve information sharing and best practices among military and civilian authorities, as well as with industry.
- Maritime Centre for the Security of Critical Underwater Infrastructure: This center, located within NATO's Allied Maritime Command, focuses on preparing for, deterring, and defending against the coercive use of energy and other hybrid tactics.
- Digital Ocean Initiative: This initiative aims to enhance Maritime Domain Awareness (MDA) through better integration of capabilities and exploration of new technologies, such as floating sensors, autonomous systems, and satellites.
- Enhanced Cooperation with Partners: NATO is strengthening cooperation with partners, including the European Union, to improve the resilience of critical infrastructure.
- EU-NATO Task Force on Resilience of Critical Infrastructure: This task force aims to identify and address security challenges related to energy, transport, digital infrastructure, and space.
- Emphasis on Technology and Innovation: NATO is exploring and developing new technologies to enhance the protection and resilience of critical infrastructure.
- Defense Innovation Accelerator for the North Atlantic (DIANA): This program focuses on energy resilience and sensing and surveillance technologies.
- Strengthening of Deterrence and Defense: NATO is strengthening its deterrence and defense capabilities to protect critical infrastructure from threats.
- Increased surveillance and reconnaissance activities: NATO is conducting more patrols and surveillance activities to monitor critical infrastructure.
- Deployment of advanced military capabilities: NATO is deploying advanced military capabilities, such as mine hunters and airborne early warning and control systems, to protect critical infrastructure.

23.2.3 Improving Crisis Response Strategies

Cyber-security is complex but important segment in security of large systems and corporations. Nowadays businesses and state facilities use in general data bases and internet, which makes them vulnerable to attacks. Existing threats in any country, or even neighboring ones, can develop uneasiness and fear among people in every field. Such fear in combination with inappropriate media content might lead to the spread of growing panic. In such environment employees become less productive, prone to mistakes and as consequence this creates security holes in the system. For example, employees with local administrator rights who override security protocols on their computers due to eagerness to obtain latest news or discussions in social media, might become victims of social engineering and let the infection spread from their computer onto the entire corporate network. Afterwards employees quite often hide the fact that their company is hit by a security incident to avoid personal responsibilities. Hiding an incident may lead to more severe consequences and increase the caused damage. Careless or uninformed staff, for example, are the second most likely cause of a serious security breach, second only to malware. In addition, in 46% of cyber-security incidents in the year 2019, careless or uniformed staff has contributed to the attack [4].

As some examples of a highly complex critical infrastructures for analysis, selected from the domain of the process industry, could be a system of power plants connected by a power grid, a network of main gas pipelines with measuring and compressor stations or a plant for the preparation of drinking water with a distribution network of pipelines. Each of those examples includes multiple units, buildings, resources and other facilities spread over a larger area. Due to the size and complexity of the system, its organized management can represent a key challenge in protecting CI [5].

Organizations, independent of their levels of complexity, can significantly enhance their crisis response strategies as a part of CIP through a combination of planning, technology, training, and collaboration. Some underlined approaches would be:

- *Develop Comprehensive Crisis Management Plans*—Organizations should create detailed crisis management plans that outline procedures for various types of incidents. These plans should include roles and responsibilities, communication protocols, resource allocation, and recovery strategies. Regularly updating and testing these plans ensures they remain relevant and effective.
- *Invest in Technology and Tools*—Leveraging advanced technologies is crucial for improving crisis response. Tools such as incident management software, communication platforms, and real-time data analytics enable organizations to respond swiftly and efficiently. Additionally, incorporating IoT devices for monitoring can provide critical insights during an incident.
- *Conduct Regular Training and Drills*—Regular training sessions and simulation drills help prepare employees for crisis scenarios. These exercises improve familiarity with crisis plans and foster teamwork and effective communication

under pressure. Organizations should involve all relevant stakeholders, including emergency services, to ensure a coordinated response.
- *Foster a Culture of Resilience*—Encouraging a culture of resilience within the organization helps employees understand the importance of preparedness. Leadership should promote open communication, encourage feedback on crisis management procedures, and recognize individuals and teams who contribute to successful crisis response efforts.
- *Establish Strong Partnerships*—Collaborating with external partners, such as local emergency services, government agencies, and industry peers, enhances an organization's crisis response capability. Establishing relationships before a crisis occurs facilitates better coordination and resource sharing during an incident.
- *Utilize Data and Analytics*—Post-incident analysis is vital for continuous improvement. Organizations should collect data on their response efforts, analyze what worked and what didn't, and make adjustments to their crisis management plans accordingly. This iterative process enhances preparedness for future events.

By implementing these strategies, organizations can enhance their resilience and efficiency in responding to crises, ultimately protecting their assets and ensuring the safety of their personnel and stakeholders. However, besides informing and training of employees it is important to equip them also with appropriate tools that would allow for the correct behavior and appropriate decision making on individual level. TOC-s tool is precisely one of such additional advanced technology segments that can be used in CIP.

23.3 Application and Results

23.3.1 The TOC-s Basic Description

In order to access TOC-s data base, a potential user (student, researcher, academic or state institution…) applies for username and password and gives some necessary personal data. After his personal data have been verified, the user obtains one of the two levels of data base access and the corresponding username and password. All passwords are encrypted with 32-bit code and stored in the base. In order to provide the protection of confident data stored in the base, two levels of data access are implemented in the TOC-s. The first level is named "blue key" and it is available for students and researches in academic institutions and research centers. The "blue key" enables the access to all open-source data stored in the base. The second level of data access, named "red key" is reserved for legal authorities, state institutions and state government. The "red key" opens the part of the TOC-s with confident data which are valuable for all legal institutions that are dealing with counterterrorism. The owner of the "red key" also has all rights and access to the open source data, as the "blue key" owner. However, there is also the third key "admin", reserved for

TOC-s data base administrator who is managing with all TOC-s data and controlling the base [3].

23.3.2 The TOC-s Database Sources

The TOC-s database is constantly updated with the reports obtained from George C. Marshal European Center for Security Studies http://www.marshallcenter.org. The Marshall Center is a leading transatlantic security and defense educational institution dedicated to teaching and nurturing an ever-expanding community of leaders that will create a more stable security environment. These provide the foundation for the College of International and Security Studies' undertakings-from international courses and research to specialized language and computer training. The College is developing a common understanding of international security and democratic defense [6].

The Program on Terrorism and Security Studies (PTSS) addresses numerous aspects of a threat that confronts nations around the globe. The five-week course features presentations by prominent civilian, military and government officials and internationally renowned scholars. It is designed for government officials, military officers and police administrators currently working in mid- and upper-level management positions of counterterrorism organizations throughout the world. The PTSS Daily is for educational purposes only and is a special service supporting counterterrorism course at The George C. Marshall European Center for Security Studies. The newsletter's is produced from the open sources media reports by GMC postgraduate students and verified by senior experts and counter terrorism officers, which is the main reason why PTSS has been chosen as the main data source for TOC-s [3].

23.3.3 The TOC-s Application and Perspectives

The work on the TOCs project is a good example of exploratory learning scenario with semi-guided elements. By extensive use of online information sources, students develop both help seek and information-searching skills. Help seek component of the learning process is additionally realized through intensive social interactions with both supervisors and colleagues via Active Collab online portal which surrounds and supports individual learning. The students have developed analytical and researchable approach to the problems and applied this skill successfully for solving different types of tasks within the TOCs project. They acquired the technique of data collecting and classification and became aware of the importance of the verification of the collected data by comparing them with other relevant and selected sources [3].

TOC-s has already been used in the purpose of Security of the Olympic Games in Beijing 2008. TOC-s has constantly been upgraded and several analytical features are to be incorporated within the database. Analytical tools will provide statistical analysis of information of global terrorist network. The analysis will examine the predictability of international terrorism in terms of the existence of trends, seasonality, and periodicity of terrorist events. Another important goal of the TOC-s team is the enhancement of an analytical-purpose system for exploring TOC-s data base. It could be used, for example, to understand the structure of different terrorist organizations with respect to particular attack type or regions of their activity. By using this system, the intelligence analysts will be able to develop hypotheses and then validate them (or not) against the data in the TOC-s information space [3].

The emphasize of the presentation of the TOC-s should be on its operability and flexibility. With customer-tailored additions to the database, user-level GUI adjustments and custom tools (optional), it would become a handy and ready available android knowledge application that can be utilized by security professionals and other CI employees alike.

23.3.4 Conclusions

The integration of advanced technologies in the field of crisis prevention and response is transforming how we protect critical infrastructure. By harnessing smart surveillance, cybersecurity innovations, resilient communication networks, and data-driven decision-making, we can build a more resilient society capable of withstanding and recovering from crises. Continued investment in these technologies and collaborative efforts in the future will be essential to safeguarding our CI against emerging risks.

Advanced technologies are pivotal in enhancing the resilience and protection of critical infrastructures against various crises. The integration of IoT, AI, GIS, drones, and blockchain technology, among others, provides unprecedented capabilities for monitoring, managing, and responding to potential threats. As we move forward, the trends of 5G, AR/VR, and predictive analytics will further shape the landscape of crisis management and infrastructure protection. By embracing these technologies, organizations and governments can better anticipate and respond to threats, mitigate risks, and ensure the continued functioning of CI and other vital systems. However, it is crucial to address the challenges and considerations associated with these technologies to ensure their safe and effective implementation. In this article TOC-s project has been provided as a real life example.

Through the TOC-s project, a team of scientists has successfully unified the perspectives of help seeking and information searching, within the context of ICT-based learning situation. An online collaboration learning environment is designed to facilitate students in group collaboration and to coordinate and monitor the learning process. Besides its educational functions, the TOC-s database has important applicative role in the national security system, as the additional powerful measure in CIP, in protecting state borders, etc. [3].

The continuous evolution of all mentioned technologies and systems underscores the importance of investment in research and development, training, and collaborative efforts among stakeholders. Building resilient infrastructures not only protects assets and resources but also fosters trust and confidence among communities, ensuring a safer future for all.

References

1. Todorovic B et al. (2016) 'Chapter 22—Contribution to enhancement of critical infrastructure re-silience in serbia, resilience and risk—methods and application in environment, cyber and social domains. In: Linkov I, Palma-Oliveira JM (eds). Proceedings of the NATO advanced research workshop on resilience-based approaches to critical infrastructure Safe-guarding, ISBN 978–94–024–1122–5 (HB), Azores, Portugal, 26–29, pp 531–551
2. Trifunovic D (2024) Intelligence and protection of key (Critical) infrastructure objects. In: Kovács TA, Nyikes Z, Berek T, Daruka N, Tóth L (eds). Critical infrastructure protection in the light of the armed conflicts. HCC 2022. Adv Sci Technol Secur Appl. Springer, Cham. https://doi.org/10.1007/978-3-031-47990-8_1
3. Stanimirović Z, Trifunović D (2011) From database to knowledge: the terrorist and organized criminal search database. Soc Sci Comput Rev 29(4):508–514. https://doi.org/10.1177/0894439310378327
4. Upal R (2020) In cybersecurity chain, humans are the weakest link as well as last line of defence. Int Def, Secur & Technol Inc. (IDST). https://idstch.com/threats/in-cybersecurity-chain-humans-are-the-weakest-link-as-well-as-last-line-of-defence/
5. Todorovic B, Trifunovic D, Todorovic M, Markovic M (2022) Operation and maintenance issues related to critical infrastructure protection in the process industry domain. In: Processing '22–35th international congress on process industry, society for process engineering within SMEITS (Serbian Union of Mechanical and Electrical Engineers and Technicians). Belgrade, Serbia
6. The Marshall Center—The George C. Marshall european center for security studies, www.marshallcenter.org, Last accessed 2025/01/03

Index

T. A. Kovács and I. Fürstner (eds.), *Critical Infrastructure Protection: Advanced Technologies for Crisis Prevention and Response*, NATO Science for Peace and Security Series C: Environmental Security, https://doi.org/10.1007/978-94-024-2308-2

MIX
Papier aus verantwortungsvollen Quellen
Paper from responsible sources
FSC® C105338

If you have any concerns about our products,
you can contact us on
ProductSafety@springernature.com

In case Publisher is established outside the EU,
the EU authorized representative is:
Springer Nature Customer Service Center GmbH
Europaplatz 3, 69115 Heidelberg, Germany

Printed by Libri Plureos GmbH
in Hamburg, Germany